Contemporary Problems in Carbonium Ion Chemistry III

Contemporary Problems in Carbonium Ion Chemistry III

Arenium Ions — Structure and Reactivity

by V. A. Koptyug

With 21 Figures and 58 Tables

AKADEMIE-VERLAG · BERLIN
1985

Die Originalausgabe erscheint im Springer-Verlag
Berlin · Heidelberg · New York · Tokyo
in der Schriftenreihe "Topics in Current Chemistry", Volume 122

Vertrieb ausschließlich für die DDR und die Sozialistischen Länder

Erschienen im Akademie-Verlag · Berlin, DDR-1086 Berlin, Leipziger Straße 3—4
Lizenznummer: 202 · 100/521/85
Printed in the German Democratic Republic
Gesamtherstellung: VEB Druckerei „Thomas Müntzer", 5820 Bad Langensalza
Umschlaggestaltung: Karl Salzbrunn
LSV 1275
Bestellnummer: 763 372 9 (6835)

11600

Table of Contents

I Introduction

Arenium ions represent a large class of organic cations—carbenium ions that have been intensively studied in recent years. Thus, benzenium ions can be regarded as derivatives of cyclohexadienyl cations (1) belonging to the group of cyclic enyl cations with the open π-electron system including also the derivatives of cyclopentenyl (2), cyclobutenyl (3), cycloheptadienyl (4) and other analogous cations

1

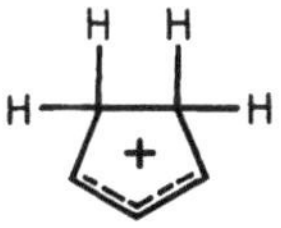

2

3

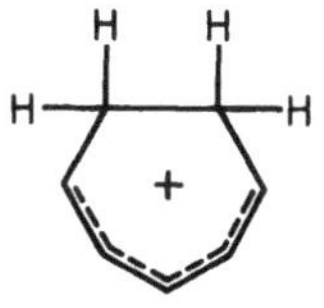

4

G. Olah proposed[1] to name the classical carbocations "carbenium ions" (cf. [2]) and those with penta-coordinated carbon atoms "carbonium ions". In keeping with his recommendations the authors dealing with the above ion groups in their English-language publications now use the term "arenium ions" more often than the one used earlier ("arenonium ions").

Arenium ions are singled out as a particular group due to their genetic relationship with aromatic compounds which owing to their properties occupy a special place in organic chemistry. Indeed, arenium ions can be classified as species actually or formally generated by the addition of an electrophile to an aromatic molecule to form a σ-bond using two electrons of the aromatic π-system:

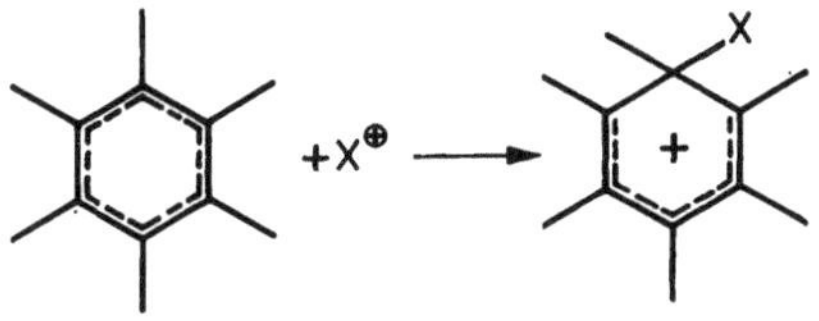

To emphasize the nature of bonding the electrophilic agents, the arenium and their salts are also called σ-complexes.

As a rule, the names of the ions corresponding with the addition of an electrophile to the molecule of an aromatic compound are built of the name of this compound followed by the suffix "-ium" and an indication of the character of the electrophile and the site of its addition [3,4]. Thus, ion (5) can be named 9-bromo-9,10-dimethylanthracenium ion, and ion (6), 2-H-mesitylenium ion.

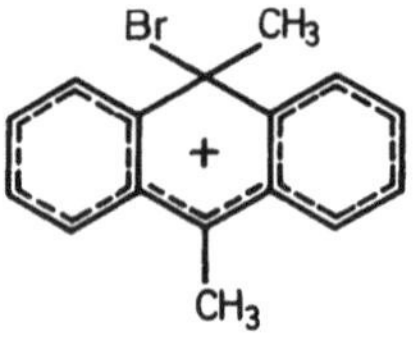

5

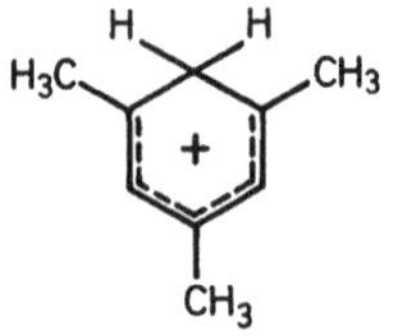

6

For benzenium ions — a somewhat different system of nomenclature is often used: the name contains the enumeration of all substituents and their positions from the ring sp^3-hybridized carbon atom. In this case ion (6) will be called 2,4,6-trimethylbenzenium ion.

Stable arenium ions are of great interest to chemists since they are analogues of intermediates in important reactions of aromatic compounds. This refers primarily to the electrophilic substitution of hydrogen in the aromatic series [5-12], as well as to acid-catalyzed transformations connected with the shifts of substituents in aromatic molecules (isomerization reactions [13,14]) and with intermolecular transfer of substituents.

With the wide application of modern physical methods in recent years much evidence has been accumulated on the structure and reactivity of arenium ions under "long life" conditions. This evidence has been partly reflected in earlier reviews [4,15-19]. In this book a fuller picture is drawn of the research on stable arenium ions and of its significance for the chemistry of aromatic compounds. This generalization is dictated by fast expansion of research on arenium ions which, in its turn, is accounted for by the striving for a deeper theoretical description of the processes in which these species take part, and for the application of the discovered regularities in solving problems of organic synthesis.

II Methods of Generating Arenium Ions in Solutions

The known methods of generating arenium ions start with the addition of electrophilic agents, in the first place the proton, to the molecules of aromatic compounds.

The data on the electrical conductivity and the character of the electronic absorption spectra allow us to conclude that in strong acids the aromatic hydrocarbons (A) behave like bases capable of adding the proton and thus forming salt-like compounds (for details see [7]):

$$A + HY \rightleftharpoons AH^+ \cdot Y^-$$

In a number of instances such salt-like compounds have been isolated and characterized. In particular, many ternary complexes of the composition $A \cdot HY \cdot MY_n$ (Y — halogen) can be considered as salts of the complex acids HMY_{n+1}, i.e. the salts $AH^+ \cdot MY^-_{n+1}$. To these belong the compounds $A \cdot HF \cdot BF_3$, $A \cdot HF \cdot PF_5$, $A \cdot HF \cdot SbF_5$, $A \cdot HCl \cdot AlCl_3$, $A \cdot HCl \cdot SbCl_5$ and $A \cdot HBr \cdot AlBr_3$. When using aluminium halogenide as Lewis-type acid one can also obtain the ternary complexes $A \cdot HCl \cdot 2\,AlCl_3$ and $A \cdot HBr \cdot 2\,AlBr_3$ which are salts of the hypothetical acids HAl_2Y_7. The formation of the salts $AH^+ \cdot Al_mY^-_{3m+1}$ with $m > 2$ is not excluded either [20]. For the complexes $A \cdot HCl \cdot 2\,GaCl_3$ see [21].

Ternary complexes of the above type are generally viscous oils or fusible solids of light yellow to red colour characterized by high specific gravity ($\sim$2 g/cm^3 [21, 22]) and high electrical conductivity [7, 23]. Their stability increases as the basicity of the hydrocarbon and the acidity of the system $HY + MY_n$ increase. In case the hydrocarbon remains the same in all the complexes the stability of the latter changes as follows (cf. [20, 22, 24–28]):

$$A \cdot HF \cdot BF_3 < A \cdot HCl \cdot AlCl_3 < A \cdot HBr \cdot AlBr_3$$

$$A \cdot HF \cdot BF_3 < A \cdot HF \cdot PF_5 < A \cdot HF \cdot SbF_5$$

$$A \cdot HCl \cdot AlCl_3 < A \cdot HCl \cdot 2\,AlCl_3 < A \cdot HBr \cdot 2\,AlBr_3$$

Thus, complexes of methylbenzenes with HF and BF_3 decompose into their components at a temperature below 0 °C while many complexes with HBr and $AlBr_3$ as well as with HF and SbF_5 are stable at room and even higher temperature.

Research of the ternary systems $A + HY + AlY_3$ [29] by phase analysis methods has shown that these complexes $A \cdot HY \cdot AlY_3$ are able to "take on" a few more molecules of an aromatic hydrocarbon not necessarily identical to the one contained in the complex [30, 31]. The additional molecules, unless they exceed the original hydrocarbon in basicity, are assumed [23, 25, 32, 33] to be bonded by the cation AH^+ due to solvation. The number of aromatic molecules likely to be added to the solvate shell of the cation is limited (to no more than 5–6), so in adding the excess aromatic hydrocarbon to the salt-like compound $AH^+ \cdot Al_mY^-_{3m+1}$ (or when HY and AlY_3 react with the excess hydrocarbon A) one observes the formation of two phases: the heavy coloured phase representing the solvate $[kA \cdot AH]^+$ $[Al_mY_{3m+1}]^-$ and the light colourless phase the excess of hydrocarbon.

The character of the interaction between salt-like compounds of the type $AH^+ \cdot Al_mY^-_{3m+1}$ and the additional molecules of aromatic hydrocarbon is certain to attract attention. There are good reasons to believe that a definite contribution to this interaction is made by the formation of charge-transfer complexes [25, 34]. This is indicated by the solvates usually having a deeper colour than the parent ternary compounds $A \cdot HY \cdot AlY_3$.

The systems $kA \cdot HY \cdot AlY_3$ ($k > 1$) containing monoalkylbenzenes as well as polymethylated and polyethylated benzenes have recently been studied by the ^{13}C- and ^1H-NMR spectroscopy [35–38].

The ability to bind the additional amount of aromatic hydrocarbons is also characteristic of complexes with other binary acid systems, for example, of the complexes $A \cdot HF \cdot BF_3$. This property of ternary complexes can be used to separate aromatic hydrocarbons from saturated ones and to separate aromatic hydrocarbons differing in their basicity (for the use of complexes with HCl and $AlCl_3$ see [39]; with HF and BF_3 see [40–47]). References recording the formation of the ternary complexes $A \cdot HY \cdot mMY_n$ and of their solvates for hydrocarbons of the benzene series are listed in Table 1.

The ternary complexes $A \cdot HY \cdot mMY_n$ are usually prepared from their components in an inert solvent or without it at low temperatures. Other ways, however, are possible as well. For instance, to avoid the inconvenience of handling HF it is proposed to obtain the complexes $A \cdot HF \cdot MF_n$ by saturating a suspension of $AgMF_{n+1}$ in aromatic hydrocarbon with hydrogen chloride and separating the silver chloride formed [26, 28]:

$$A + AgMF_{n+1} + HCl \rightarrow AH^+ \cdot MF_{n+1} + AgCl\downarrow .$$

In the early 1960's it was described [20, 24, 55–57] that salt-like compounds of aromatic hydrocarbons are σ-complexes, i.e. their cations AH^+ possess the structure of arenium ions. This conclusion was first based on indirect arguments ensuing from the analysis of the AH^+-cation electronic absorption spectra (in particular, from the similarity of the spectra of anthracene and 1,1-diphenylethylene solutions in conc. H_2SO_4 [58]). It also results from the linear dependence of the logarithms of the relative stability constants of $A \cdot HF \cdot BF_3$ complexes on those of the rate constants of electrophilic substitution reaction of the hydrocarbons A [56]. Direct proof of this point of view was obtained from studies into the $A \cdot HY \cdot mMY_n$ complexes and the solutions of aromatic hydrocarbons or their derivatives in various acids (HF, HF + BF_3, HSO_3F and others) by the nuclear magnetic resonance measurements of Dutch investigators [59–61].

NMR spectroscopy has opened unique possibilities of observing the formation of arenium ions in solutions and studying their structure and reactivity. The data

Table 1. Ternary Complexes $A \cdot HY \cdot mMY_n$ and their Solvates

Hydrocarbons (A)	Complexes
C_6H_6	$4A \cdot HBr \cdot 2AlBr_3$ [48]; $6A \cdot HBr \cdot 2AlBr_3$ [33,49] (cf. [24]);
$CH_3C_6H_5$	$A \cdot HF \cdot BF_3$ [26]; $xA \cdot HCl \cdot AlCl_3$ [20,32];
	$6A \cdot HCl \cdot 2AlCl_3$ [32] (cf. [20,31]); $A \cdot HBr \cdot AlBr_3$ [24];
	$A \cdot HBr \cdot 2AlBr_3$ [24,50]; $6A \cdot HBr \cdot 2AlBr_3$ [24,49] (cf. [51]).
1,3-$(CH_3)_2C_6H_4$	$A \cdot HF \cdot BF_3$ [26]; $A \cdot HF \cdot PF_5$ [28]; $A \cdot HF \cdot SbF_5$ [28];
	$xA \cdot HBr \cdot AlBr_3$ [24,25,52]; $A \cdot HBr \cdot 2AlBr_3$ [27];
	$xA \cdot HBr \cdot 2AlBr_3$ [24,25,52].
1,2-$(CH_3)_2C_6H_4$	$A \cdot HCl \cdot 2AlCl_3$ [31].
1,4-$(CH_3)_2C_6H_4$	$3A \cdot HBr \cdot AlBr_3$ [52]; $3A \cdot HBr \cdot 2AlBr_3$ [52].
1,3,5-$(CH_3)_3C_6H_3$	$A \cdot HF \cdot BF_3$ [26]; $A \cdot HF \cdot PF_5$ [28]; $A \cdot HF \cdot SbF_5$ [28];
	$4A \cdot HCl \cdot 2AlCl_3$ [23,53]; $2A \cdot HCl \cdot AlCl_3$ [23]; $A \cdot HCl \cdot 2AlCl_3$ [31,53]
	$4A \cdot HCl \cdot 2AlCl_3$ [31]; $A \cdot HCl \cdot 2GaCl_3$ [21];
	$A \cdot HBr \cdot AlBr_3$ [53] (cf. [24]); $A \cdot HBr \cdot 2AlBr_3$ [22,54];
	$3A \cdot HBr \cdot 2AlBr_3$ [51] (cf. [24]).
1,2,4-$(CH_3)_3C_6H_3$	$3A \cdot HBr \cdot 2AlBr_3$ [51].
1,3,5-$(C_2H_5)_3C_6H_3$	$A \cdot HCl \cdot AlCl_3$ [30]; $A \cdot 3C_6H_6 \cdot HCl \cdot AlCl_3$ [30];
	$A \cdot HCl \cdot 2AlCl_3$ [29,30]; $A \cdot 6C_6H_6 \cdot HCl \cdot 2AlCl_3$ [30];
	$2A \cdot HCl \cdot 2AlCl_3$ [51]; $A \cdot HBr \cdot 2AlBr_3$ [54]; $2A \cdot HBr \cdot 2AlBr_3$ [51];
1,3,5-$(i\text{-}C_3H_7)_3C_6H_3$	$A \cdot HCl \cdot AlCl_3$ [31]; $A \cdot HCl \cdot 2AlCl_3$ [31]; $2A \cdot HCl \cdot 2AlCl_3$ [29].
1,2,3,5-$(CH_3)_4C_6H_2$	$A \cdot HF \cdot BF_3$ [26].
1,2,4,5-$(CH_3)_4C_6H_2$	$A \cdot HF \cdot SbF_5$ [28].
$(CH_3)_5C_6H$	$A \cdot HF \cdot PF_5$ [28]; $A \cdot HF \cdot SbF_5$ [28]; $A \cdot HBr \cdot 2AlBr_3$ [27].
$(CH_3)_6C_6$	$A \cdot HF \cdot PF_5$ [28]; $A \cdot HF \cdot SbF_5$ [28]; $A \cdot HCl \cdot 2AlCl_3$ [53].
	$A \cdot HBr \cdot 2AlBr_3$ [47].

obtained in doing so are discussed in the following sections. Here only are some general points:

Molecules of aromatic compounds normally contain several non-equivalent ring carbon atoms, hence in general the protonation must lead to an equilibrium mixture of isomeric arenium ions. As a rule, however, the equilibrium is sharply shifted to the side of one or two isomers; only the most stable ions are detected by the NMR spectra. For example, when anthracene is protonated the 9-H-anthracenium ion is the only one observed.

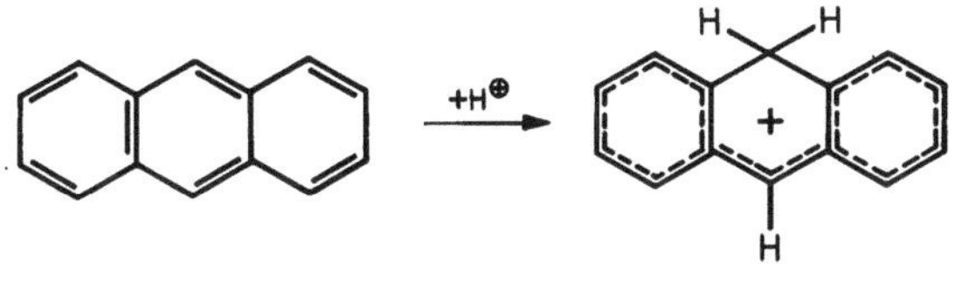

This is easily understood since the energy required to localize the electron pair at the 9-position of anthracene is much lower than for the other positions [11]. Using this criterion one can predict the sites of priority attachment of the proton for other polycyclic aromatic hydrocarbons [4,11].

When substituents in aromatic compounds are able to take part in the delocalization of the positive charge of the π-system via hyperconjugation (alkyls, $-CH_2MR_n$) or conjugation (OR, NR_2, Hal), the arenium ion isomers in which the substituents are located at the carbon atoms carrying larger positive charges turn out to be more stable. Thus, in the case of 9-methylanthracene the proton is attached at the 10-position [4,62].

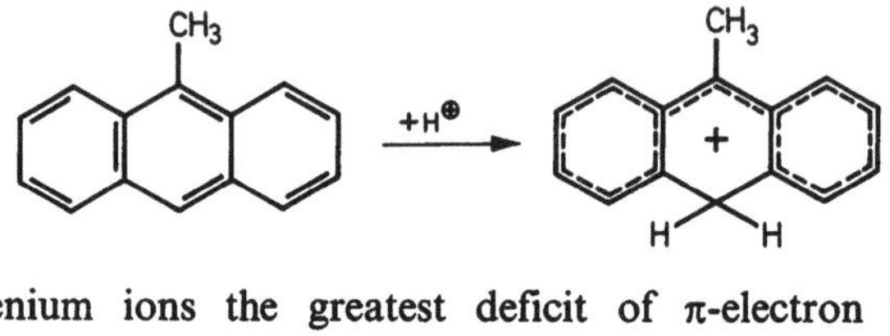

In benzenium ions the greatest deficit of π-electron density, according to molecular-orbital calculations, manifests itself at the 2-, 4-, and 6-positions. Accordingly, in protonating mesitylene, out of two possible isomers, (6) and (7), only 2,4,6-trimethylbenzenium ion (6) is detected by the PMR method

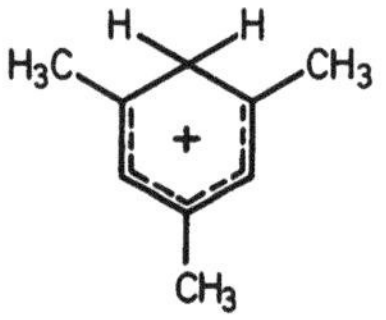

6

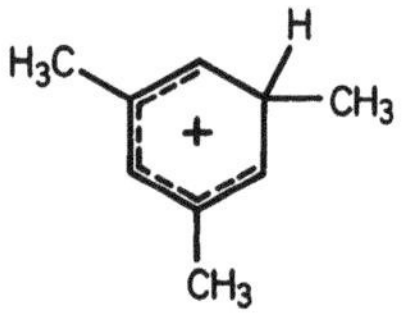

7

On the whole the effect of substituents on the relative stability of isomeric arenium ions (for details see Sect. IV, 1) is described in the same terms as those used to explain the influence of substituents on the orientation and relative rates of electrophilic aromatic substitution. However, the isomeric composition of electrophilic substitution products is often controlled by kinetic factors while the equilibrium composition of isomeric arenium ions formed in aromatic compound protonation is determined by thermodynamic equilibrium. Therefore, no quantitative agreement may be observed between the relative hydrogen substitution rates at different positions of this compound and the ratio of equilibrium concentrations of the respective arenium ions formed in protonating the same compound even under identical conditions (cf. Sect. IV, 7).

When monosubstituted derivatives of benzene with electron-releasing substituents are protonated under conditions ensuring their full conversion into benzenium ions the NMR spectra usually show the formation of only the ion corresponding to the attachment of the proton at the para position:

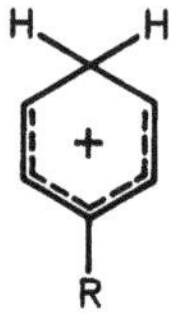

R = CH_3 [63–66], C_2H_5 [66], $CH_2Si(CH_3)_3$ [66]
F [67], HO [68,69], CH_3O [34,63,68,70,71]

As a rule, the detection of the isomer corresponding to the proton attachment at the ortho position is hindered by the fact that the isomer is formed in smaller amounts and under the ordinary conditions of spectrum recording its signals, due to proton exchange, may merge with those of the principal isomer. This can be exemplified by the ortho protonation of toluene and ethylbenzene [72].

Electron-withdrawing substituents of -M type must diminish the basicity of the aromatic ring, naturally hinders direct observation of arenium ions due to the reduction of their stability. Besides, the -M type substituents are basic centres[1] themselves, so in similar cases the proton is often attached to the substituent rather than to the aromatic ring. This has been observed, for instance, upon protonation of acetophenone, benzophenone and their derivatives [74–76], benzoic acid [74], nitrobenzene [77] and nitromesitylene [74]. The formation of arenium ions from compounds containing -M type substituents can only be observed when simultaneously the molecule contains strong electron-releasing substituents. An example is the formation of ion (8) when 2,4,6-trihydroxybenzoic acid is dissolved in 70% $HClO_4$ or HSO_3F [68].

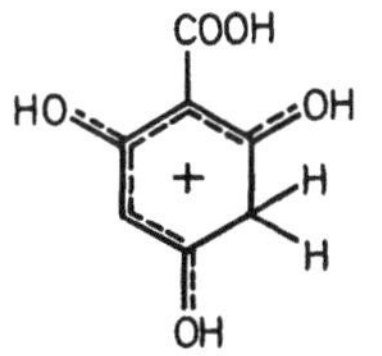

8

For a similar protonation of 2,4,6-trimethylbenzoic acid see [78].

The basicity of a compound to be protonated imposes certain restrictions on the choice of an acid system to be used for generation and direct observation of the appropriate arenium ion. For example, hexamethylbenzene is completely protonated in 100% H_2SO_4. To observe the unsubstituted benzenium ion by the NMR method one of the strongest acid systems, $HF—SbF_5$, was necessary.

In these strong acid systems one can often observe diprotonation of aromatic compounds. Thus, bimesityl and decamethylbiphenyl in $HSO_3F—SbF_5$ attach the protons at 3- and 3'-positions [79] while 1,2,3,5,6,7-hexamethyl- and octamethylnaphthalenes, at the 1- and 5-positions [80, 81]; from octamethylnaphthalene two isomeric ions are formed corresponding to cis- and trans-diprotonation:

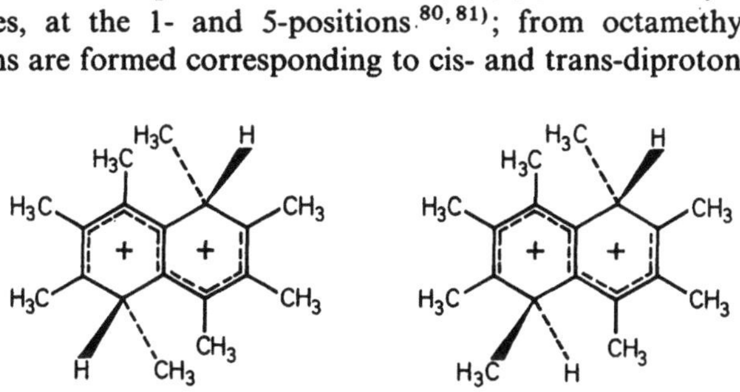

1 The basicity of different functional groups can be judged by the values of pK_a of aliphatic derivatives: CH_3NO_2 (−11.9), CH_3CN (−11), CH_3COCH_3 (−7.2), $CH_3COOC_2H_5$ (−6.5), CH_3SH (−6.8), CH_3OH (−2.0), CH_3NH_2 (+10.6) [73].

Diprotonation has also been observed for some di- and trimethylnaphthalenes in the $HF—SbF_5$ system [82]; depending on the arrangement of the methyl groups one can observe the 1,5-, 1,8-, and 1,6-attachment of the protons.

Data on the relative basicity of a number of aromatic hydrocarbons and their alkyl derivatives were obtained by the Dutch researchers [83].[2] They measured the coefficients of hydrocarbon distribution between an inert solvent (n-hexane) and liquid hydrogen fluoride, containing additives of inorganic fluorides or saturated with BF_3. The protonation of aromatic hydrocarbons in HF is described by the equation:

$$A + 2\,HF \rightleftarrows AH^+ + HF_2^-$$

and characterized by the equilibrium constant K_B:

$$K_B = \frac{[AH^+] \cdot [HF_2^-]}{[A]} \cdot \frac{f_\pm^2}{f_A},$$

where [x] are molar concentrations of respective particles and f are their activity coefficients.

Decreasing the acidity by adding NaF to strong bases and increasing it by introducing BF_3 to weak bases have created a range of values of K_B from 10^{-5} to 10^6. Some of the data are presented in Table 2 [83] (for more complete tables see [4]).

In recent years some experimental data have been accumulated on the basicity of aromatic compounds in the gas phase; they are of great importance, in particular, for refining the parameters of quantum chemical calculations in which it is not yet possible to take account of solvation effects. The gross basicity of aromatic compounds in the gas phase is determined by the method of ion-molecule reactions with mass-spectrometric registration (methylbenzenes [84]) or by that of ion-cyclotron resonance (monoalkylbenzenes [85], methylbenzenes [85][3] trideuteromethylbenzenes [86], halogenbenzenes [87], monosubstituted benzenes with donor and acceptor substituents [88].[4]

Recent publications calculate the basicity of aromatic compounds and the electronic structure of the respective arenium ions by quantum chemical methods in different approximations — by semi-empirical methods MO LCAO (methylbenzenes [92–96]), CNDO, CNDO/2 and CNDO/2FK (benzene [97–100], toluene and other monoalkylbenzenes [98, 100–102], anisole [98], a series of monosubstituted benzenes [103], polymethylbenzenes [100, 104], monomethylnaphthalenes [102, 105] and polycyclic aromatic hydrocarbons [102]); INDO (benzene [99], cresols [106]); MINDO-2′ and MINDO-3 (benzene [99, 107, 108], toluene [108]); by nonempirical (ab initio) methods using the basis STO-3G (benzene [107, 109–111], monoalkylbenzenes [85] a series of monosubstituted benzenes [88, 89], methylbenzenes [112], cresols [106], methylanisoles [113], fluorobenzene

2 Various experimental estimations of the relative basicity of aromatic compounds and the results of quantum chemical calculations are considered in the reviews [4, 7, 11, 16].

3 p-Xylene appears to be subject to ipso-protonation [86].

4 Papers [89, 90] present evidence showing aniline in the gas phase to be protonated at the nitrogen atom rather than at the ring. Phenol [91], just as expected [88], is protonated at the ring.

Table 2. Basicity Constants of Aromatic Hydrocarbons (HF, 0 °C [4,83])

Compounds	log K_B	Compounds	log K_B
Benzene	−9.2	Toluene	−6.3
Biphenyl	−5.5	p-Xylene	−5.7
Triphenylene	−4.6	o-Xylene	−5.3
Naphthalene	−4.0	m-Xylene	−3.2
Phenanthrene	−3.5	Pseudocumene(1,2,4)	−2.9
Chrysene	−1.7	Hemimellitene(1,2,3)	−2.8
Pyrene	+2.1	Durene(1,2,4,5)	−2.2
Anthracene	+3.8	Prehnitene(1,2,3,4)	−1.9
Perylene	+4.4	Mesitylene(1,3,5)	−0.4
Naphthacene	+5.8	Isodurene(1,2,3,5)	+0.1
1-Methylnaphthalene	−1.7	Pentamethylbenzene	+0.4
2-Methylnaphthalene	−1.4	Hexamethylbenzene	+1.4
9-Methylanthracene	+5.7	Hexaethylbenzene	+2.0

[111,114]), the basis 4-31G (benzene [109], fluorobenzene [114]), the basis CGTF (benzene [115], toluene [116]) and other bases (benzene [117,118], fluorobenzene [117,118], chlorobenzene [118]). The nonempirical calculations yield the results which agree rather well with the experimental data for the gas phase [85,88,112,115]. The calculated proton affinities of individual positions of substituted benzenes correlate with σ^+-constants of the substituents [88,119].

The degree of protonation for aromatic compounds in acid systems may be estimated from the data available for some compounds on the values of the acidity function H_0 at which they are half converted into arenium ions:

pentamethylbenzene (log K_B = 0.1)	$H_0 = -9.3(CF_3COOH{-}H_2O{-}BF_3)$ [4]
hexamethylbenzene (log K_B = 1.4)	$H_0 = -9.1(90.5\%\ H_2SO_4)$ [120,121]
benzopyrene (log K_B = 6.5)	$H_0 = -6.9(HF{-}C_2H_5OH)$ [4]
1,3,5-trihydroxybenzene	$H_0 = -3.8$(aqueous $HClO_4$) [122]
1,3,5-triaminobenzene	pH = 5.5(aqueous acids) [123,124]

Intensive development of research on protonation of aromatic and other organic compounds for generating carbocations has led to the search for new acid systems with extremely high protonating ability. Particularly strong acids were $HSO_3F{-}SbF_5$ and $HF{-}SbF_5$ with a different ratio of the components ("magic acids", or "superacids" [125]). The quantitative estimation of the protonating ability of very strong acids presents difficulties, so the data reported by different authors are somewhat different. The values of the acidity function H_0 for the superacids together with the data for some other acids are given in Table 3.

In order to work with acid solutions at as low temperatures as possible (for example when recording the NMR spectra of unstable ions) they are commonly diluted with SO_2, SO_2ClF or SO_2F_2. Thus the $HSO_3F{-}SbF_5$ system allows us to work at temperatures down to −120 °C [136] and the system $HF{-}SbF_5{-}SO_2ClF{-}SO_2F_2$, at as low a temperature as −134 °C [139] (the freezing point of HSO_3F itself is −89 °C [127]). Since the above diluents are very weak bases, the decrease in the acidity of the system after dilution is insignificant [136].

Table 3. H_0-values for Some Strongly Acidic Media

Acidic Media	$-H_0$	Acidic Media		$-H_0$
H_2SO_4	11.9 [126]; 12.1 [127]	$H_2SO_4-SO_3$	(3:1)	13.6 [126]
HSO_3F	13.9 [127]; 15.1 [128]; 14.7 [129]	$H_2SO_4-SO_3$	(1:1)	14.4 [126]
HSO_3Cl	12.8 [130]; 13.8 [126]	HSO_3F-SbF_5	(1:1)	17.5 [136]; ~25 [137]
HF	10.1 [131]; 9.7 [132]	$HF-SbF_5$	(9:1)	20 [138]
CF_3SO_3H	13 [133]; 14.1 [129]	$HF-SbF_5$	(1:1)	>20 [138]
CF_3COOH	3.05 [134]; 2.7 [135]			

The nature of complex acid systems has been investigated by the methods of vibrational spectroscopy and nuclear magnetic resonance. See, for example: $HF-SbF_5$ [140], HSO_3F-SO_3 [141], $HSO_3F-SO_3-SO_2ClF$ [141], HSO_3F-SbF_5 [125, 127, 141-143], $HSO_3F-SbF_5-SO_2$ [125, 127], $HSO_3F-SbF_5-SO_2ClF$ [142], $SbF_5-SO_2-SO_2FX$ (X=F, Cl), $H_2SO_4-SbF_5$ [142], $CF_3COOH-SbF_5$ [144].

When generating arenium ions by the protonation of aromatic compounds some acid systems, along with protonation, cause some other substrate transformations. H_2SO_4, oleum, HSO_3F and systems with these acids may sulphonate aromatic compounds; in the systems $HF-SbF_5-SO_2$ [139-145] and $HSO_3F-SbF_5-SO_2$ [67, 145, 146] aromatic hydrocarbons yield sulphinic acids. In some instances the PMR signals from sulphonation and sulphination products of aromatic compounds have been assigned to arenium ions [28, 65, 147, 148]. In sulphuric acid the polycyclic aromatic hydrocarbons and benzene derivatives with several strong electron-releasing substituents, for example, p-dimethoxybenzene, are oxidized to form radical cations (see Sect. IV.6). SbF_5 is a strong oxidizer, so working with the $A-HX-SbF_5$ systems [28] it is necessary to dissolve aromatic compounds in a previously prepared mixture of HX and SbF_5 in which the fast protonation of the aromatic compound suppresses the oxidizing transformations.

In the protonation of aromatic compounds most of the arenium ions have two hydrogen atoms at the ring sp^3-hybridized atom:

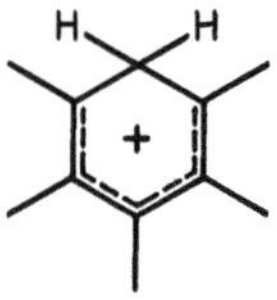

Ions of this type correspond to the intermediates of the electrophilic substitution reaction of one hydrogen atom for the other.

Similarly, aromatic compounds react with deuteroacids to yield salts of arenium ions which are models of the intermediates in isotopic hydrogen exchange reactions [26, 27, 50].

In a few instances the proton is known to be attached to the carbon atom bearing an alkyl group, e.g. the protonation of [2,2]paracyclophane (9) and that of [2,2]metaparacyclophane (10) in the systems $HSO_3F-SO_2ClF-CH_2Cl_2$ and HCl—

$AlCl_3$—SO_2 at low temperatures [149], resulting in the formation of the ions (11) and (12).

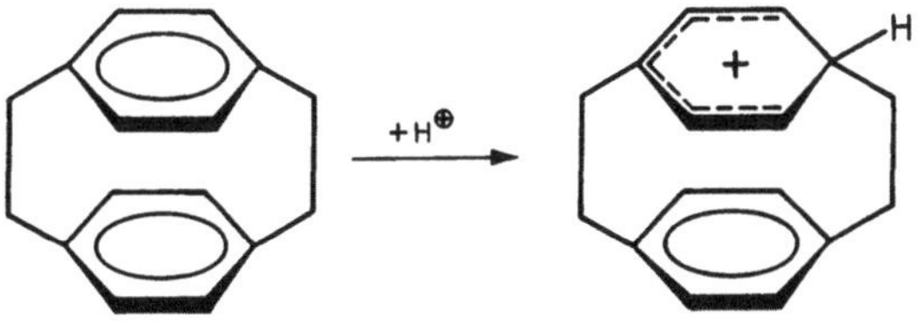

9 11

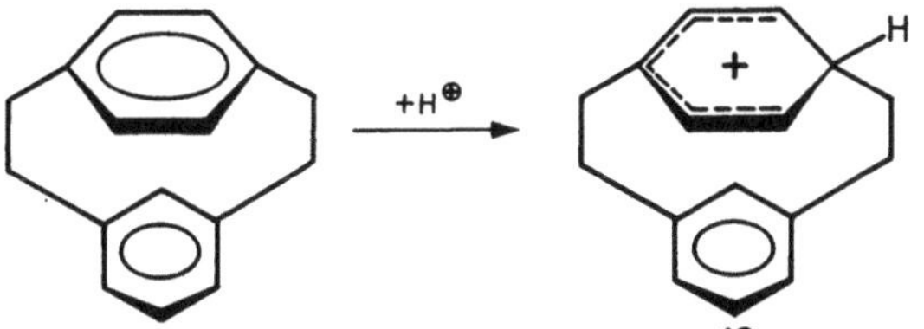

10 12

The preference of attaching the proton to the substituted carbon atom (note that p-xylene is protonated at the 2-position) in these cases is due to the transition of the carbon atom into the sp^3-hybridized state accompanied by a decrease in angular strain.

The formation of ions containing hydrogen and an alkyl group at the ring sp^3-hybridized carbon atom is also observed in the protonation of fully alkylated aromatic compounds (hexaalkylbenzenes [59, 60, 62], octamethylnaphthalene [150] as well as 9,10-dialkylanthracenes and their analogues [59, 60, 62]). Arenium ions of this kind, say, ions (13), (14)

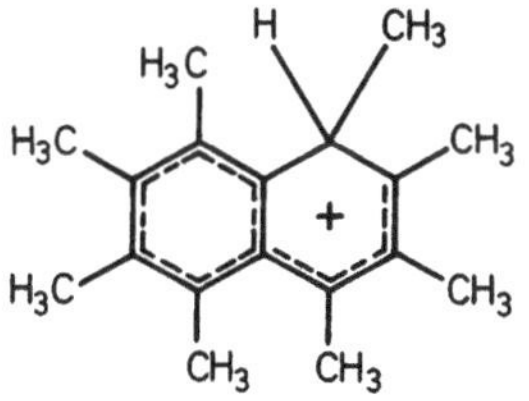

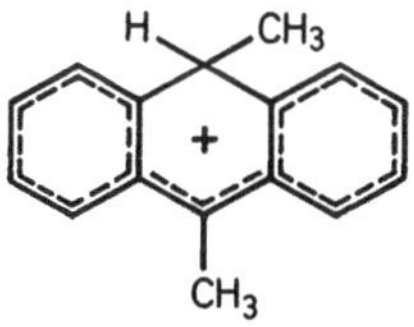

13 14

and (15), correspond to the intermediates in the reactions of electrophilic alkylation of aromatic compounds and can also be generated by interaction of a less alkylated aromatic compound with an alkylhalide and a sufficiently strong Lewis-type acid. For example [28]:

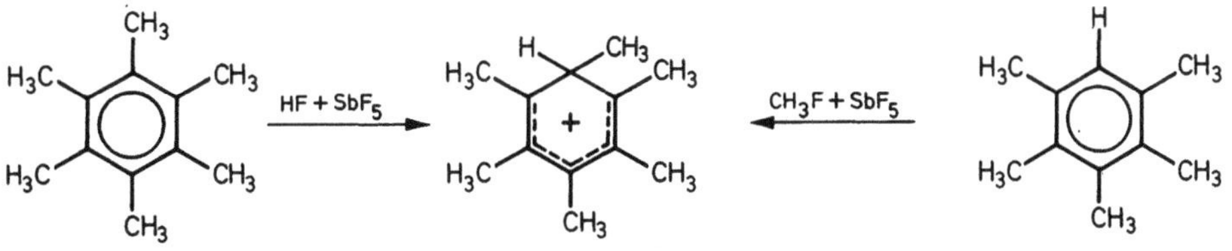

15

Similarly, completely alkylated aromatic compounds interact with alkylating agents in the presence of the Lewis-type acid to yield completely substituted arenium ions, for example, benzenium ions of the types (16) and (17) [3,151–157].

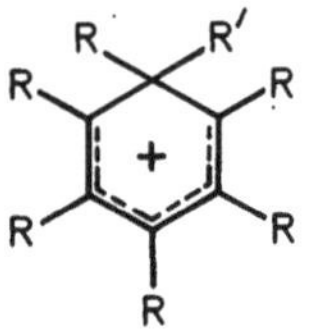

16 $R = CH_3$
a $R' = CH_3$ *b* $R' = C_2H_5$
c $R' = CH_2Cl$ *d* $R' = CHCl_2$

17 $R = R' = C_2H_5$

A number of ternary complexes of mono-, di- and trialkylbenzenes with halogen alkyls and Lewis-type acids have been described: $A \cdot AlkF \cdot BF_3$ [158], $A \cdot C_2H_5Br \cdot 2\,AlBr_3$ [54], $nA \cdot YCH_2Y \cdot 2\,AlCl_3$ [159]. G. Olah and S. Kuhn [158] assumed these complexes to be salts of benzenium ions containing, like the ions (13)–(15), a hydrogen atom and an alkyl group at the 1-position. There are, however, a number of doubts to this point. If in the ternary system A—AlkY—MY_n the alkyl group is attached to the aromatic ring to form an arenium ion of the (18) type the latter must be quickly rearranged, as a result of the hydrogen shift (see Sec. IV.1 and IV.2), into a more stable ion of the (19) type. When the $A \cdot AlkY \cdot MY_n$ complexes are prepared under the same conditions at a very low temperature which would inhibit the rearrangement of the (18) type ion, the interaction of the components seems to be confined to the formation of π-complexes (for the A—AlkF—BF_3 systems see [160–163]).

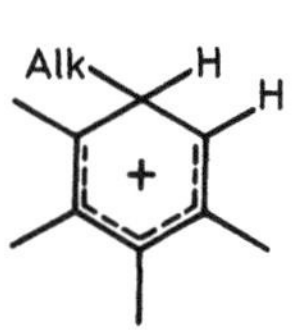

18

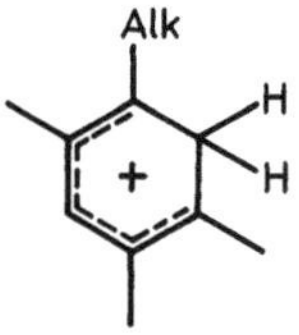

19

Attempts have been repeated to reveal, by physical methods or by separating as salts of aromatic compounds, the formation of arenium ions from electrophilic reagents for nitration, sulphonation, acylation etc.

G. Olah and co-workers [164, 165] obtained, through a reaction of benzotrifluoride with NO_2F and BF_3 at -100 °C, a solid yellow salt the cation of which they considered to be the 1-H-1-nitro-3-trifluoromethylbenzenium ion (20). Later on, however, it was suggested [74] the cation of this salt is 3-nitrobenzotrifluoride protonated at the oxygen of the nitro group (21) (see also [166]).

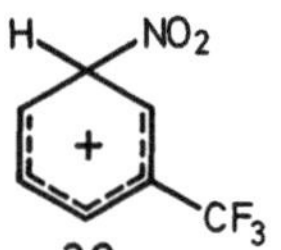

20

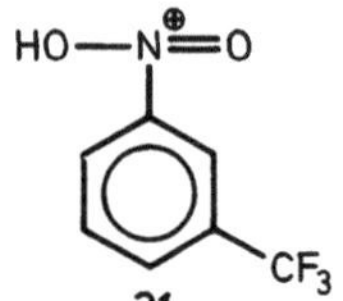

21

When the interaction of pentamethylbenzene with nitronium tetrafluoroborate in polar solvents was studied a coloured adduct was observed [167] which, judging by the way it was obtained, should have been a salt of 1-H-1-nitro-2,3,4,5,6-pentamethylbenzenium ion. But it proved to be a π-complex of pentamethylbenzene and a nitrosonium cation, whose salts are usually present in small amounts in nitronium salts [168].

The difficulty of fixing arenium ions with the hydrogen and a strongly electronegative substituent at the ring sp^3-hybridized carbon atom, is due to the ease with which the proton migrates. The ions from attachment of the nitronium cation and other heteroatomic electrophilic particles to the substituted carbon atom of the aromatic ring are much more stable. Indeed, hexamethylbenzene, by the action of electrophilic agents in strong proton acids at low temperatures generates a number of 1-X-1,2,3,4,5,6-hexamethylbenzenium ions (22). Thus, the interaction of hexamethylbenzene with nitric acid in HSO_3F [169] or with nitronium tetrafluoroborate in HSO_3F [169] and in the $HSO_3F—SO_2$ system [166] forms the 1-nitro-1,2,3,4,5,6-hexamethylbenzenium ion (22a) [170]. The ion (22b) modelling the intermediate σ-complex of sulphonation results from the interaction of hexamethylbenzene with sulphur trioxide in HSO_3F [171], and the ion (22c) from that with Cl_2 in the system $HSO_3F—SbF_5—SO_2ClF$ [166]. Interestingly, the presence of a strong acid protonating hexamethylbenzene to form the ion (15) does not inhibit the formation of 1-X-1,2,3,4,5,6-hexamethylbenzenium ions. This seems to be due to the presence of very small amounts of free hexamethylbenzene in equilibrium with the protonated form.

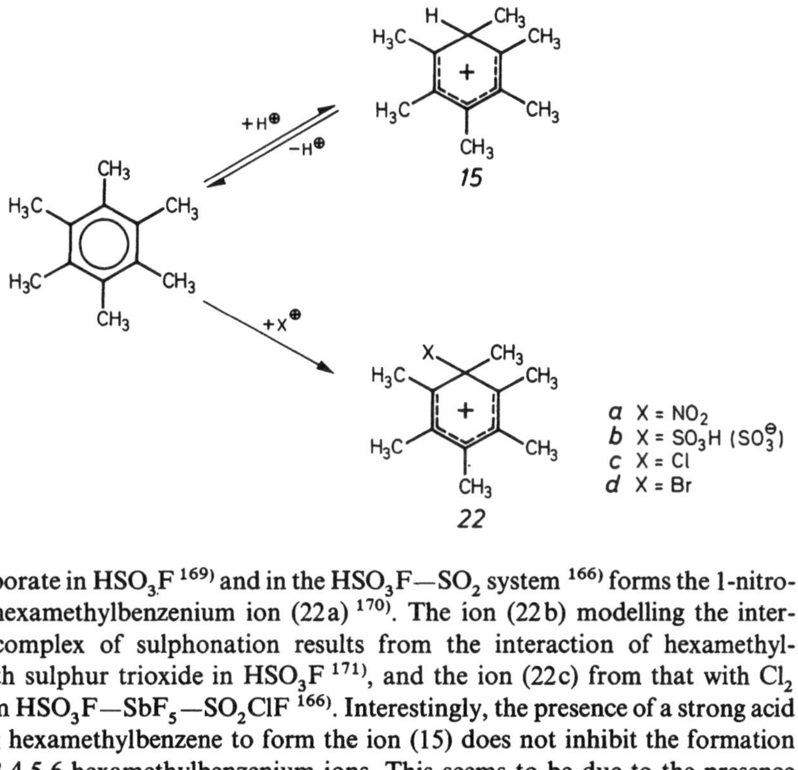

The ion (22c) is also formed from hexamethylbenzene with chlorine and $AlCl_3$ in CH_2Cl_2 at −90 °C [172]. The same method has been used to generate the 1-bromo-1,2,3,4,5,6-hexamethylbenzenium ion (22d).

The formation of arenium ions from heteroatomic electrophilic reagents was observed by the NMR method for other polysubstituted aromatic compounds as well. Thus, from pentamethylphenol, pentamethylanisole, pentamethylhalogenobenzenes, pentamethylnitrobenzene, pentamethylbenzamide and trifluoromesitylene, the following ions have been generated [166, 169, 173–176]:

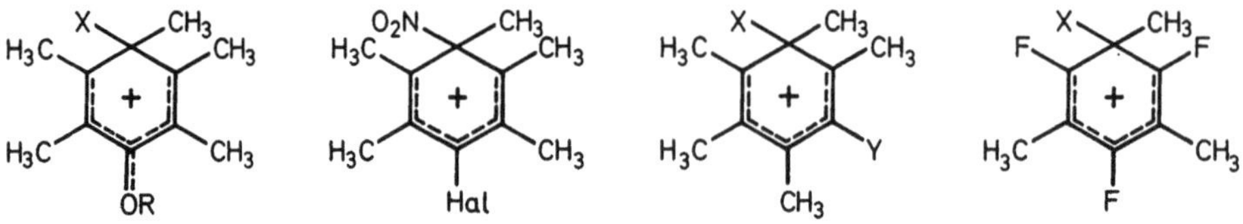

X = NO_2, Cl, SO_2Y X = NO_2, Cl; Y = NO_2, = $CONR_2$ X = NO_2, Cl

The addition of Cl^+ to the carbon atom bearing a CH_3 group was demonstrated by NMR when 9,10-dimethylphenanthrene and 1,2-dimethylacenaphthylene interacted with chlorine in strong acids [177].

Hexafluorobenzene, octafluoronaphthalene and a number of other polyfluorinated aromatic compounds reacted with chlorine in SO_2ClF at 60 °C in the presence of SbF_5 to give the ions corresponding with the addition of Cl^+ to the carbon atom bonded with the fluorine atom [178–180]:

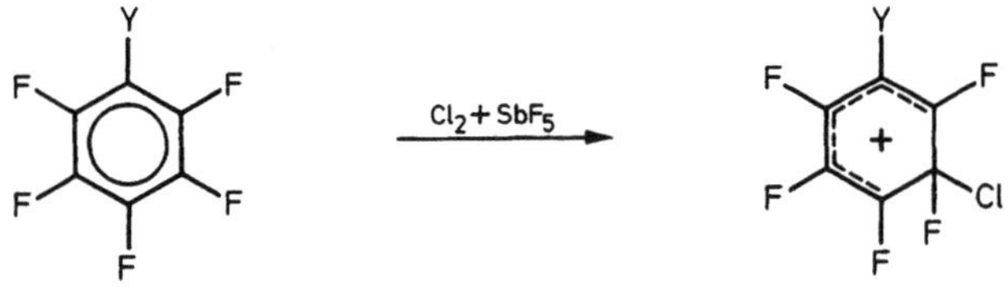

Y = F, CH_3, C_6F_5

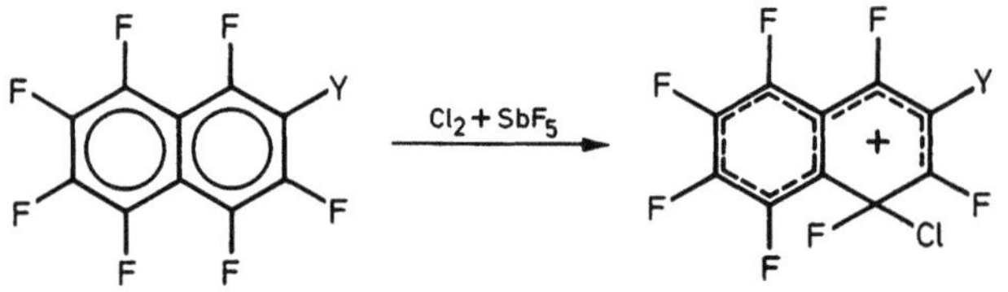

Y = H, F, CH_3

Similarly, by a reaction with CH_3F in the presence of SbF_5 one generates ions corresponding with the addition of the CH_3^+ cation to polyfluoroaromatic compounds [19, 181]. Judging by the reaction products from polyfluorinated aromatic compounds and nitrating agents, the nitronium cation is also capable of adding to the carbon atom linked with fluorine; however, no arenium ions with F and NO_2 at the ring sp^3-hybridized carbon atom have so far been observed, seemingly due to their further conversions in the presence of SbF_5 [19].

Of great interest is the possibility of fixing the arenium ions from the addition of a heteroatomic electrophilic particle to the unsubstituted carbon atom. As mentioned above, it is difficult to observe such ions since they decompose eliminating the proton. The detachment of the proton from σ-complexes in reactions of hydrogen electrophilic substitution, however, is known to be hindered sterically if the normal arrangement of the incoming substituent in the plane of the aromatic ring of the product is prevented. This occurs in a lower substitution rate of deuterium or tritium than that of protium. A kinetic isotope effect was observed, for example, in bromination of 1,3,5-tri-t-butylbenzene [182] (see also [183, 184]), nitration of 1,3,5-tri-t-butyl-2-methylbenzene with a nitrating mixture in nitromethane ($k_H/k_D = 3.7$) and in nitration of anthracene with nitronium tetrafluoroborate in sulpholane ($k_H/k_D = 2.6$) or acetonitrile ($k_H/k_D = 6.1$) [185].

One can hope to stop the reaction of hydrogen electrophilic substitution in such cases at the stage of arenium ion formation. Anthracene and nitronium tetrafluoroborate reacted in sulpholane to form a complex which was converted into 9-nitroanthracene [186]. It was assumed that this complex is the tetrafluoroborate of the 9-nitroanthracenium ion (23). However, more recent data on reactions of methylbenzenes with nitronium tetrafluoroborate [168] do not support this conclusion.

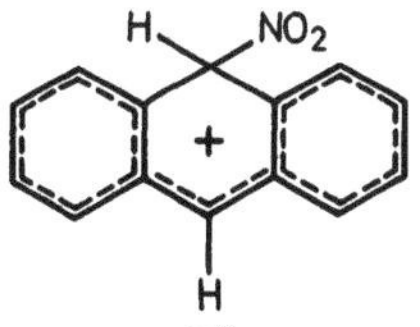

23

Pentamethylbenzene is shown [173] to react with nitric acid in HSO_3F at —80 °C to yield 1-nitro-1,2,3,4,6-pentamethylbenzenium ion (24) which, when the temperature rises to —60 °C, turns readily into pentamethylnitrobenzene as a result of nitrogroup shift and proton detachment. The PMR spectra have so far failed to show any intermediate formation of the ion (25).

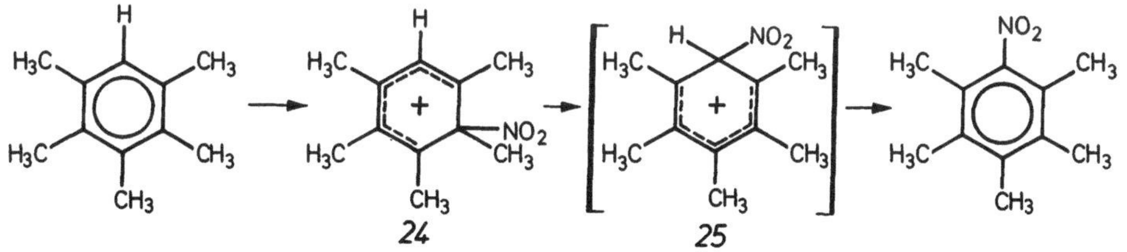

In studying the reaction between pentamethylbenzene and mercury trifluoroacetate in CF_3COOH by UV- and PMR spectroscopy [187] indications were obtained of an equilibrium mixture of the ions (26) and (27)

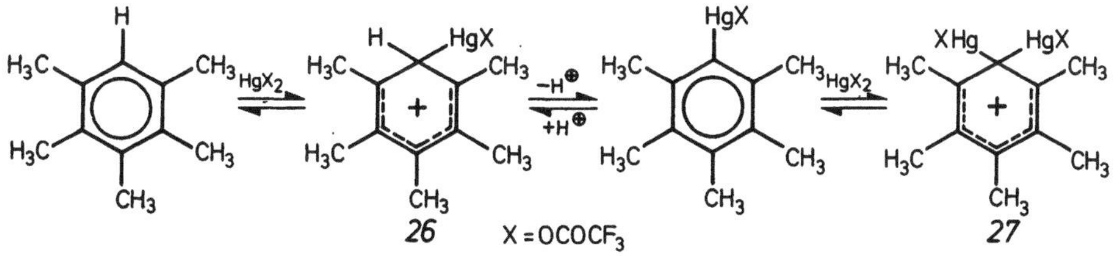

Since, however, no splitting resulting from the interaction with the isotope ^{199}Hg was observed for the signal of a single proton, the authors [187] do not exclude a different interpretation of the spectroscopic data.

Nor did the analysis of the values of $J_{^{13}C-^{1}H}$ and the position of signals in the NMR-^{13}C spectra allow one to draw unequivocal conclusions on the character of the complexes resulting from the reaction of benzene derivatives with $Hg(OCOCF_3)_2$ [188]. The authors [188] consider the data obtained to conform with the formation of σ- and π-complexes undergoing quick interconversions.

For the studies of aromatic complexes of hydrocarbon with $Hg_2(AsF_6)_2$ see [189].

In analyzing the IR spectra of the binary systems methylated benzene-$GaCl_3$ recorded at —200 °C bipolar σ-complexes of the type (28) were concluded to have

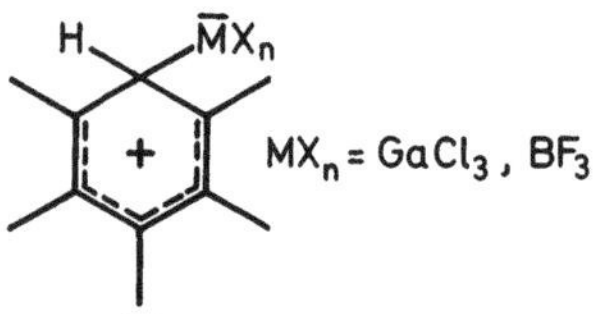

28

formed [190]. The formation of similar complexes was also assumed in discussing the electronic absorption spectra of the systems anthracene-BF_3 [191], anthracene-$AlCl_3$ [192], anthracene-$ZnCl_2$ [193], tetracene-BF_3 [191] (see [16, 194]). The measurement of the dipole moments for the complexes of mesitylene with aluminium bromide and gallium chloride and the analysis of the PMR spectra, however, permit one to believe [195] that the interaction of aromatic hydrocarbons with Lewis-type acids is confined to the formation of π-complexes.

Interesting results have been obtained from the reactions between 1,3,5-tris-(dialkylamino)benzenes and electrophilic reagents. The combination of the strong electron-releasing influence of dialkylamino groups and the steric hindrance created by them to the entry of the substituting group into the ring plane when the proton breaks off, made it possible to obtain the salts of the stable benzenium ions (29) corresponding to the intermediates in electrophilic substitution of hydrogen by alkyl, acyl, sulphonyl, rodan groups or halogens [196–200].

29

X = CH_3, C_2H_5, $(CH_3)_2CH$, $(C_6H_5)_2CH$, Cl, Br, CH_3CO, C_6H_5CO, CH_3SO_2, $ArSO_2$, SCN

The protonation of 4-X-2,3,5,6-tetramethylphenols (X = Br [201], SO_2F [202]) and the reactions between the donors of X^+ and durenol in strong acids (X = Br, SO_3H [202])

generate a number of 1-H-1-X-4-hydroxy-2,3,5,6-tetramethylbenzenium ions:

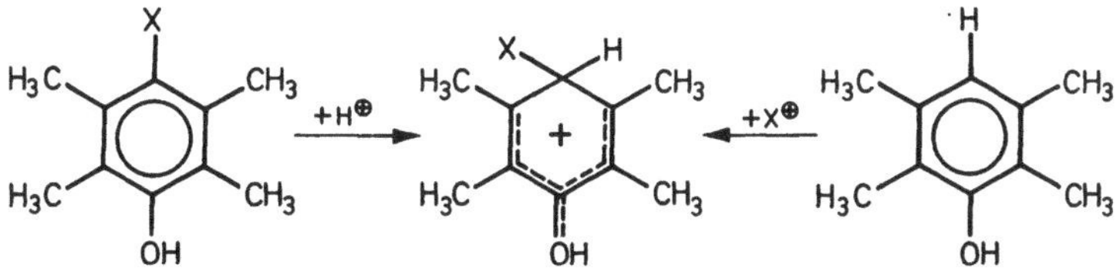

Similarly, the protonation of a number of substituted 1- and 2-naphthols and their methyl ethers resulted in the formation of ions corresponding to σ-complexes of electrophilic substitution reactions:

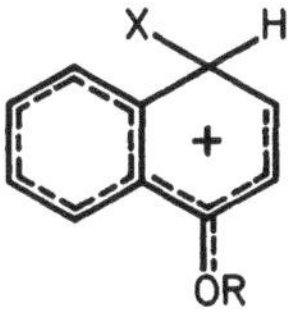

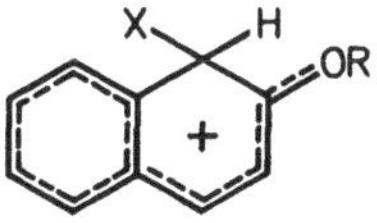

X = CH_3, R = H [203)]
X = Br, R = H [203, 204)]

X = Br, R = H, CH_3 [204, 205)]
X = Cl, R = CH_3 [204, 205)]
X = SO_3H, R = H, CH_3 [206)]

Rather unexpected is the communication [180)] that in the reaction of 1-H-heptafluoronaphthalene with chlorine in SO_2ClF in the presence of SbF_5 at —80 °C the NMR spectrum indicates the formation of the 1-H-1-chloroheptafluoronaphthalenium ion:

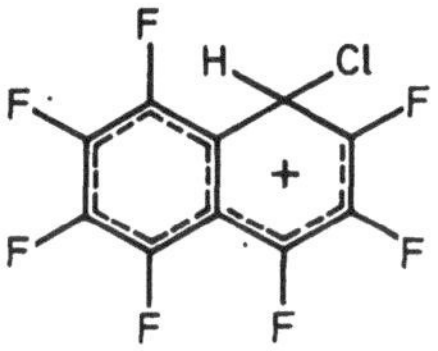

when the solution is poured into CH_3OH, a proton splits off to form 1-chloroheptafluoronaphthalene. For pentafluorobenzene such an ion could not be identified — in this case chlorpentafluorobenzene is formed even in the acid system.

Besides the above-considered interactions of aromatic compounds with electrophilic agents to generate arenium ions, there are a number of other ways. In 1958 E. Döering et al. [3)] showed 4-methylene-1,1,2,3,5,6-hexamethylcyclohexa-2,5-diene (30) to be a very strong base (pK_a = 1.38 [207)]), and, unlike polymethylbenzenes, to be readily solved in hydrochloric acid yielding the heptamethylbenzenium ion (16a).

30 *16a*

This procedure was later used to generate a number of other 1-R-1,2,3,4,5,6-hexamethylbenzenium ions (R=C_2H_5, CH_2Cl [152,153], C_6H_5 [208]), heptaethylbenzenium ion [151,157] (cf. [153]), 4-β-X-ethyl-1,1,2,3,5,6-hexamethylbenzenium ions with X = $\overset{+}{N}(CH_3)_3$, $\overset{+}{P}(C_6H_5)_3$ and other groups [209], polymethylated ions of the naphthalene series [210], 9-R-9,10-dimethylphenanthrenium ions (R = CH_3 [211], C_2H_5 [212], aryl [213]), 9,9,10-triethylphenanthrenium ion [214] (see also [215–217]) and 9,9,10-trimethylanthracenium ion [153]. 9,9-Dimethyl-10-methylene-9,10-dihydrophenanthrene is half protonated in 36% H_2SO_4 [207].

Similarly, the protonation of substituted cyclohexadienones in strong acids (conc. H_2SO_4, HSO_3Cl, HSO_3F, HCl + $AlCl_3$ etc.) results in hydroxybenzenium ions:

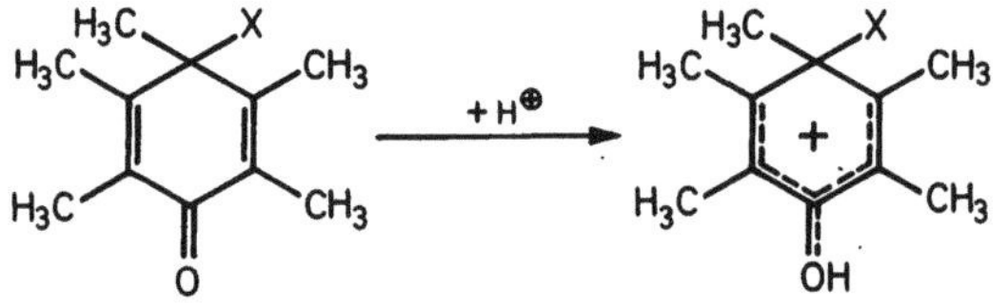

31

X = CH_3 [218,219]; F, Cl, Br [220]; OH, OCH_3 [221]; $OCOCH_3$ [222]; NO_2 [223]

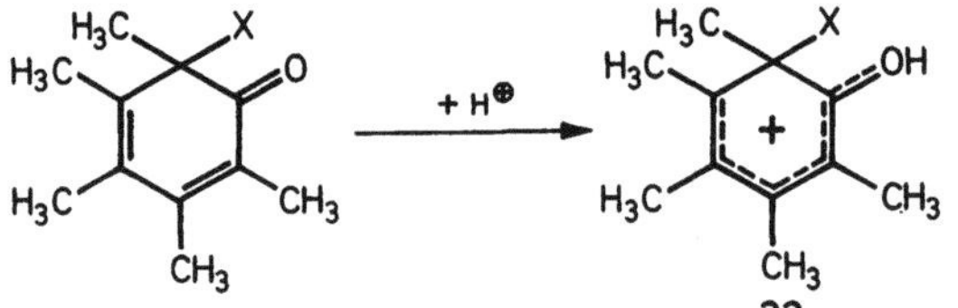

32

X = CH_3 [218,219]; $OCOCH_3$ [222]; NO_2 [223]

As mentioned earlier, some of the ions (31) can be obtained in a different way — by adding electrophiles to pentamethylphenol.

There are indications that cyclohexadienones are not Hammett bases [224–226] and that in many cases their protonation is better described by the amide acidity function H_A [226,227]. Therefore the basicity of cyclohexadienones is generally characterized either by the value of pK_a in the scale H_A or by that of $(H_0)_{1/2}$ — the value of the Hammett acidity function at which the cyclohexadienone is half converted into a conjugate acid — the hydroxybenzenium ion. The available data on the basicity of cyclohexa-2,5-dienones are summarized in Table 4. In [227] a comparison is drawn between the basicity of cyclohexa-2,5-dienones and that of cyclohexa-2-enones.

According to [232] the basicity of alkyl-substituted cyclohexa-2,5-dienones is described by the following equation:

$$-pK\,(25\,°C) = 2.66 \pm 0.12 + 2.32\,\Sigma\,\sigma^+$$

the σ^+ ortho constants being used for the substituents at 3- and 5-positions, while the σ^+ meta constants, for those at 2- and 6-positions. The character of the alkyl

Table 4. Basicity of Cyclohexa-2,5-dienones

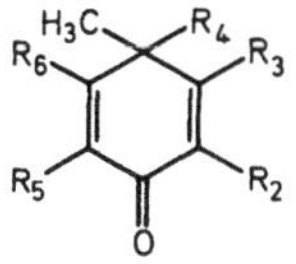

Substituents					pK_a^A	$(H_0)_{1/2}$
R_4	R_3	R_5	R_2	R_6		
CH_3	H	H	H	H	−2.37 [226, 227]	−3.8 [228]; −3.66 [225] −3.15 [227]; −3.24 [229]
C_2H_5	H	H	H	H	−2.26 [230]	−2.96 [230]
n-C_3H_7	H	H	H	H	−2.43 [230]	−3.32 [229, 230]
CH_3	H	H	H	H		−6.20 [229]
$CHCl_2$	H	H	H	H		−5.52 [224, 225]
CCl_3	H	H	H	H		−6.12 [231]
OCH_3	H	H	H	H		−6.02 [231]
CH_3	CH_3	H	H	H	−2.01 [226, 227]	−2.44 [227]
CH_3	C_2H_5	H	H	H	−1.92 [226] −1.97 [227]	−2.39 [227]
CH_3	H	H	CH_3	H	−2.7 [232]	−3.27 [232]; −3.36 [229]
CH_3	CH_3	CH_3	H	H	−1.38 [226, 227]	−1.56 [227]
$CHCl_2$	CH_3	CH_3	H	H	−2.31 [226, 227]	−3.3 [224]; −2.99 [227]
CH_3	CH_3	H	H	CH_3	−1.86 [227]	−2.25 [227]
CH_3	H	H	CH_3	CH_3	−4.2 [232]	−4.20 [232]; −4.13 [229]
CH_3	CH_3	CH_3	CH_3	CH_3 [a]		−2 [219]

[a] For isomeric hexamethylcyclohexa-2,4-dienone $(H_0)_{1/2} \approx -2$ [219]

substituent located at the sp^3-hybridized carbon atom does not essentially affect the value of pK. If both positions ortho to the oxygen function are occupied by the substituents, the calculated value of pK must be decreased by 1.2 units due to the steric inhibition to the OH group in the ring plane; for one substituent at the above positions the correction amounts to 0.3 units (allowance for the statistical factor).

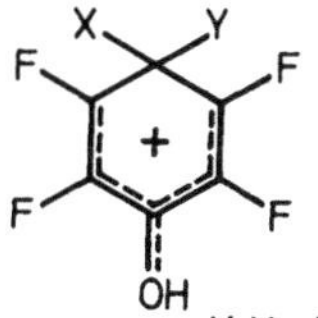

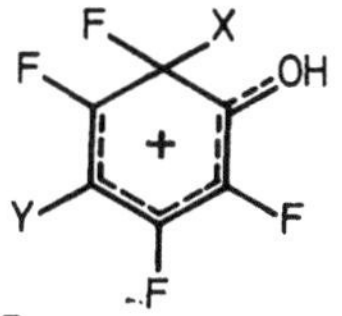

X,Y = F, F; F, Cl; F, Br;
CH_3, Cl; CH_3, Br

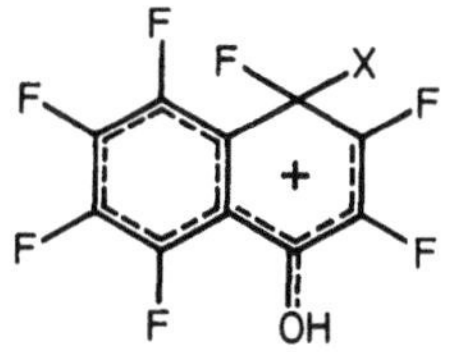

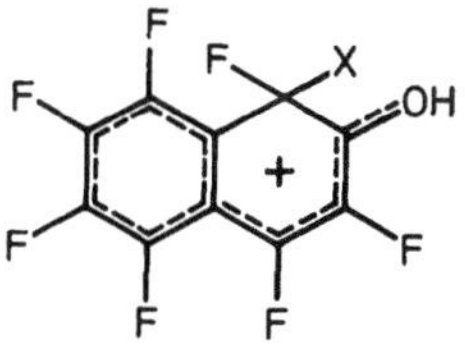

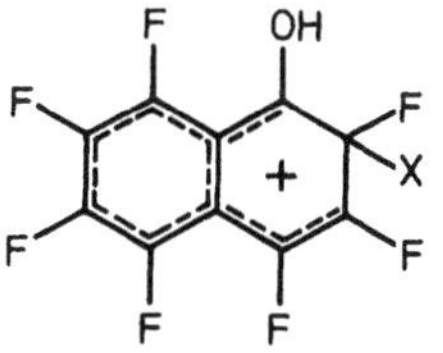

X = F, Cl, Br

By the protonation of polyfluorinated cyclohexadienones in the $HSO_3F—SbF_5$ system a large number of polyfluorinated hydroxyarenium ions have been generated [233–235].

Carbonyl derivatives of dihydroaromatic compounds can be used to generate polycyclic hydroxyarenium ions, e.g. [59,60]:

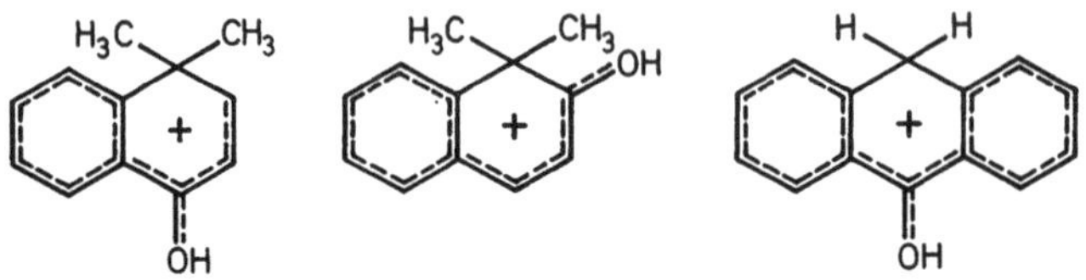

Another way of generating arenium ions based on dihydroaromatic compounds is to break off, with a pair of electrons, a substituent located at one of the ring sp^3-hybridized carbon atoms. G. Olah et al. [236] have used this method to obtain tetrafluoroborate of the methylbenzenium ion by the action of $AgBF_4$ on the bromination product of 1-methylcyclohexa-1,4-diene at —60 °C.

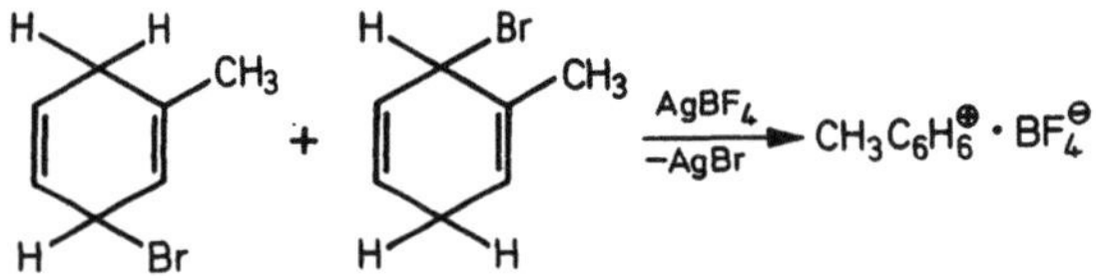

Polyfluorinated cyclohexadienes reacted with SbF_5 to form hexafluoroantimonates of the heptafluorobenzenium ion (Y = F) and close analogues [237–239]:

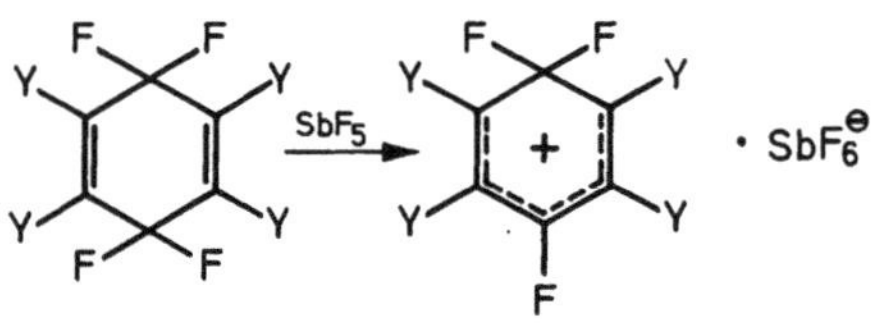

The same procedure was used to generate perfluoro-α-naphthalenium ion [237,238]:

its analogues [239], as well as the 9,9,10-trifluoroanthracenium ion and its derivatives [240,241].

If an eliminated substituent in a dihydroaromatic compound is an OR group, then the respective arenium ion can be generated by dissolving this compound in a strong protonic acid. Thus, the dissolution of 4-hydroxy-4-R-hexamethylcyclohexa-2,5-

dienes in HSO_3F at —70:—90 °C yields 4-R-1,1,2,3,5,6-hexamethylbenzenium ions.

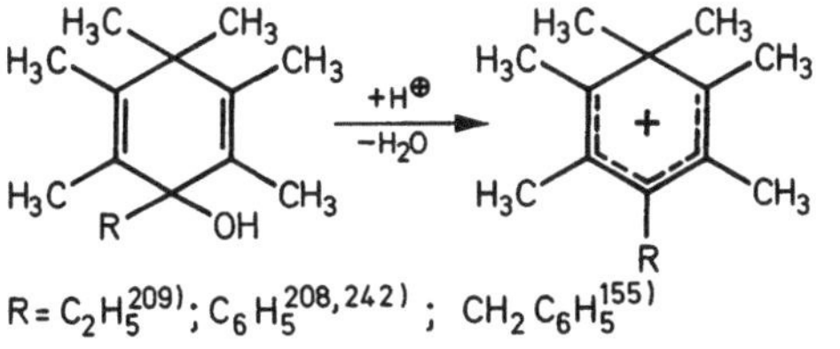

Similarly, 1,4-dimethoxyhexamethylcyclohexa-2,5-diene in HSO_3F at —75 °C yields the 1-methoxy-1,2,3,4,5,6-hexamethylbenzenium ion [243]:

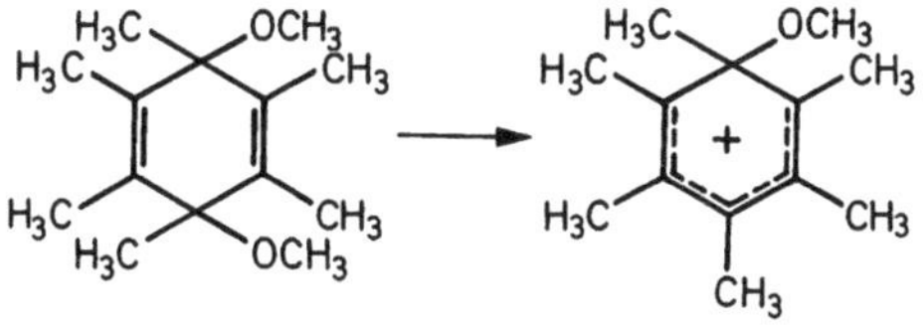

while 1-hydroxy-6-benzyl-1,2,3,4,5,6-hexamethylcyclohexa-2,4-diene gives the 1-benzyl-1,2,3,4,5,6-hexamethylbenzenium ion [244]:

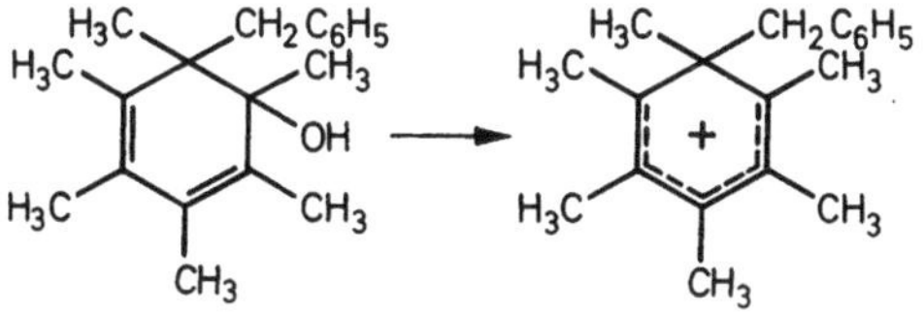

The leaving group may also be a hydrogen atom (elimination of the hydride-ion under the action of $SbCl_5$ [245], carbocation salts [246] etc.):

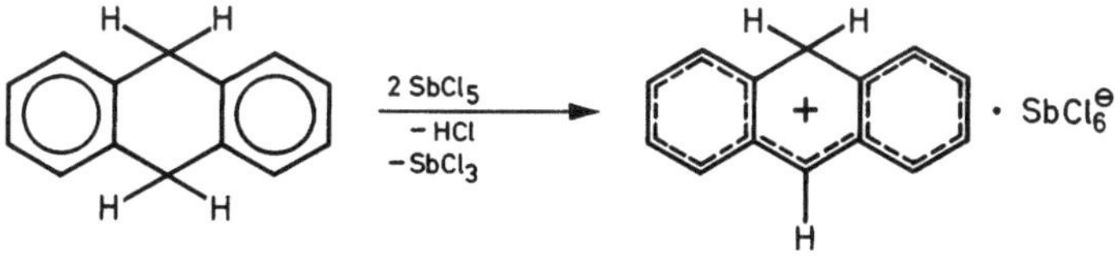

and an aromatic molecule [247]:

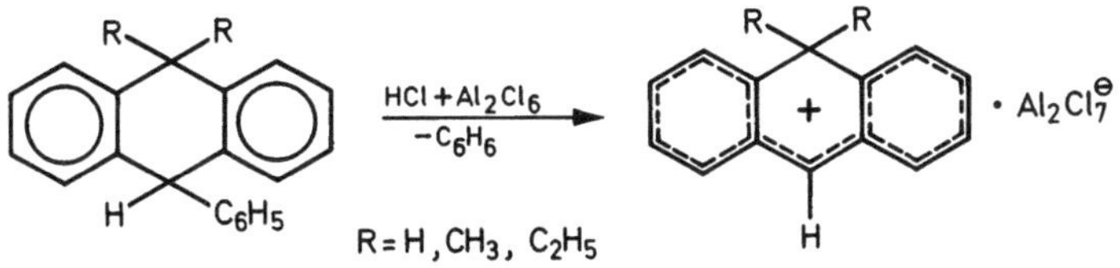

Arenium ions can be formed as a result of rearrangement of other, less stable carbocations. Thus, the 9-t-butylfluorenyl cation is generated by dissolving the corresponding chloride at −110 °C in a mixture of SbF_5 and SO_2ClF or by other methods, then as the temperature rises to −70 °C, it is smoothly converted into the 9,9,10-trimethylphenanthrenium ion [248–250] (see also [177,251,252]).

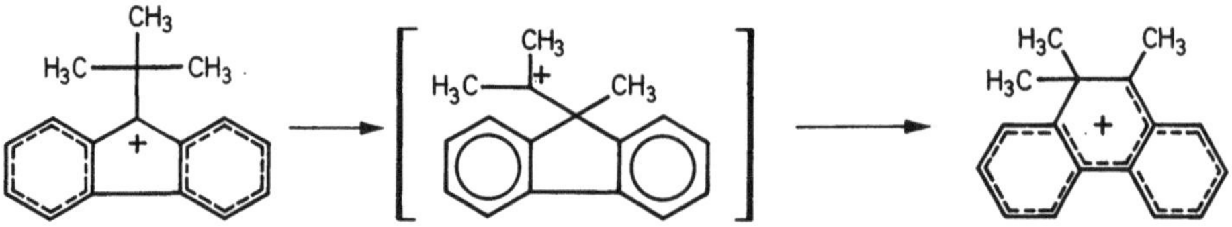

Another example is formation of benzenium ions from their valence isomers – bicyclo[3,1,0]hexenyl cations:

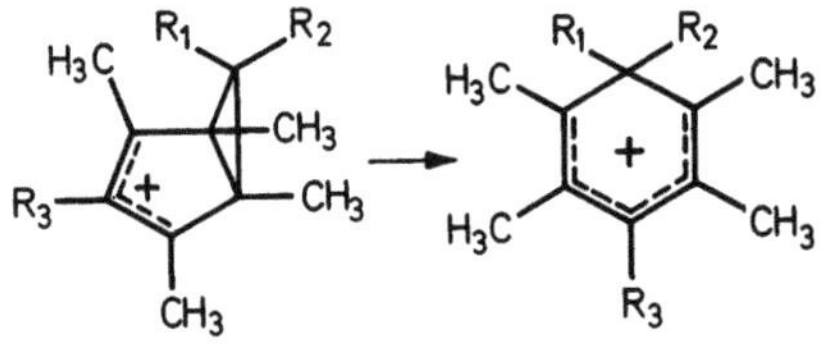

$R_1 = R_2 = H$, $R_3 = CH_3$ [253,254]; $R_1 = R_2 = R_3 = H$ [255]; R_1 and R_2 = H and CH_3, $R_3 = CH_3$ [253,254,256]; $R_1 = R_2 = R_3 = CH_3$ [253,254]; R_1 and R_2 = CH_3 and C_2H_5, $R_3 = CH_3$ [254]; R_1 and R_2 = CH_3 and CH_2Cl, $R_3 = CH_3$ [254].

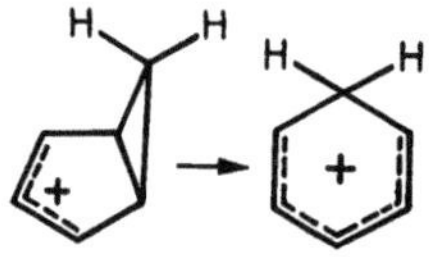 257)

and bicyclo[2,1,1]hexenyl cations:

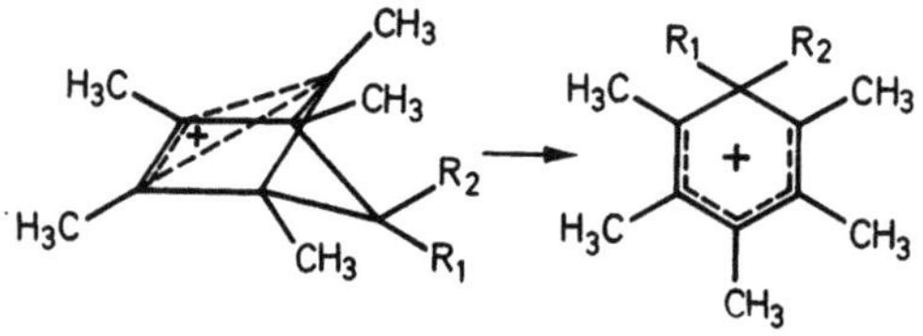

R_1 and R_2 = H and CH_3 [258]; R_1 = Cl or Br, $R_2 = CH_3$ [172,259]

The 2-isopropyl-5-methyl-4-hydroxybenzenium ion has been obtained [260] from the following transformations:

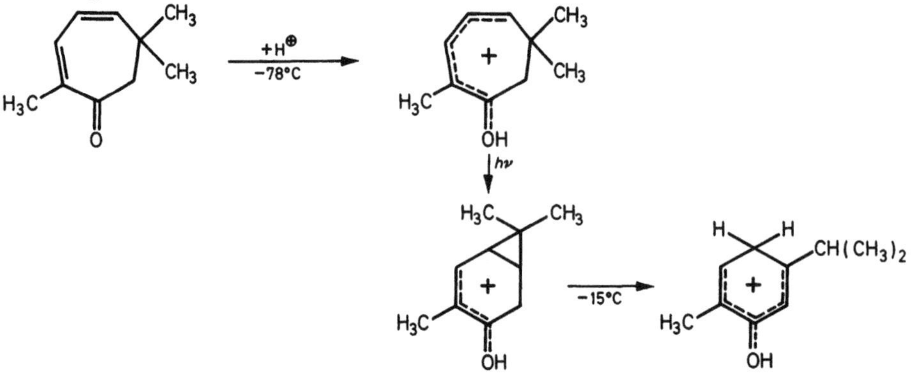

The isomeric transformations of arenium ions themselves due to 1,2-shifts of the substituents can also be used to obtain new ions. Thus, the 1-methoxy-1,2,3,4,5,6-hexamethylbenzenium ion formed by dissolving 1,4-dimethoxyhexamethylcyclohexa-2,5-diene in HSO_3F at —75 °C is wholly converted, as the temperature rises to —30 °C, into the 6-methoxy-1,1,2,3,4,5-hexamethylbenzenium ion which at —10 °C is rearranged into the 4-methoxy-1,1,2,3,5,6-hexamethylbenzenium ion [243].

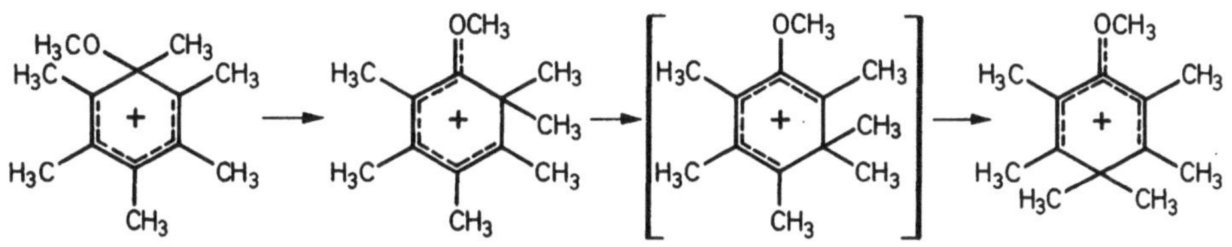

Other examples of isomeric transformations of arenium ions and data on their regularities can be found in Sect. IV.

Rather interesting is the possibility of transition from benzyl-type cations to benzenium ions. The benzyl cations can, just as other carbenium ions, act as hydride-anion acceptors turning, in so doing, into an aromatic hydrocarbons which, in a strong acid, are protonated to form the benzenium ions. Similar transformations occurred in the case of the 2,3,4,5,6-pentamethylbenzyl cation, the hydride-anion donors being the aliphatic hydrocarbons, 9,10-dihydroanthracene and even the molecular hydrogen [242, 261–264]:

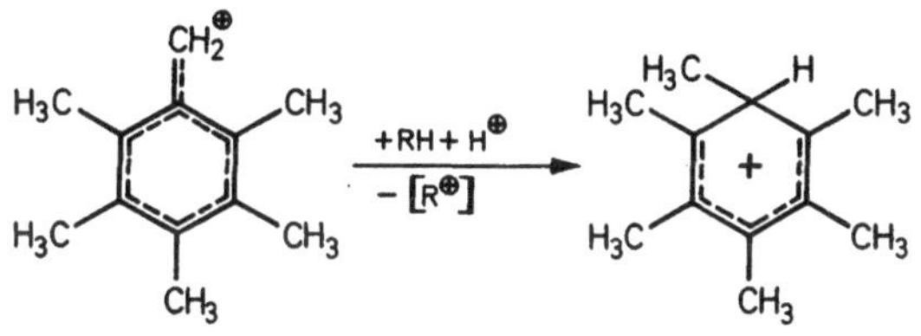

Analogous conversions are described for the 4-X-2,3,5,6-tetramethylbenzyl cations (X = H, CH_3, F, Br, OCH_3, NO_2) [263] (cf. [265]).

For the 4-hydroxy-2,3,5,6-tetramethylbenzyl cation which can be regarded as the conjugate acid of 2,3,5,6-tetramethyl-p-methylenequinone bromine is added to form the 1-bromo-1-bromomethyl-4-hydroxy-2,3,5,6-tetramethylbenzenium ion [220].

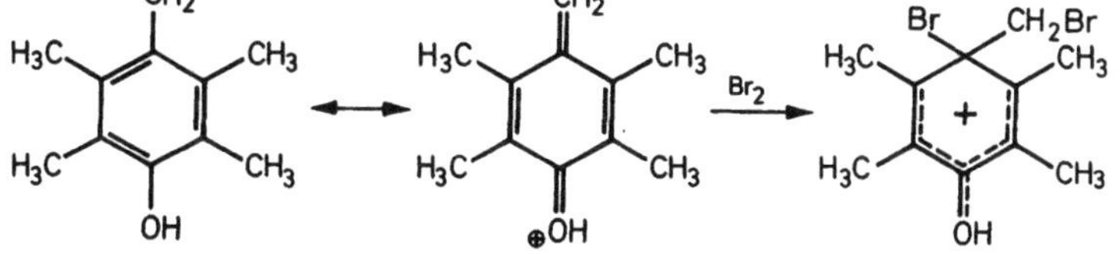

It would be of interest to study the possibility of adding reagents of other types to hydroxybenzyl cations.

Oxidation of 1,3,5-tris(dialkylamino)benzenes [266,267] yielded salts of dications (*33*) which appear to be the products of radical cation dimerization (*34*):

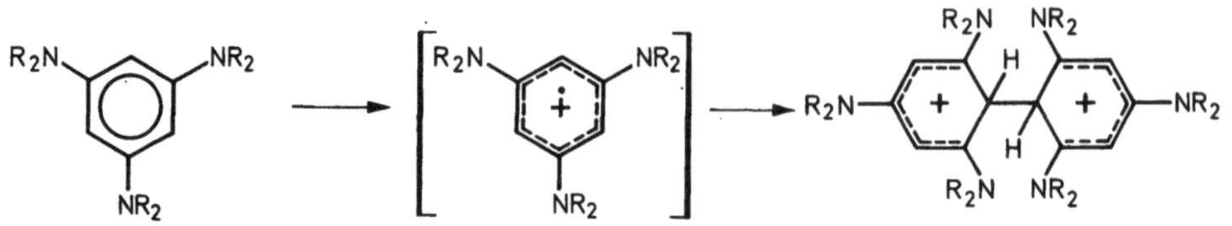

34 *33*

Thus at present we possess a wide choice of ways and means of generating arenium ions in solutions, among them ions containing at the ring sp^3-hybridized carbon atom an alkyl group and a heteroatomic substituent of the type NO_2, Hal, SO_2Y, OR etc. Ions of this type are intermediates in reactions of polyalkylated aromatic compounds with electrophilic reagents and can be regarded as structural analogues of σ-complexes formed in the course of electrophilic hydrogen substitution reactions in the aromatic series. The further development of the methods of generating and observing arenium ions will make it possible to proceed to a more extensive research into the ions of the latter type — ions having at the ring sp^3-hybridized carbon atom a hydrogen and a heteroatomic group.

III Structure of Arenium Ions According to Physical Data

The most reliable information on the structure of species of a new type is provided by X-ray analysis. It has so far been little applied to arenium ions. One of the few examples is the determination of the structural parameters of the tetrachloroaluminate of the heptamethylbenzenium ion [268]. The anion of this salt is a nearly regular tetrahedron with the Al—Cl bond equal to 2.12 Å.[5] Five sp^2-hybridized carbon atoms of the cation are lying in the same plane while the sp^3-hybridized atom C, is deflected from it by 0.07 Å (the angle between the planes C_2, C_4, C_6 and C_2, C_1, C_6 is 5.4°). The mean values of the bond lengths (in Å) are given in the following diagramme:

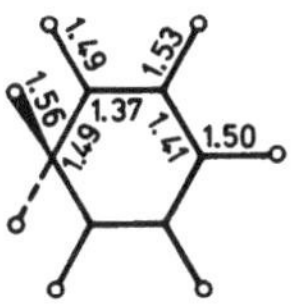

The bonds $C_{2(6)}$—CH_3 and C_4—CH_3 are shorter than $C_{3(5)}$—CH_3 which seems to reflect the predominant localization of the positive charge at the 2-, 4- and 6-positions. The length of the C_1—CH_3 bonds is close to that usually observed in the bonds between sp^3-hybridized carbons.

The nonplanar character of the cyclohexadienyl framework has also been noted for 2,4,6-triaminobenzenium ions (*29*): the angle between the planes C_2 C_4, C_6 and C_2, C_1, C_6 amounts to about 15° [198,199].

At present the principal method of studying arenium ions in solutions is NMR. Interesting evidence is also derived from the analysis of vibrational and electronic spectra. The data on the structure of arenium ions obtained using these methods, as well as the spectral characteristics of the ions, are now discussed.

1 Proton Magnetic Resonance Spectroscopy

NMR spectroscopy allows one to obtain information on arenium ions generated in solutions, but also to follow their transformations. Of particular importance is the NMR spectroscopy for studying degenerate rearrangements of arenium ions. This

5 The X-ray data and the analysis of the signal width in the NMR-^{13}C spectrum of a solid salt sample [269] show the anion $AlCl_4^-$ chlorine atoms to be located somewhat closer to the positively charged sp^2-hybridized carbons than to the atom C_1.

aspect will be discussed in detail in Section IV.2. When an ion undergoes any fast reversible intra-or intermolecular transformations, the shape of the spectrum strongly depends on the temperature, and in some cases (if there is an intermolecular proton transfer) on the acidity of the medium. The discussion here will therefore be confined to the NMR data related to the conditions under which the reversible transformations of arenium ions are "frozen" by a drop of temperature or inhibited by an increase in acidity.

When comparing proton chemical shifts (δ scale) it should be born in mind that some workers used external standards introducing no corrections for the difference in the magnetic susceptibilities of the specimen and the standard. It is more reliable to compare the shifts obtained by means of internal standards.

Listed below are the abbreviated notations of the main standards and the values of their chemical shifts (ppm) which were accepted by most workers in determining the position of signals in the PMR spectra of arenium ion salts:

TMS	— tetramethylsilane	0.00
CH	— cyclohexane	1.43
TMA	— tetramethylammonium-cation	3.20
TBC	— t-butyl cation	3.93
CH_2Br_2	— methylene dibromide	4.94
CH_2Cl_2	— methylene dichloride	5.33
C_6H_6	— benzene	7.27

The data taken from the publications in which different values of chemical standard shifts were used have been recalculated in making up the Tables of this survey. The abbreviations "int." and "ext." in the Tables stand for "internal standard" and "external standard", respectively.

A Unsubstituted, as well as Alkyl- and Aryl-Substituted Arenium Ions

As noted earlier, the Dutch investigators [59–61] first studied the formation of arenium ions from aromatic hydrocarbons in liquid HF in the presence of BF_3. Consider, e.g., the PMR spectra of hexamethylbenzene and mesitylene solutions in the $HF—BF_3$ system (Fig. 1) recorded at temperatures providing the "freezing" of exchange processes.

The PMR spectrum of a hexamethylbenzene solution contains, apart from the HF signal not shown in Fig. 1a, five signals (at 4.00; 2.77; 2.60; 2.32 and 1.60 ppm) with a 1:3:6:6:3 intensity ratio. The relative intensities indicate that the last four signals belong to CH_3-groups; the first one belongs to a single hydrogen atom which, to judge by the quadruplet splitting ($J = 6.8$ cps), is adjacent to the CH_3 group giving a doublet signal at 1.60 ppm. Thus, when hexamethylbenzene is dissolved in the $HF—BF_3$ acid system, the structural fragment $>CH-CH_3$ is observed. It is natural to conclude this fragment to belong to the 1,2,3,4,5,6-hexamethylbenzenium ion (*15*). The signal at 2.70 ppm corresponds to the 4-CH_3 group of this ion and those at 2.60 and 2.72 ppm, to the two pairs of CH_3 groups occupying 2(6)- and 3(5)-positions.

These two signals can be finally assigned by turning to the spectrum of the ion formed by mesitylene protonation. The proton can be attached to the mesitylene

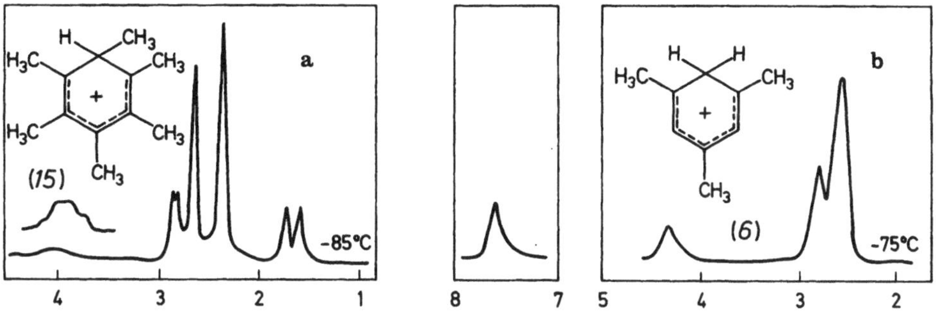

Fig. 1. PMR spectra of hexamethylbenzene **a** and mesitylene **b** in HF saturated by BF_3 [4,61]

molecule in two ways: to the unsubstituted ring carbon atom to form the 2,4,6-trimethylbenzenium ion (6) and to the carbon atom bonded with the CH_3 group to form the 1-H-1,3,5-trimethylbenzenium ion (*7*). The absence of the signals characteristic for $>CH-CH_3$ in Fig. 1.**b** indicates that only (*6*) is formed. It is logical to assign the signal in the lowest field (7.67 ppm) to single hydrogen atoms accommodated in the charged pentadienyl part of the ion at the C_3 and C_5 atoms while the 4.34 ppm signal belongs to the ring CH_2 group and those at 2.77 and 2.57, to 4- and 2(6)-CH_3 groups.

Comparison of the CH_3 groups resonance for the hexamethylbenzenium and mesitylenium ions leads to the conclusion that the signal at 2.32 ppm observed only in the spectrum of the first ion corresponds to the 3(5)-CH_3 groups.

Data on the chemical shifts of other methyl-substituted benzenium ions generated in various acid systems are summarized in Table 5. To judge by the values obtained using internal standards, the proton chemical shifts of methylbenzenium ions are relatively little dependent on the nature of the anion or the solvent, the signal of different types of protons being usually observed in the following ranges:

$C_1(R)-CH_3$	1.5–1.7	$C_{2(6)}-CH_3$	2.6–2.9	$C_{2(6)}-H$	9.0–9.4
$C_1(CH_3)-H$	4.0–4.1	$C_{3(5)}-CH_3$	2.3–2.5	$C_{3(5)}-H$	7.6–8.2
$C_1{<}^{H}_{H}$	4.4–5.0[6]	C_4-CH_3	2.8–3.1	C_4-H	8.6–9.2

Data on the PMR spectra of 2,4,6-alkyldimethylbenzenium ions (alkyls-CH_2CH_3, $CH(CH_3)_2$, $C(CH_3)_3$ and cyclopropyl) are given in [280] and the proton chemical shifts of some 4-alkylbenzenium ions are presented in Table 6. A series of ethylbenzenium ions (from mono- to heptaethyl-substituted) [157] is also described.

The PMR spectrum of the unsubstituted benzenium ion has only been recorded for benzene solutions in particularly strong acid systems $HF-SbF_5-SO_2ClF-SO_2F_2$ [65,139] and $HSO_3F-SbF_5-SO_2ClF-SO_2F_2$ [282] (the earlier data for the $HF-SbF_5-SO_2$ [10,28,213] are erroneous). The chemical shifts of this ion measured at

6 For the effect of the 2- and 4-substituents different from alkyl groups on the position of the CH_2 group signal see Sec. III-1B, C, D.

Table 5. PMR Data of Methylbenzenanium Ions

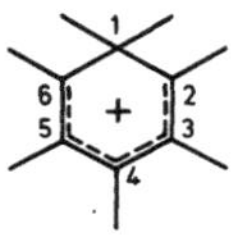

Position of methyl groups	Acid system	Temp. °C	Chemical shifts (ppm) of methyl or ring (in parentheses) protons				Standard	Ref.
			C_1	$C_{2,6}$	$C_{3,5}$	C_4		
4-CH_3	HF—SbF_5—SO_2ClF	—100	(4.93)	(9.00)	(8.03)	3.05	CH_2Cl_2, int	[64]
	HF—SbF_5—SO_2ClF		...	(9.43)	(8.49)	3.48	TMS, ext	[63]
	HF—SbF_5—SO_2ClF	—100	(5.05)	(9.38)	(8.38)	3.30	TMS, ext	[65]
	HSO_3F—SO_2		(5.05)	(9.43)	(8.38)			[188]
2,4-$(CH_3)_2$	HF—SbF_5	—45	(4.67)	2.88 (8.68)	(7.83, 7.93)	(3.00)	TMA, int	[270]
	HF—SbF_5—SO_2	—71	(4.6)	2.7 (8.4)	(7.8)	(2.9)	TMS, ext	[10, 28, 63]
	HSO_3F—SbF_5	—50	(5.6)	3.7 (9.7)	(8.8, 8.9)	(3.9)	TMS, ext	[271]
2,5-$(CH_3)_2$	HF—BF_3	—120	(4.92)	3.00 (8.8)	2.60 (8.06)	(9.2)	TMA, int	[270]
	HF—SbF_5—SO_2	—65	(4.7)	2.4 (7.8)	2.2 (7.6)	(8.2)	TMS, ext	[10, 28]
3,4-$(CH_3)_2$	HF—BF_3	—125	(4.96)	(8.8)	2.60 (8.18)	3.10	TMA, int	[270]
	HF—SbF_5—SO_2	—70	(4.4)	(8.3)	2.2 (7.6)	2.4	TMS, ext	[10, 28]
2,3,4-$(CH_3)_3$	HF—BF_3	—90	(4.66)	2.85 (8.57)	2.47 (7.87)	2.99	TMA, int	[270]
	HF—SbF_5—SO_2	—62	(4.6)	2.2 (8.3)	2.0 (7.6)	2.6	TMS, ext	[10, 28]
2,4,5-$(CH_3)_3$	HF—SbF_5	—30	(4.67)	2.84 (8.43)	2.47 (7.87)	2.92	TMA, int	[270]
	HF—SbF_5—SO_2	—60	(4.6)	2.3 (8.3)	2.1 (7.8)	2.6	TMS, ext	[10, 28]
2,4,6-$(CH_3)_3$	HF—SbF_5	+20	(4.49)	2.74	(7.63)	2.87	TMA, int	[270]
	HF—SbF_5	—60	(4.53)	2.74	(7.63)	2.89	TMA, int	[270]
	HF—BF_3	—60	(4.43)	2.76	(7.64)	2.91	TMA, int	[270]
	HF—BF_3	—75	(4.34)	2.57	(7.67)	2.77	B, int	[60, 61]
	HBr—2 $AlBr_3$	+25	(4.61)	2.80	(7.64)	2.93	CH, int	[22, 27, 272]

Table 5. (continued)

Position of methyl groups	Acid system	Temp. °C	Chemical shifts (ppm) of methyl or ring (in parentheses) protons				Standard	Ref.
			C_1	$C_{2,6}$	$C_{3,5}$	C_4		
	HBr—2 $AlBr_3$—$(CHCl_2)_2$	—40	(4.66)	2.77	(7.60)	2.90	CH, int	273)
	HCl—2 $AlCl_3$—CH_2Cl_2	—45	(4.58)	2.80	(7.66)	2.94	CH_2Cl_2, int	273)
	HF—SbF_5—SO_2	—80	(4.6)	2.8	(7.7)	2.9	TMS, ext	10, 28, 63)
	HSO_3F	—77	(4.5)	2.8	(7.7)	2.9	TMS, ext	271)
	HCl—$AlCl_3$—$CDCl_3$	—30	(5.27)	3.36	(8.30)	3.44	TMS, ext	274)
2,3,4,5-$(CH_3)_4$	HF—BF_3	—84	(4.59)	2.83 (8.35)	2.52	2.94	TMA, int	270, 275)
	HF—SbF_5—SO_2	—63	(4.4)	2.4 (7.9)	2.2		TMS, ext	10, 28)
2,3,4,6-$(CH_3)_4$	HF—SbF_5	—20	(4.56)	2.70	2.37 (7.67)	2.86	TMA, int	270)
	HF—SbF_5—SO_2	—66	(4.4)	2.5	2.2 (7.4)	2.7	TMS, ext	10, 28)
	HF—BF_3	...	...	2.60	2.32 (7.67)	2.77	B, int	60)
	HCl—$AlCl_3$—$CDCl_3$	—30	(5.32)	3.28	2.88 (8.28)	3.38	TMS, ext	276)
2,3,5,6-$(CH_3)_4$	HF—BF_3	—97	(4.95)	2.75	2.47	8.6	TMA, int	270)
	HF—BF_3	—85	...	2.74	2.34	8.59	B, int	60)
	HF—SbF_5—SO_2	—68	(4.7)	2.8	2.5	7.8	TMS, ext	10, 28)
	HSO_3F	—88	(5.0)	2.9	2.6	8.9	TMS, ext	271)
2,3,4,5,6-$(CH_3)_5$	HF—SbF_5	+20	(4.59)	2.68	2.39	2.84	TMA, int	270)
	HF—SbF_5	—20	(4.59)	2.68	2.40	2.86	TMA, int	270)
	HF—SbF_5—SO_2	—68	(4.3)	2.4	2.1	2.5	TMS, ext	10, 28)
	HF—BF_3	—60	(4.50)	2.69	2.41	2.87	TMA, int	270)
	HF—BF_3	—60	(4.47)	2.60	2.32	2.77	B, int	59, 60, 92)
	HBr—2 $AlBr_3$—C_6H_{12}	+25	(4.55)	2.64	2.30	2.75	CH, int	27)
	HBr—2 $AlBr_3$—C_6F_6	+25	(4.55)	2.70	2.38	2.85	CH, int	27)
	HSO_3F	—33	(4.46)	2.52	2.24	2.69	CH_2Cl_2, int	277)
	HSO_3F	—49	(4.7)	2.8	2.5	3.0	TMS, ext	271)
	HSO_3F	—85	(4.53)	2.58	2.29	2.74	CH_2Cl_2, int	173)

Table 5. (continued)

Position of methyl groups	Acid system	Temp. °C	Chemical shifts (ppm) of methyl or ring (in parentheses) protons				Standard	Ref.
			C_1	$C_{2,6}$	$C_{3,5}$	C_4		
1,2,3,4,5,6-$(CH_3)_6$	$HF—BF_3$	−83	1.68 (4.08)	2.67	2.39	2.83	TMA, int	270)
	$HF—BF_3$	−85	1.60 (4.00)	2.60	2.32	2.77	B, int	60)
	$HF—SbF_5—SO_2$	−62	1.9 (4.2)	2.8	2.6	3.0	TMS, ext	10, 28)
	HSO_3F	−70	1.90 (4.30)	2.85	2.55	3.00	TMA, int	263)
	HSO_3F	−85	1.8 (4.2)	2.8	2.5	2.9	TMS, ext	271)
	$HSO_3F—SO_2$	−80	1.87 (4.30)	2.88	2.58	3.04	TMS, ext	174)
	$HSO_3F—SbF_5—SO_2$	−96	1.69 (4.22)	2.72	2.41	2.88	TMA, int	278)
$(CH_3)_7$	$HCl—2\ AlCl_3—C_6F_6$	+20	1.66	2.73	2.47	2.92	CH, int	154, 279)
	CF_3COOH	+20	1.54	2.62	2.36	2.82	CH, int	279)
	HSO_3Cl	+20	1.54	2.64	2.35	2.84	TMA, int	279)
	HSO_3F	...	1.44	2.55	2.28	2.75	CH_2Cl_2, int	253)
	HCl—aq	+20	1.47	2.55	2.26	2.77	TMA, int	279)

Table 6. PMR Data of 4-Alkylbenzenium Ions ($HF—SbF_5—SO_2ClF$, —100 °C, TMS, ext.) [65]

Alkyl	Chemical shifts (ppm)			
	protons of alkyl groups	ring protons at C_1	$C_{2,6}$	$C_{3,5}$
CH_3	3.30 (CH_3)	5.05	9.38	8.40
C_2H_5	1.73 (CH_2), 3.60 (CH_3)	5.13	9.44	8.47
n-C_3H_7	1.37 (CH_2), 2.07 (CH_2), 3.40 (CH_3)	5.02	9.30	8.30
i-C_3H_7 [a]	1.63 (CH), 3.70 (CH_3)	5.07	9.37	8.39

[a] Data for n-, sec-, t- and neo-C_5H_{11} see [65], cyclopentyl and cyclohexyl [281]

—130 ÷ —140 °C relative to the TMS as external standard are listed below:

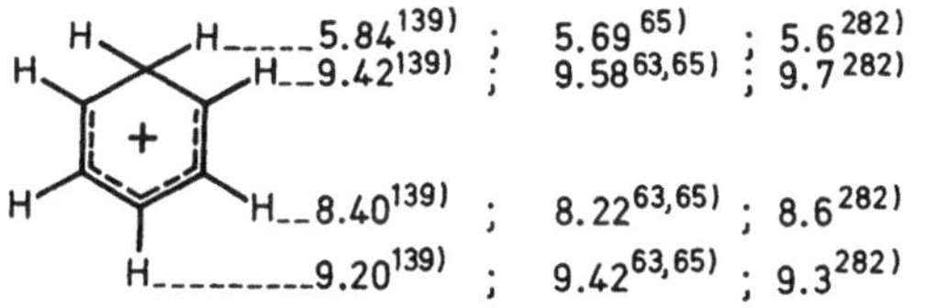

Important information on the structure of arenium ions is provided by the spin-spin coupling constants of different types of protons. For the vicinal protons located in the pentadienyl part of alkylbenzenium ions the spin coupling constant is 7–9 cps [270]. Since long-range interactions for this kind of protons are usually insignificant[7], the hydrogen atoms at 2- and 3-, or at 5- and 6-positions give in the PMR spectra of para-substituted benzenium ions the characteristic AB-type signals. Fig. 2, e.g. shows the spectrum of a 4-methylbenzenium ion.

An unexpectedly strong spin coupling (3.5—4.0 cps [65,270]) is observed between the protons of the ring CH_2 group and those of the 4-CH_3 group, i.e. via 6 bonds. The value of the spin coupling constant of the protons of the ring CH_2 group and the 2(6)-CH_3 groups is much smaller (0.5–1.5 cps [4,92]) while for the 3(5)-CH_3 groups it is close to zero. The interaction constant for the proton of the CH_2 group and for the ortho-proton is about 2 cps [283].

These regularities are widely used in the elucidation of the structure of methyl-substituted arenium ions; they make it easy to identify the signal of the CH_3 group occupying the para-position relative to the ring CH_2 group. Usually this signal has the form of a rather well resolved triplet (see Fig. 2).

The doublet splitting of the 4-CH_3-group signal in 1-H-1,2,3,4,5,6-hexamethylbenzenium ion is slightly smaller (J = 2.0 cps, see Fig. 1) than the triplet splitting of the signal of the CH_3 groups in the position para to the ring CH_2 group of the other methylbenzenium ions. A similar decrease of the long-range coupling constant is observed for 10-H-9,10-dimethylanthracenium ion (0.9 cps) [62] as compared with the 10-H-9-methylanthracenium ion (1.8 cps).

7 Cf., however, the data for hydroxybenzenium ions.

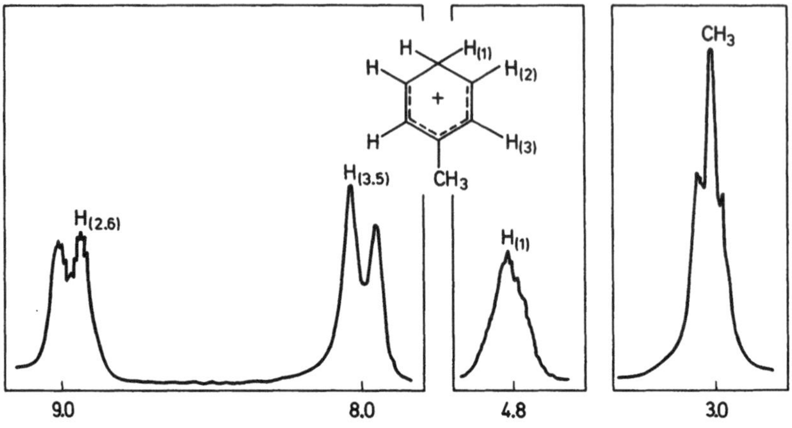

Fig. 2. PMR spectra of hexafluorantimonate of the 4-methylbenzenium ion in SO_2ClF at -100 °C, 100 MHz [64,65]

Of special interest are the reasons for the strong spin-spin coupling between the protons of the para-CH_3 group and those of the ring C(R)H fragment. It seems to be a reflection of the effective hyperconjugative interaction between the C—H bond of the C(R)H fragment and the π-electron system of the ion. In 1-H-1,2,3,4,5,6-hexamethylbenzenium and 10-H-9,10-dimethylanthracenium ions the spatial interaction of the CH_3 group at the sp^3-hybridized carbon atom with the ortho-positioned substituents leads to an increase in the deviation angle of the C_{sp^3}—CH_3 bond from the plane of the unsaturated part of the ion and, as a consequence, to a decrease in the deviation angle of the C_{sp^3}—H bond. In this case the efficiency of hyperconjugation drops reflecting a smaller constant of long-range spin-spin coupling and a smaller degree of deshielding the protons of the C(R)—H fragment. Indeed, the signal of the single hydrogen of 1-H-1,2,3,4,5,6-hexamethylbenzenium ion is shifted by $\sim$0.4 ppm [270] in a high field relative to the usual signals of the ring CH_2 groups of methylbenzenium ions. The calculation analysis of long-range spin-spin couplings in arenium ions is given in [92].

As a rule, the splitting of the ring CH_2 group signals of methylbenzenium ions cannot be observed due to the overlap of interactions with other types of protons.

For the methyl-substituted benzenium ions the down-field shift of signals observed for hydrogen atoms and CH_3 groups remaining at the sp^2-hybridized carbons can be attributed to partial positive charges at these carbon atoms. An attempt was made [60,92] to estimate the distribution of the positive charge in methylbenzenium ions from their proton shifts. The approach used is essentially as follows. The proton attachment to a carbon atom of the benzene ring must lead to disturbance of the π-electron system and hence to changes in the anisotropy of magnetic susceptibility which manifests itself [284] in the deshielding the protons located in (or near) the plane of the benzene nucleus and in the excessive shielding of the protons appearing in the zone above the ring. Ring current interruptions must therefore cause some up-field shifts for the signals of protons located near the ring plane, i.e. offset partially the down-field signal shift which is due to positive charges at the ring carbons. Believing the effect of the anisotropy of magnetic susceptibility for the protons of the CH_3 group in the protonated benzene ring to be comparable to nonaromatic unsaturated systems C. MacLean and E. Mackor [60,92] assumed the i-CH_3 group signal shift for the methyl-substituted benzenium ion relative to the one of the CH_3 group at the double carbon-carbon bond (~ 1.60 ppm) to be proportional to the partial positive charge q_i at the ring atom C_i, for instance, in the 2,3,4,5,6-pentamethylbenzenium ion:

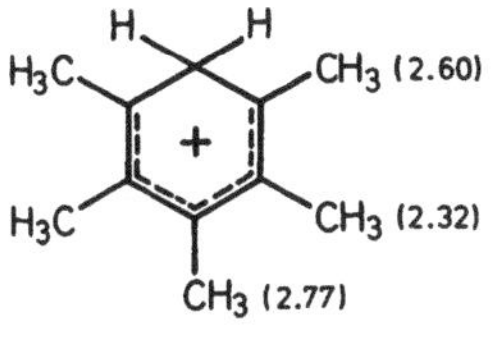

$$\Delta_2 = 1.00 \qquad q_{2(6)} = \Delta_2/\Sigma\Delta = +0.22$$
$$\Delta_3 = 0.72 \qquad q_{3(5)} = \Delta_3/\Sigma\Delta = +0.16$$
$$\Delta_4 = 1.17 \qquad q_4 = \Delta_4/\Sigma\Delta = +0.25$$
$$\Sigma\Delta = 2\Delta_2 + 2\Delta_3 + \Delta_4 = 4.61$$

The thus estimated distribution of the positive charge[8] is unexpectedly uniform. The molecular-orbital calculations [4,60,92,93,285] result in the deficiency of π-electron density, being much more localized at atoms C_2, C_4 and C_6. This was later confirmed by the analysis of the carbon-13 magnetic resonance spectra of methylbenzenium ions (see Sect. III.2): cf. also the results of the X-ray analysis of the heptamethylbenzenium ion tetrachloraluminate (p. 25).

The different factors contributing to the chemical shifts of arenium ion protons are important enough to be studied more closely. As noted above C. MacLean and E. Mackor [60,92] believed the anisotropy of magnetic susceptibility for the protonated benzene ring to be comparable to that of usual unsaturated systems. An attempt was made [273,286] to verify experimentally this assumption by the protonation of 2,4,6,2',4',6'-hexamethylbiphenyl (*35*) and other model compounds.

Due to spatial interaction of the CH_3 groups located at 2-, 2'-, 6- and 6'-positions the benzene rings of this hydrocarbon are turned at 90° to each other; so the ortho-CH_3 groups of one ring are placed over the plane of the other. Consequently, the signal of the 2-, 2', 6- and 6'-CH_3 groups shifts compared to that of the 4- and 4'-CH_3 groups to the high field by 0.46 ppm [276,286]. The calculations [276] according to C. Johnson and E. Bovey [287] confirmed the difference in the chemical shifts of the two types of CH_3 groups of hydrocarbon (*35*) to be wholly due to the anisotropy of the benzene ring magnetic susceptibility.

The protonation of 2,4,6,2',4',6'-hexamethylbiphenyl (HCl + 2 $AlCl_3$ in CH_2Cl_2, at -50 °C [286]) yields ion (*36*)[9] with the following chemical shifts of its protons:

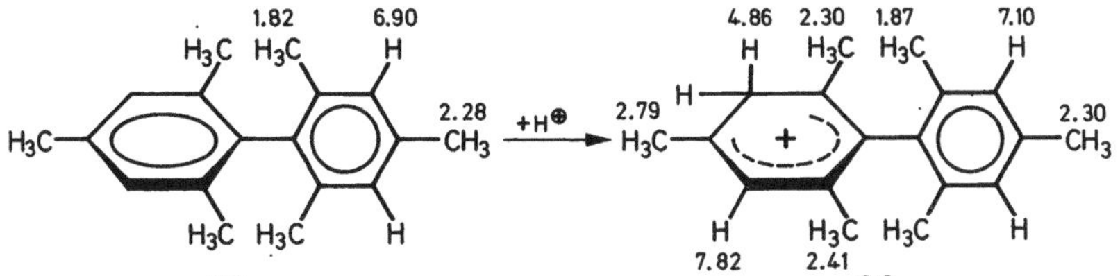

35 *36*

8 A similar result was obtained by calculating the chemical shifts of the ring protons of incompletely substituted benzenium ions [60,92].

9 Unsubstituted biphenyl is protonated at the 4-position [288].

The positions of CH_3 groups signals of the protonated ring of ion (*36*) agree well with the chemical shifts of the mesitylenium ion signals (see Table 5) considering that the signals of the two CH_3 groups of ion (*36*) adjacent to the unprotonated ring must be shifted up-field, due to the reasons mentioned, by ~0.5 ppm. The position of the unprotonated ring ortho-CH_3 group signal differs little from that of the corresponding signal in the spectrum of the parent hydrocarbon. The ortho-CH_3 group signal has been observed to change as little on protonation of one of the decamethylbiphenyl rings (from 1.74 to 1.81 ppm [286])).[10] The suggestion that the down-field shift of ortho-CH_3 group signal due to the decreasing ring-current of the neighbouring protonated benzene ring is largely offset by a reverse shift due to the electric fields of positive charge does not explain this unexpected fact. Indeed, the calculations for the ortho-CH_3 groups of the unprotonated (*36*) ion ring according to Buckingham-Musher [290,291] have shown the expected total electric field effect of the partial positive charges of the protonated ring is not great (~0.1 ppm), and it must cause a shift of the signal to the low, rather than high field [272].

The above data, therefore, make it possible to conclude that the contribution of anisotropy of magnetic susceptibility to the shielding of the protons over the benzene ring changes comparatively little upon the protonation of this ring.[11]

This conclusion can also be extended to the protons located near the plane of the protonated ring, in particular, to those of the CH_3 groups of methylbenzenium ions. Indeed, estimation of anisotropic effects according to the semiempirical method suggested by J. Goldstein and G. Reddy [292,293] indicates [273] the ring anisotropy of magnetic susceptibility on the chemical shift of the CH_3 group bonded with this ring for mesitylene and mesitylenium ion being comparable in value; it equals ~0.7 ppm.

Thus the anisotropic effects of the protonated benzene ring are not so simple as the Dutch scientists thought them to be. The available data point to a comparatively small change in the effect for the benzene ring upon its protonation. This may be due to the effective participation of the C—H bonds in the "aliphatic" ring fragments of benzenium ions in the σ,π-conjugation with the charged pentadienyl system of an ion due to which the protonation does not break the ring chain of conjugation completely. The importance of hyperconjugation was emphasized in the analysis of the relative stability of arenium ions in the molecular orbital method [285,294,295].

From these considerations on the magnetic effect of the benzenium ion ring system it is easy to explain the chemical shift anomalies of the CH_2- and CH_3-fragments of alkyl groups located in benzenium ions at the ring sp^3-hybridized carbon atoms. Thus, as compared to the 1-ethyl-1,2,3,5,6-pentamethyl-4-methylenecyclohexa-2,5-diene (*37*) in 1-ethyl-1,2,3,4,5,6-hexamethylbenzenium ion (*16b*) and its analogues the signal of the CH_2 fragment of the ethyl group undergoes a down-field shift by more than 1 ppm; the signal of the CH_3 fragment shifts, on the contrary, slightly in the

10 By contrast upon protonation of the 14 π-electron aromatic system of trans-15,16-dihydropyrene the signal of the CH_3 groups located above and below the plane of the π-system is drastically shifted to the low field, the shift value being close to the expected one for the complete ring current interruption [289].

11 Interesting substrates for verifying this conclusion may be protonated cyclophanes [149].

opposite direction. Ions of this kind seem to favour, for sterical reasons (cf. [62,157]), conformation (*39*) in which the CH_3 fragment of the ethyl group is located above the

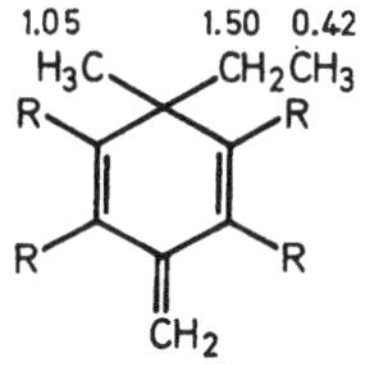

37
$R = CH_3$ [153]
cf. [296]

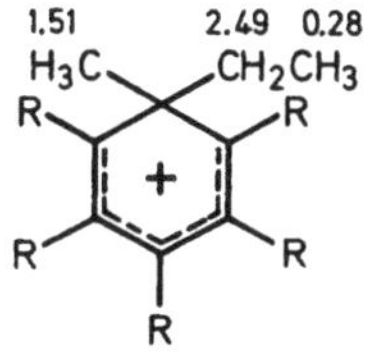

16b
$R = CH_3$ [254] [12]

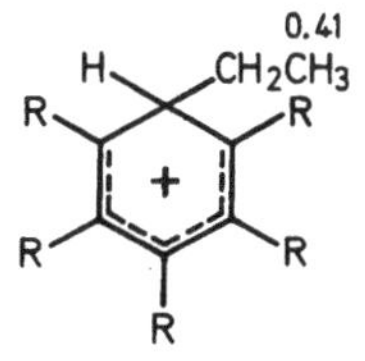

38
$R = C_2H_5$ [62]
cf. [156,157]

ring in the zone of the shielding of the π-system.

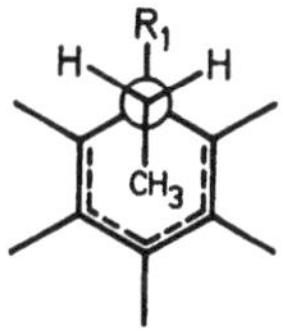

39

For the CH_3 fragment this effect offsets the deshielding influence of the positive charge of the ion. By contrast, the protons of the CH_2 fragment of the ethyl group are placed near, but outside, the ring plane and are subjected to the deshielding magnetic influence of the ring ion system (the direction of the effect of the positive charge [290, 291] is the same). No wonder, therefore, that the down-field shift of the ethyl group CH_2 fragment (~1 ppm) proves to be by far greater than that of the CH_3 group ~0.6 ppm located at the same ring carbon and capable of turning around the $C_1—CH_3$.

To determine the effect of the positive charge on the changes in chemical shifts in benzenium ions at sp^2-hybridized carbons these changes should be counted off from the position of protons and proton-containing groups in the spectra of aromatic hydrocarbons rather than from that of the signals in the appropriate unsaturated compounds.

Another point should also be noted. C. MacLean and E. Mackor [60, 92] studied the effect of the positive charge on the chemical shifts of benzenium ions considering only the charge on the carbon bonded with the proton or CH_3 group and assuming the electric fields of the other ring carbon atom charges to be negligible [4]. In [273], however, convincing reasons are given in favour of taking this additional factor into account. Through strict consideration of the effect of ring C_j-atom positive charges on the deshielding of the proton-bearing group located at the C_i atom is difficult, with single protons at the ring sp^2-hybridized carbons of benzenium ions it is possible to use the semiempirical relationship obtained [297] (cf. [298]) from the

---

12 See also [152,153,155].

analysis of the proton chemical shifts of π-electron systems with a known charge distribution ($C_5H_5^-$, C_6H_6, $C_7H_7^+$):

$$-\Delta\delta_{H_i} \approx 7.1q_i + 1.4(q_{i+1} + q_{i-1}), \text{ppm},$$

where q_{i+1} and q_{i-1} are the charges at the carbons adjacent to the C_i atom.

Using this relation and the data on the chemical shifts of the protons of benzene (7.27 ppm) and those occupying in the unsubstituted ion 2-, 3- and 4-positions (9,58, 8.22 and 9.42 ppm [63,65]) one can readily determine the values of $q_{2(6)}$, $q_{3(5)}$ and q_4 (q_1 for a first approximation is taken to equal zero). The thus found distribution of the positive charge in a benzenium ion sharply differs from that obtained in terms of C. MacLean and E. Mackor[13].

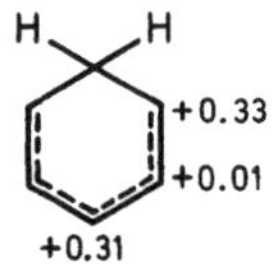

It is in good agreement with the results of the NMR-^{13}C spectroscopy (see Sect. III.2). Therefore we can believe that the suggested estimation of various factors contributing to the proton chemical shifts of benzenium ions is sufficiently reasonable.

Since, as shown in Sect. III.2, the values of the deficiency of π-electron density at sp^2-hybridized carbons linearly depend on the chemical shifts of these atoms in the NMR-^{13}C spectra the above relation can be rewritten in the form

$$\Delta\delta_{H_i} = a\,\Delta\delta_{C_i} + b(\Delta\delta_{C_{i+1}} + \Delta\delta_{C_{i-1}})$$

where Δδ are the differences of the chemical shifts of respective atoms in the spectra of ions and their aromatic precursors.

This relation proved [299] to describe well the chemical shifts of ring protons ($a = 4.03 \cdot 10^{-2}$ and $b = 0.69 \cdot 10^{-2}$) and those of CH_3 groups bonded with C_i atoms ($a = 1.07 \cdot 10^{-2}$ and $b = 0.25 \cdot 10^{-2}$) in the series of methylbenzenium ions ($r = 0.988$, $s = 0.075$).

As mentioned earlier for monoprotonated bimesityl and decamethylbiphenyl the magnetic anisotropy of the aryl residue makes a marked contribution to the chemical shifts of cyclohexadienyl fragment protons. This effect must also account for other aryl-substituted benzenium ions. Significant are the differences in the chemical shifts of the CH_3 groups of the heptamethylbenzenium ion and of three phenylhexamethylbenzenium ions [208] (see also [292]).

13 If chemical shift changes are to be counted off from positions typical of the olefin hydrogen (5.40 ppm [60,92]) then the C. MacLean and E. Mackor approach yields for the unsubstituted benzenium ion roughly the same charge distribution as the one given earlier for the 2,3,4,5,6-pentamethylbenzenium ion (p. 33).

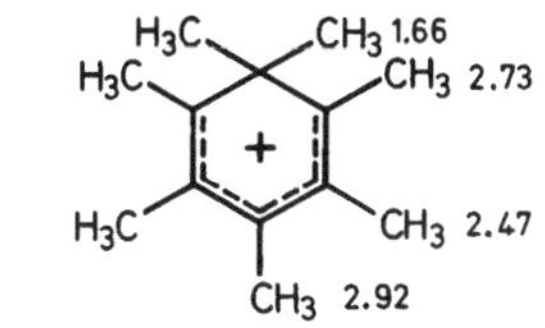

Anion = $Al_2Cl_7^{\ominus}$

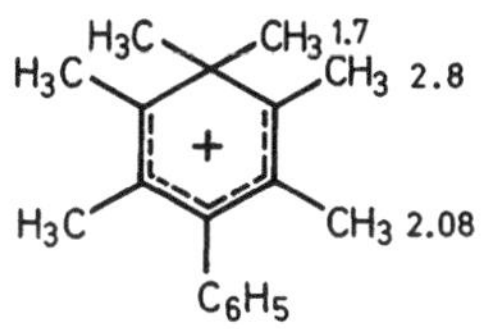

Anion = $SO_3F^{\ominus}$

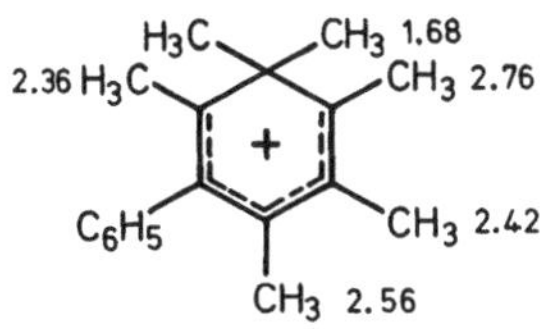

Anion = $BF_4^{\ominus}$

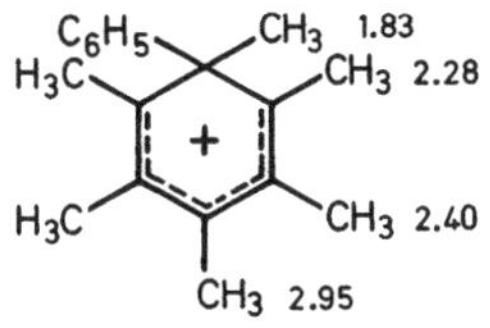

Anion = $BF_4^{\ominus}$

Naphthalene and its methylated derivatives are generally protonated at one of α-positions.[14] The PMR spectra of the respective ions are presented in Table 7. The most characteristic by their position are the signals H^4 (9.3–9.6), H^2 (8.9–9.3), 4-CH_3 (3.1–3.6) and 2-CH_3 (2.8–3.2 ppm). The introduction of a CH_3 group to 4-position was observed to cause a down-field shift of the H^5 signal by ~0.5 ppm [81, 301]. The spin-spin coupling constants for the protonated ring of 1-H-naphthalenium ions have the following values [300, 301].

H^1—H^2	1.5–2.5 cps	H^1-4-CH_3	2.5–3 cps
H^1—H^4	1 –2 cps	H^2—H^3	8 –9 cps
H^1-1-CH_3	7.5–8 cps	H^3—H^4	7 –8 cps

The influence of the position of CH_3 groups on the relative stability of methyl-naphthalenium ions is discussed in Sect. IV.2C.

For the diprotonation of polymethylnaphthalenes see [80, 81, 303].

Acenaphthene is protonated at the unsubstituted α-position to form ion (*40a*) [301].

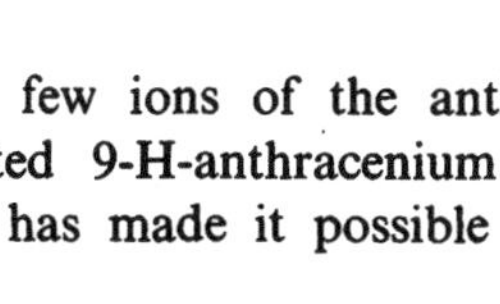

40 a

Quite a few ions of the anthracenium series have been described. For the unsubstituted 9-H-anthracenium ion the recording of the PMR spectrum at 300 MHz has made it possible to determine the chemical shifts of all protons

14 β-Protonation has been noted for monohalogenonaphthalenes [300] and 4-halogen-1-naphthols [203, 204]. A representative of β-series ions is also the 1,2,2,3,4-pentamethylnaphthalenium ion [210]. See also data on protonation of 1,4-dimethylnaphthalene [300, 301].

Table 7. PMR Data of Methyl-substituted 1-Naphthalenium Ions

Substituents	Acid system	Temp. °C	Chemical shifts (ppm) of	
			1	2
—	$HF-SbF_5-SO_2ClF$	−78	(5.27)	(9.51)
$4-CH_3$	$HF-SbF_5-SO_2ClF$	−60	(5.20)	(9.26)
$2-CH_3$	$HF-SbF_5-SO_2ClF$	−60	(5.24)	3.26
$4,8-(CH_3)_2$	$HF-SbF_5-SO_2ClF$	−40	(4.95)	(9.36)
	$HSO_3F-SbF_5-SO_2ClF$	−80	(4.76)	(9.18)
	HSO_3F	−40	(4.32)	(9.16)
$2,6-(CH_3)_2$	$HF-SbF_5-SO_2ClF$	−60	(5.16)	3.17
	$HSO_3F-SbF_5-SO_2ClF$	−80	(5.05)	3.09
$2,3-(CH_3)_2$	$HF-SbF_5-SO_2ClF$	−50	(5.28)	3.14
	$HSO_3F-SbF_5-SO_2ClF$	−80	(5.16)	3.04
$2,4-(CH_3)_2$	$HSO_3F-SbF_5-SO_2ClF$	−40	(4.95)	3.06
	$HSO_3F-SbF_5-SO_2ClF$	−80	(4.90)	3.02
$4,7-(CH_3)_2$	$HSO_3F-SbF_5-SO_2ClF$	−80	(4.87)	(8.90)
$2,8-(CH_3)_2$	$HSO_3F-SbF_5-SO_2ClF$	−80	(4.83)	3.13
$4,6-(CH_3)_2$	$HSO_3F-SbF_5-SO_2ClF$	−80	(4.94)	(9.07)
$4,5-(CH_3)_2$	$HSO_3F-SbF_5-SO_2ClF$	−80	(4.95)	(8.89)
$3,4-(CH_3)_2$	$HSO_3F-SbF_5-SO_2ClF$	−80	(4.95)	(8.93)
$2,7-(CH_3)_2$	$HSO_3F-SbF_5-SO_2ClF$	−80	(4.93)	3.01
$5,8-(CH_3)_2$	$HSO_3F-SbF_5-SO_2ClF$	−95	(4.88)	(9.47)
$1,2,4-(CH_3)_3$	HSO_3F	−60	1.77 (4.57)	2.92
$1,4,8-(CH_3)_3$	HSO_3F-SO_2ClF	−80	1.61	
$4,5,8-(CH_3)_3$	HSO_3F-SO_2ClF	−80	(4.60)	
$2,4,8-(CH_3)_3$	HSO_3F-SO_2ClF	−80	(4.61)	2.99
$1,2,3,4-(CH_3)_4$	HSO_3F	−20	1.80 (4.63)	2.94
$1,1,2,3,4-(CH_3)_5$	HSO_3F	+28	1.82	2.93
$1,2,3,4,7-(CH_3)_5$	HSO_3F	−60	1.73 (4.52)	2.80
$1,2,3,4,6,7-(CH_3)_6$	HSO_3F	−30	1.70 (4.44)	2.79
$1,2,3,4,5,8-(CH_3)_6$	CF_3COOH	+20	1.49 (4.54)	2.84
$1,2,3,4,7,8-(CH_3)_6$	CF_3COOH	+20	1.45 (4.57)	2.75
$2,3,4,6,7,8-(CH_3)_6$	HSO_3F	−40	(4.57)	2.88
$(CH_3)_8$	HSO_3F	−30	1.50 (4.65)	2.75
	CF_3COOH	+20	1.37 (4.49)	2.65

($HF-SbF_5-SO_2ClF$, −55 °C, TMS, ext)[282]:

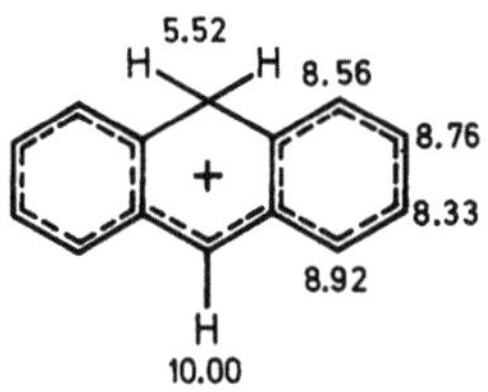

The data on anthracenium ions containing substituents in the 9- and 10-positions are collected in Table 8. The proton chemical shifts of anthracenium ions are influenced by the partial transfer of the positive charge to the side rings and by the effect of magnetic anisotropy of the entire ring system. To illustrate the latter note an

methyl group and ring (in parentheses) protons						Standard	Ref.
3	4	5	6	7	8		
(8.4)	(9.81)	(8.4)	(8.4)	(8.4)	(8.4)	TMS, ext	300)
(8.4)	3.62	(8.92)	(8.4)	(8.4)	(8.4)	TMS, ext	81, 300)
(8.3)	(9.63)	(8.3)	(8.3)	(8.3)	(8.3)	TMS, ext	300)
(8.4)	3.70	(8.93)	(8.4)	(8.4)	2.94	TMS, ext	81, 300)
(8.18)	3.50	(8.74)	(8.00)	(8.24)	2.75	TMS, ext	301)
(8.2)	3.50	(8.72)	(7.96)	(8.2)	2.72	CH_2Cl_2, int	81)
(8.2)	(9.44)	(8.2)	2.76	(8.2)	(8.2)	TMS, ext	300)
(8.02)	(9.37)	(8.21)	2.68	(8.10)	(8.10)	TMS, ext	301)
2.95	(9.43)	(8.4)	(8.4)	(8.4)	(8.4)	TMS, ext	300)
2.64	(9.29)	(8.2)	(8.2)	(8.2)	(8.2)	TMS, ext	301)
(8.1)	3.60	(8.1)	(8.1)	(8.1)	(8.1)	TMS, ext	300)
(8.00)	3.36	(8.68)	(8.14)	(8.14)	(8.14)	TMS, ext	301)
(7.92)	3.39	(8.67)	(7.98)	2.82	(7.98)	TMS, ext	301)
(8.04)	(9.42)	(8.30)	(7.90)	(8.16)	2.70	TMS, ext	301)
(8.09)	3.44	(8.57)	2.72	(8.09)	(8.09)	TMS, ext	301)
(8.00)	3.50	3.10	(8.00)	(8.00)	(8.00)	TMS, ext	301)
2.68	3.39	(8.84)	(8.17)	(8.17)	(8.17)	TMS, ext	301)
(7.88)	(9.28)	(8.29)	(7.83)	2.79	(7.96)	TMS, ext	301)
...	(10.00)	...	...	...	...	TMS, ext	301)
(7.91)	3.25	(8.66)	(8.1)	(8.1)	(8.1)	CH_2Cl_2, int	81)
	3.37	(8.65)				CH_2Cl_2, int	302)
	3.47	3.07			2.63	CH_2Cl_2, int	302)
(7.95)	3.30	(8.55)	(7.88)	(8.01)	2.64	TMA, int	82)
2.55	3.28	(8.60)	(8)	(8)	(8)	CH_2Cl_2, int	81)
2.59	3.32						210)
2.46	3.16	(8.52)	(7.70)	2.67	(7.85)	CH_2Cl_2, int	81)
2.4–2.6	3.17	(8.32)	2.4–2.6	2.4–2.6	(7.77)	CH_2Cl_2, int	81)
2.4–2.7	3.15	2.4–2.7	(7.64)	(7.64)	2.4–2.7	TMA, int	150)
2.4–2.6	3.09	(8.35)	(7.64)	2.4–2.6	2.4–2.6	TMA, int	150)
2.5	3.20	(8.34)	2.5	2.5	2.5	CH_2Cl_2, int	81)
2.4–2.6	3.15	2.4–2.6	2.4–2.6	2.4–2.6	2.4–2.6	CH_2Cl_2, int	81)
2.4–2.6	3.07	2.4–2.6	2.4–2.6	2.4–2.6	2.4–2.6	TMA, int	150)

abnormally large up-field shift for a CH_3 fragment of ethyl groups if these are located at the sp^3-hybridized meso-atom of the anthracenium ion (0.2 ppm [62, 247]). For steric reasons the most advantageous conformation seems to be the one in which a CH_3 group gets into the shielding zone of the π-systems of the side (cf. data on derivatives of 9-ethyl-9,10-dihydroanthracene [247, 305]) and central rings (see p. 34). The preference of the above conformation follows likewise from the analysis of the spin-spin coupling constants of the fragment $>CHCH_2CH_3$ [62].

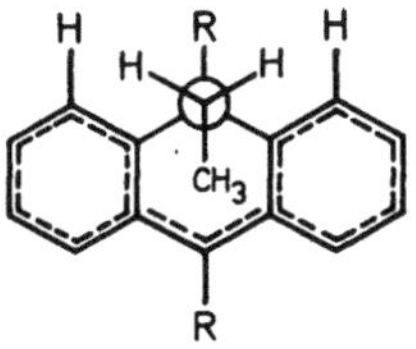

Table 8. PMR Data of Anthracenium Ions

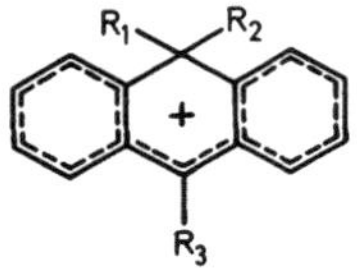

Substituents			Acid system	Temp. °C	Chemical shifts (ppm) of substituent or ring (in parentheses) protons		Standard	Ref.
R_1	R_2	R_3			$R_{1,2}$	R_3		
H	H	H	$HF—BF_3$	−35	(5.15)	(9.92)	TMA, int	[4]
			$HF—SbF_5—SO_2ClF$	−55	(5.52)	(10.00)	TMS, ext	[282]
			$SbCl_3—AlCl_3$	+100	(4.9)	(9.6)	TMA, int	[304]
H	H	CH_3	$CF_3COOH—H_2O—BF_3$	+20	(4.95)	3.68	TMA, int	[4]
			$HF—BF_3$	−20	(5.07[a])	3.75	TMA, int	[62]
H	H	C_2H_5	$HF—BF_3$	−20	(5.02)	1.75, 4.16	TMA, int	[62]
H	H	Cl	$HSO_3F—SO_2ClF$	−60	(5.20)		TMS, ext	[282]
H	H	Br	$HSO_3F—SO_2ClF$	−60	(5.18)		TMS, ext	[282]
H	CH_3	CH_3	$HCl—2\,AlCl_3—C_6H_5Cl$	+20	1.82 (4.78)	3.64	TMA, int	[278]
			$CF_3COOH—H_2O—BF_3$	+20	1.67 (4.8)	3.70	TMA, int	[4, 59, 60]
			$HF—BF_3$	−20	1.71 (4.94)	3.71[b]	TMA, int	[62]
			$SbCl_3—AlCl_3$	+100	1.6 (4.7)	3.6	TMA, int	[304]
H	CH_3	C_2H_5	$HF—BF_3$	−20	1.64 (4.8)	1.68, 4.01	TMA, int	[62]
H	C_2H_5	CH_3	$HF—BF_3$	−20	0.19, 2.41 (4.85)	3.62[c]	TMA, int	[62]
CH_3	CH_3	H	$HCl—2\,AlCl_3—CHCl_3$	+25	1.90	(9.95)	CH, int	[247]
CH_3	CH_3	CH_3	CF_3COOH	+25	2.00	3.84	CH_2Cl_2, int	[153]
C_2H_5	C_2H_5	H	$HCl—2\,AlCl_3—CHCl_3$	+25	0.20, 2.30	(9.90)	CH, int	[247]

[a] quartet, J = 1.8 Hz; [b] doublet, J = 0.9 Hz; [c] doublet, J = 1.2 Hz.

Interestingly, 1,4,5,8,9-pentamethylanthracene protonates at the C_9 atom to yield ion (*40b*) rather than at the 10-position [306]. This seems to be favoured by a decreasing steric interaction, on protonation at the C_9 atom, of the CH_3 groups at the 1-, 8- and 9-positions.

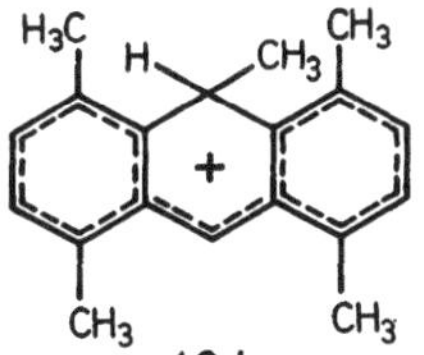

40 b

The chemical shifts of the alkyl groups of 9- and 10-alkylsubstituted phenanthrenium ions are listed in Table 9. The multiplet signals of the phenanthrenium ion side ring protons is given in [309]. For more detailed information the reader is referred to [177, 211, 215–217, 252, 308, 310–312].

The PMR spectra of 9-p-X-phenyl-9,10-dimethylphenanthrenium ions recorded at low temperatures allow us to conclude that the rotation of the aryl group around the C_9—C_{ar} is restricted and the aryl group is located in the preferable conformation so that two pairs of adjacent protons of this residue turn out to be inequivalent [309].

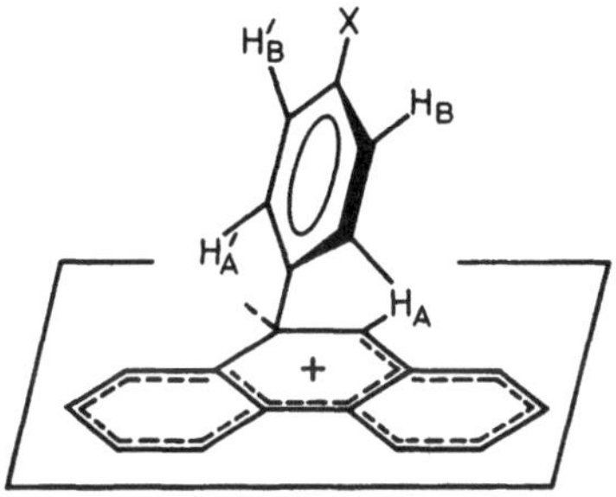

Thus, for X = CF_3 two pairs of doublets correspond to the p-X-phenyl group — at 6.38 and 7.51 ppm and at 8.06 and 8.24 ppm; the signals at higher field are assigned to the H_A and H_B protons which get into the shielding zone of the π-system in the phenanthrenium ion.

The preference of this conformation (the barrier hindering the rotation of the p-XC_6H_4 — at X = Cl, CH_3, CF_3 amounts to ~8.5 kcal/mol., cf. also [307, 312]) is assumed [313] to be due, at least partially, to the donor-acceptor interaction between the π-system of the aryl residue and the carbenium centre at C_{10}; in the electronic absorption spectra of the 9-p-X-phenyl-9,10-dimethylphenanthrenium ions the "charge transfer bands" have been revealed whose position correlates well with the ionization potentials of respective X-substituted benzenes. Judging by the NMR-^{13}C spectra, however, the extent of this interaction in the main state of ions is insignificant — the chemical shift of the C_{10} atom nearly remains unchanged as the X substituent varies [303].

Table 9. PMR Data of Phenanthrenium Ions

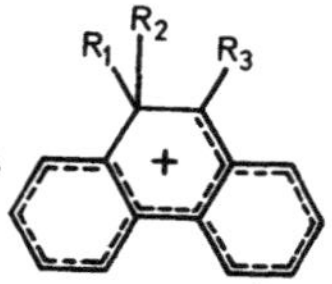

Substituents			Acid system	Temp. °C	Chemical shifts (ppm) of substituent protons[a]			Ref.
R_1	R_2	R_3			R_1	R_2	R_3	
CH_3	CH_3	CH_3	$HSO_3F—SO_2ClF$	—80	1.88	1.88	3.47	[307]
			$HSO_3F—SO_2ClF—SO_2F_2$	—110	1.74	1.74	3.44	[308]
			$HCl—2\ AlCl_3—CH_2Cl_2$	—71	1.85	1.85	3.54	[211, 248, 250]
CH_3	C_2H_5	CH_3	$HCl—2\ AlCl_3—CH_2Cl_2$	—83	1.82	0.12, 2.79	3.49	[212]
CH_3	C_6H_5	CH_3	$HSO_3F—SO_2ClF$	—100	2.13		3.13	[211, 309]
			$HCl—2\ AlCl_3—CH_2Cl_2$	—95	2.05		3.11	[211, 309]
C_2H_5	C_2H_5	CH_3	$HCl—2\ AlCl_3—CH_2Cl_2$	—92	0.04, 2.76	0.04, 2.76	3.47	[215]
C_2H_5	C_2H_5	C_2H_5	$HCl—2\ AlCl_3—CH_2Cl_2$	—84	0.10, 2.79	0.10, 2.79	1.64, 3.75	[214]

[a] standard: CH_2Cl_2, int

The most characteristic proton chemical shifts for the 2-H-biphenylenium [314] and 2-H-acenaphthylenium ions [315] are as follows:

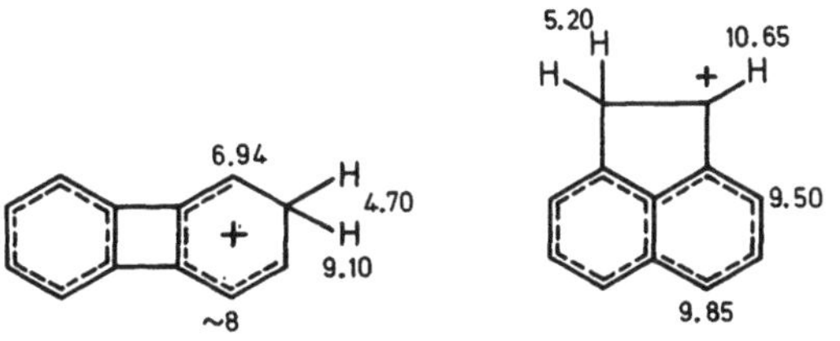

A series of 1- and 2-substituted acenaphthylenium ions has been generated by removing OH^- in strong acids from respective acenaphthenols [315, 316].

The PMR spectra of the biphenylene and ancephthylene ions fail to reveal the spin-spin coupling of the CH_3 protons with those of neighbouring carbons [315, 316].

For the PMR spectra of the arenium ions formed by the protonation of 7-methyl- and 7,12-dimethylbenz[a]anthracene, pyrene, benz[a]pyrene, naphthacene and other polycyclic systems see [304, 317, 318].

B Hydroxyarenium Ions and Their O-Derivatives[15]

Arenium ions having an OR group (R = H, Alk, Ar[16]) as substituent at one of the ring sp^2-hybridized carbons are usually generated by protonating the aromatic compounds. As a rule, the proton is attached para to an oxygen. However, with other electron-releasing substituents at the ring and especially if the para position is occupied, the ortho protonation may be predominant. Thus, the dissolution of 3,4-dimethylphenol in HSO_3F at −50 °C yields a mixture of ions (*41*) and (*42*) in a ratio 9:1 [319] or 11:1 [320][17]:

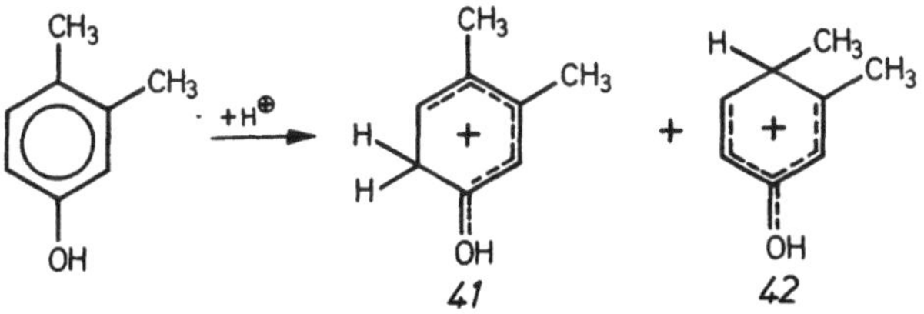

41 *42*

while the protonation of p-cresol with HBr and $AlBr_3$ at −70 °C yields, according to PMR, ion (*43*) alone [69]:

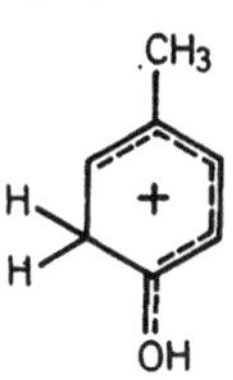

43

---

15 For the protonation of ArSR see [321, 322].

16 The protonation of ArOAr compounds has been studied on diphenyl ether [68] and xanthene [323].

17 According to [324], also 3 % of an ion corresponding to protonation at the 2-position is formed.

Table 10. PMR Data of 2-Hydroxybenzenium Ions and their O-Derivatives 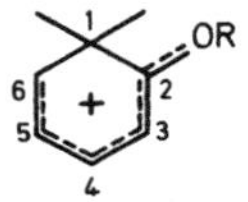

Substituents	Acid system	Temp. °C	Chemical shifts (ppm) of substituent and ring (in parentheses) protons						Standard	Ref.
			C_1	C_3	C_4	C_5	C_6	OR		
1,1-$(CH_3)_2$; R=H	HSO_3F	−50	1.74	(7.22)	(8.68)	(7.22)	(7.72)		TMA, int	324, 325)
1-CH_3-1-$CHCl_2$; R=H	H_2SO_4	+35	1.82	(7.25)	(8.60)	(7.37)	(7.87)		TMS, ext	326)
1,3-$(CH_3)_2$-1-$CHCl_2$; R=H	H_2SO_4	+35	1.83	2.42	(8.50)	(7.37)	(7.84)		TMS, ext	326)
1,4-$(CH_3)_2$-1-$CHCl_2$; R=H	H_2SO_4	+35	1.80	(7.09)	2.82	(7.35)	(7.79)		TMS, ext	326)
1,5-$(CH_3)_2$-1-$CHCl_2$; R=H	H_2SO_4	+35	1.80	(7.27)	(8.51)	2.42	(7.69)		TMS, ext	326)
1,6-$(CH_3)_2$-1-$CHCl_2$; R=H	H_2SO_4	+35	1.87	(7.14)	(8.57)	(7.25)	2.72		TMS, ext	326)
5-CH_3; R=H	HSO_3F	−50	(4.46)	(7.23)	(8.64)	2.36	(7.45)		TMA, int	324)
	HSO_3F—SbF_5	−60	(4.45)	(7.23)	(8.64)	2.36	(7.50)		TMA, int	324)
	HSO_3F—SbF_5	−60	(4.47)	(7.32)	(8.70)	2.37	(7.60)		TMA, int	320, 327)
	HBr—2 $AlBr_3$—CS_2	−70	(4.59)	(7.30)	(8.67)	2.38	(7.50)	10.34	CH_2Br_2, ext	69)
5-CH_3; R=CH_3	HSO_3F	−60	(4.38)	(7.40)	(8.55)	2.38	(7.40)			320, 328)
3,5-$(CH_3)_2$; R=H	HSO_3F	−50	(4.41)	2.36	(8.47)	2.36	(7.34)		TMA, int	324)
	HSO_3F	−50	(4.36)	2.34	(8.44)	2.34	(7.40)		TMA, int	320)
3,5-$(CH_3)_2$; R=CH_3	HSO_3F	−50	(4.27)	2.34	(8.32)	2.34	(7.32)	4.53	TMA, int	324)
4,5-$(CH_3)_2$; R=H	HSO_3F	−50	(4.51)	(7.15)	2.77	2.36	(7.42)		TMA, int	319, 325)
	HSO_3F	−50	(4.30)	(7.18)	2.77	2.34	(7.43)		TMA, int	324)
	HSO_3F	−50	(4.28)	(7.18)	2.76	2.34	(7.44)		TMA, int	320)
	HSO_3F—SbF_5	−60	(4.26)	(7.14)	2.78	2.35	(7.40)		TMA, int	327)
5,6-$(CH_3)_2$; R=H	HSO_3F	−50	(4.33)		(8.58)		2.47		TMA, int	324)
3,4,5-$(CH_3)_3$; R=H	HSO_3F	−50	(4.22)	2.32	2.72		(7.28)		TMA, int	324)
	HSO_3F	−50	(4.25)	2.31	2.72	2.31	(7.30)		TMA, int	320)
3,5,6-$(CH_3)_3$; R=H	HSO_3F	−50	(4.34)	2.31	(8.40)	2.31	2.40		TMA, int	324)
	HSO_3F	−50	(4.35)	2.40	(8.44)	2.40	. . .		TMA, int	320)
4,5,6-$(CH_3)_3$; R=H	HSO_3F	−50	(3.87)	(6.65)	2.35	1.87	2.07		TMS, ext	319)
	HSO_3F	−50	(4.26)	(7.00)	2.74	2.26	2.48		TMA, int	324)
	HSO_3F	−50	(4.25)	(7.02)	2.74	2.35	2.47		TMA, int	320)

$R=CH_3$	HSO_3F	−64	(3.85)	(6.75)	2.42	1.92	2.12	4.17	TMS, ext	319)
$R=COCH_3$	HSO_3F	−54	(3.92)	(6.70)	2.41	1.92	2.13	2.50	TMS, ext	319)
1,3,4,6-$(CH_3)_4$; $R=CH_3$	HSO_3F	−70	1.66	2.16	2.68	(6.9)	2.44	4.64	TMA, int	202)
3,4,5,6-$(CH_3)_4$; $R=H$	HSO_3F	−50	(4.19)	...	2.56	2.23	...		TMA, int	320)
1,3,4,5,6-$(CH_3)_5$; $R=CH_3$	HSO_3F	−50	1.60 (4.15)	2.22	2.64	2.22	2.41	4.58	TMA, int	175, 265)
	CF_3SO_3H	+20	1.67 (4.06)	2.22	2.62	2.22	2.40	4.50	TMA, int	265)
1,1,3,4,5,6-$(CH_3)_6$; $R=H$	HSO_3F	−50	1.62	2.27	2.72	2.27	2.39		TMA, int	324)
	HSO_3Cl	−75	1.63	2.28	2.72	2.28	2.39	9.89		218)
	HSO_3F		1.52	2.17	2.62	2.17	2.29		...	332)
	$HCl-2\,AlCl_3-CH_2Cl_2$	−70	1.50	2.1–2.2	2.60	2.1–2.2	2.28		CH_2Cl_2, int	219)
	H_2SO_4 (75%)	+20	1.56	2.21	2.66	2.21	2.32		TMA, int	329)
	CF_3SO_3H	+10	1.61	2.26	2.69	2.26	2.37		TMA, int	330)
$R=CH_3$	HSO_3F	−20	1.58	2.43	2.78	2.28	2.43	4.65	TMA, int	243)
1,3,4,6-$(CH_3)_4$-5-Br; $R=H$	HSO_3F-SO_2ClF	−120	1.84	...	2.95		2.52		TMA, int	201)
$R=CH_3$	HSO_3F	0	1.72	2.30	2.84		2.56	4.66	TMA, int	201)
1,1,3,4,6-$(CH_3)_5$-5-F; $R=H$	HSO_3F	−30	1.77	2.36	2.77		2.36		TMA, int	331)
1,1,4,5,6-$(CH_3)_5$-3-NO_2; $R=H$	HSO_3F	−50	1.85		3.07	2.38	2.61		CH_2Cl_2, int	173)
1,1,3,5,6-$(CH_3)_5$-4-NO_2; $R=H$	HSO_3F	−50	1.63	1.97		2.00	2.37	11.5	CH_2Cl_2, int	173)
1,1,3,4,5-$(CH_3)_5$-6-NO_2; $R=H$	HSO_3F	−50	1.68	2.09	2.67	2.23		11.4	CH_2Cl_2, int	173)
1-C_2H_5-1,3,4,5,6-$(CH_3)_5$; $R=H$	HSO_3F-SO_2ClF	−20	1.66	2.33	2.77	2.33	2.43		TMA, int	330)
1-C_6H_5-1,3,4,5,6-$(CH_3)_5$; $R=H$	HSO_3F-SO_2ClF	−20	1.99	2.38	2.84	2.38	2.14		TMA, int	330)
	CF_3SO_3H	+10	1.93	2.28	2.79	2.28	2.04		TMA, int	330)

Other examples of ortho protonation in substituted phenols and their O-derivatives are found in Table 10.

The indication of the meta protonation of 2,4-dimethylanisole in HSO_3F [333] seems to be erroneous [320, 324]. However, 2,4,6-trimethylphenol in $HSO_3F—SbF_5$ at —50 °C is subjected to meta-C-protonation; this may be due to modification of the oxygen function by protonation or coordination with SbF_5 [324].

Quite a few hydroxyarenium ions have been generated by the protonation of substituted cyclohexadienones and their polycyclic analogues. In this way one can obtain ions both with the para and with the ortho position of the hydroxy group relative to the ring sp^3-hybridized carbon:

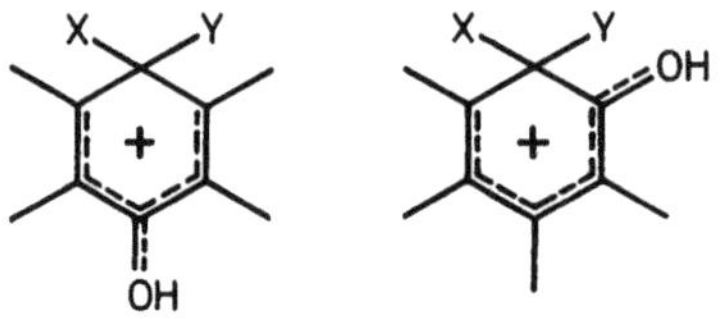

neither of the X or Y substituents being a hydrogen. For X = H or/and Y = H the precursor of the hydroxyarenium ion is the tautomeric form of phenol. One example described is the protonation of anthrone to form a 9-H-10-hydroxyanthracenium ion [59, 60]:

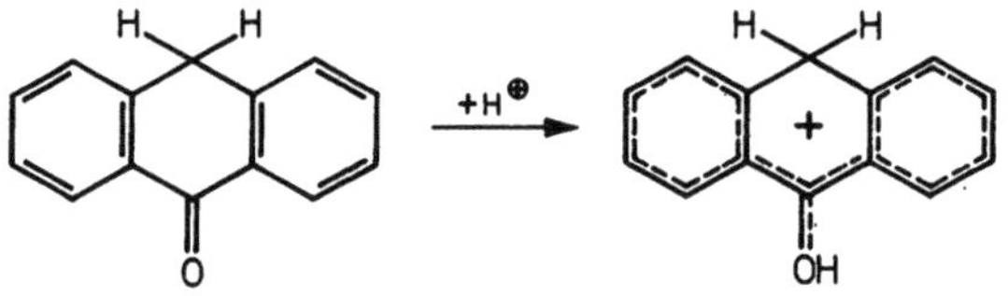

Many phenols interact with aluminium halides to yield complexes in tautomeric ketoforms; these complexes, as shown by the NMR spectra, are close analogues of hydroxyarenium ions (X = Cl or Br).

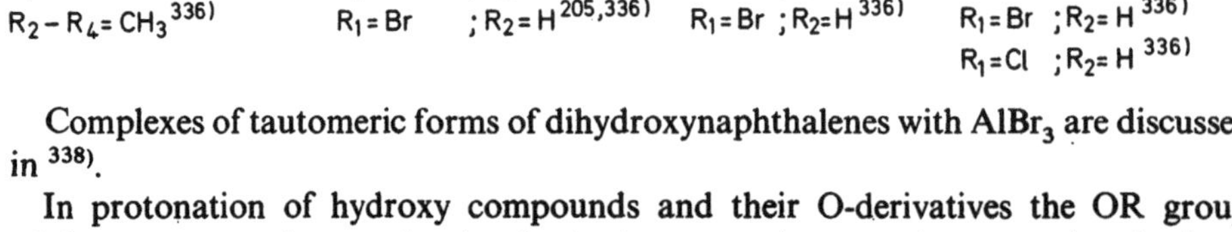

R_1 = H or CH_3 [334, 335]
R_1 = Br;
$R_2 - R_4 = CH_3$ [336]

R_1 = H, CH_3; R_2 = H [334]
R_1 = H ; R_2 = Ar [337]
R_1 = Br ; R_2 = H [205, 336]

R_1 = H ; R_2 = H, CH_3 [334]
R_1 = CH_3; R_2 = H [337]
R_1 = Br ; R_2 = H [336]

R_1 = CH_3 ; R_2 = H [337]
R_1 = Ar ; R_2 = H [337]
R_1 = Br ; R_2 = H [336]
R_1 = Cl ; R_2 = H [336]

Complexes of tautomeric forms of dihydroxynaphthalenes with $AlBr_3$ are discussed in [338].

In protonation of hydroxy compounds and their O-derivatives the OR group of these compounds can also be the basic centre; in general protonation leads to an equilibrium of arenium and oxonium ions. Equilibrium between these cations

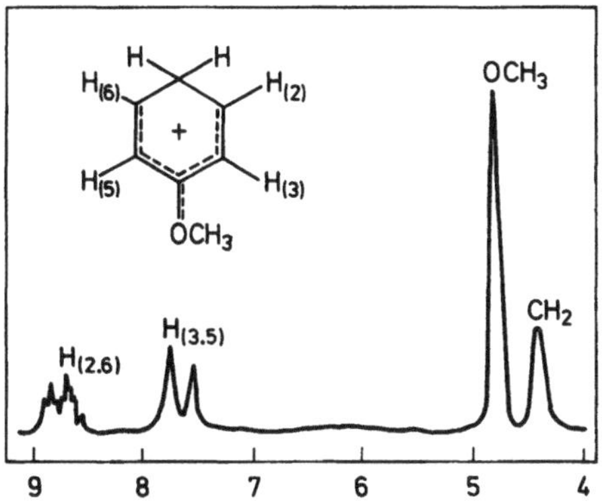

Fig. 3. PMR spectrum of of anisole in HSO_3F containing 7.5 mol.% of SbF_5 at +25 °C [68]

depends on the structure of the compound to be protonated as well as on the acid medium and on temperature. The effect of these factors is often illustrated by the protonation of anisole and p-methylanisole.

The PMR spectrum of anisole solution in the $HSO_3F—SbF_5$[18] recorded at +25 °C is given in Fig. 3. This spectrum corresponds to the structure of the 4-methoxybenzenium ion (*44*), its shape does not change until the temperature drops to —90 °C. When anisole is protonated in the $HF—BF_3$ at —80 °C the PMR spectrum shows, besides the signals of this ion, that of aromatic protons at 7.7 ppm (C_6H_5) as well as a doublet and a quartet of protons of the $—O^+(H)CH_3$ fragment at 5.0 and 12.8 ppm (J = 2.7 cps) [70], respectively, which evidently belong to the oxonium cation (*45*).

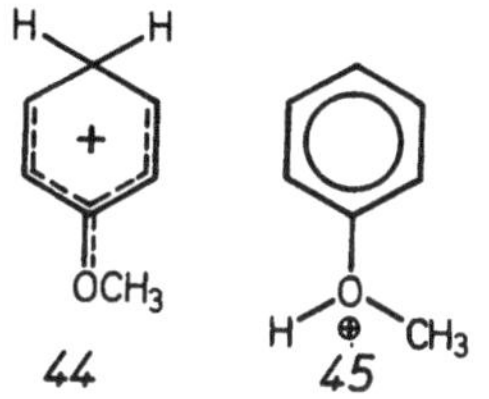

44 *45*

The observed ratio of the ions (*44*) and (*45*) is 3:2 at —80 °C and increases to 50:1 as the temperature rises to 0 °C [70].

The difference in the ratios of the ions (*44*) and (*45*) for the two acid systems [70] points to specific solvation, the change of this specific interaction on the transfer from $HF—BF_3$ to $HSO_3F—SbF_5$ causing a shift of equilibrium to the benzenium ion.

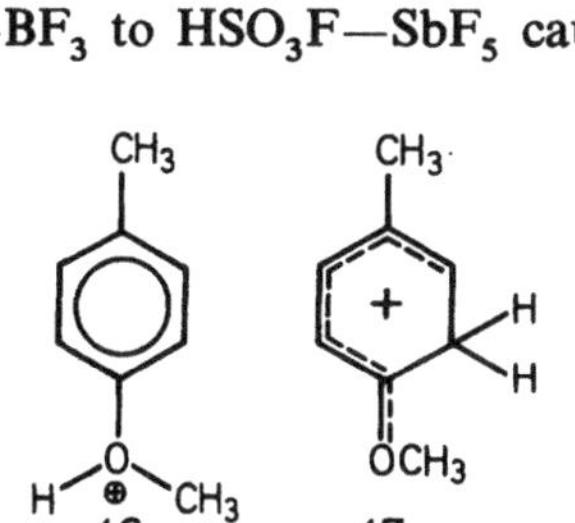

46 *47*

18 In the absence of SbF_5 the anisole in HSO_3F is subject to fast sulphonation [148] already at 0 °C.

In HF—BF_3 at —80 °C p-methylanisole is practically completely protonated at oxygen to form cation (*46*) [70]. In HSO_3F—SbF_5 [327, 328, 339] appreciable amounts of benzenium ion (*47*) are observed; its quota at —40 °C grows fast with the increasing content of SbF_5. With 20 wt% of SbF_5 this ion becomes predominant. For the ratio of the O- and C-protonation of p-cresol see [320, 324], for that of halogenphenol see [340].

The content of SbF_5 affects the dependence of the ratio of the ions (*46*) and (*47*) on temperature. When in HSO_3F, in the absence of SbF_5, the concentration of ion (*46*) is much higher than that of ion (*47*) the change of protonation temperature from —85 to —22 °C causes no change in the ratio of the ions formed (78:22). Apparently in HSO_3F the equilibrium enthalpies of the C- and O-protonation of p-methylanisole are about the same and the predominance of oxonium cation (*46*) is attributed to entropy factors. The smaller equilibrium concentration of benzenium ion (*47*) indicates that its solvation in HSO_3F requires a better ordering of surrounding molecules.

With 30 wt% of SbF_5 in the acid system an increase of protonation temperature from —82 to —42 °C diminishes the quota of the oxonium cation (from 54% to 30%). This was interpreted [327, 328, 340] as an indication of a somewhat higher exothermicity of the O-protonation as compared with C-protonation.

The equilibrium ratio of O- and C-protonated forms depends on the acid medium and temperature. However, the observed ratio of ion concentrations not necessarily corresponds with that of equilibrium. The relative rates of the O- and C-protonation (kinetic control) of phenols and their ethers may fail to correspond with the relative stability of the O- and C-protonated forms (thermodynamic control). A more detailed study of the protonation of p-methylanisole in the HSO_3F—SbF_5 system containing 30% of SbF_5 has shown ions (*46*) and (*47*) to be formed at —70 °C in a ratio of ~42:58 with a subsequent fall of the oxonium cation content to 30% in a few hours [201]. This implies that initially the O- and C-protonated p-methylanisoles were formed in a nonequilibrium ratio.

An even more complicated picture is observed in the protonation of pentamethylanisole [265]. In HSO_3F—SO_2ClF (1:5) at —85 °C it is mostly protonated to an oxonium cation. Upon a slight rise in temperature it turns into para and ortho C-protonated forms with the former one predominating; at —50 °C the O- and para C-protonated forms are rapidly and wholly converted into the ortho C-protonated form. Initially the oxonium cation yields mostly the para C-protonated form though thermodynamically it is less stable than the ion corresponding with ortho C-protonation. The acceleration of the above conversions in the strong acid system HSO_3F—SO_2ClF—SbF_5 lead to the conclusion that they are carried out through a stage of diprotonation (cf. O- and meta C-diprotonation of 2,4,6-trimethylphenol [324]):

The higher thermodynamic stability of the ortho C-protonated form of pentamethylanisole (see also 4-bromo-2,3,5,6-tetramethylanisole [201]) is due to the fact that this form, as distinct from the para C-protonated one, allows the OCH_3 group to be located in the ring plane and hence to be effectively involved in the delocalization of the positive charge. In the absence of sterical hindrances the stabilizing effect of the OCH_3 group, as remarked above, manifests itself stronger from the para position.

O-protonated and isomeric C-protonated forms in nonequilibrium ratios have been

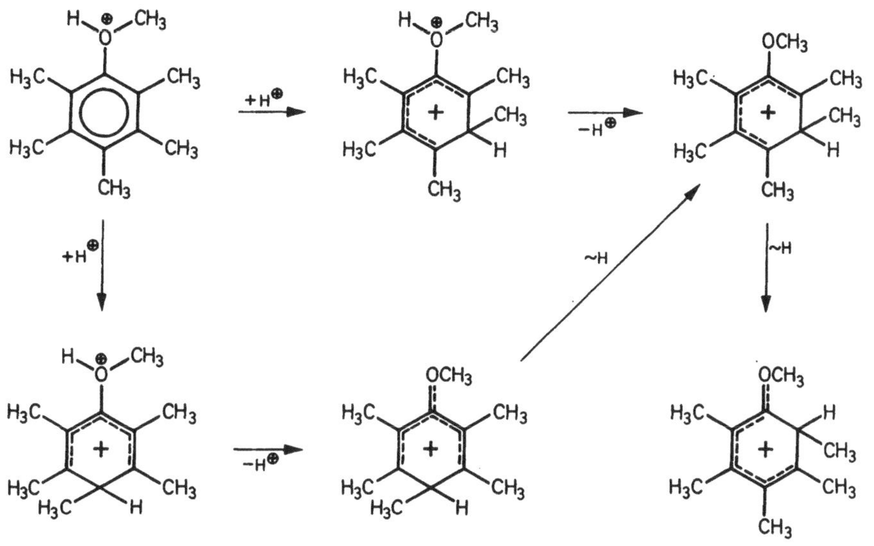

observed in other cases as well. Thus, 4-bromo-2,3,5,6-tetramethylphenol in HSO_3F—SO_2ClF at a temperature below —90 °C is mostly protonated at the ortho position while thermodynamically the more stable ion is the one corresponding with para protonation [201]. The protonation of the corresponding methyl ether in HSO_3F at —50 ÷ —70 °C is also controlled by kinetic factors. In this case the proton is first attached to an oxygen; when the system is equilibrated, ortho and para C-protonated forms emerge, the state of equilibrium between the O- and C-protonated forms highly depending on temperature [201]. In the case of 1-naphthol and its 4-bromo- and 4-methylderivative in HSO_3F the proton has been captured at the 2-position with its subsequent transfer to the 4-position [203,204].

The dissolution of m-chloro- and m-bromoanisoles in the HSO_3F—SbF_5—SO_2ClF at —80 °C yields, besides the benzenium ions, small amounts (5–10%) of the respective oxonium cations which, as the temperature rises, are converted irreversibly into C-protonated forms [340].

The higher rate of O-protonation than that of C-protonation at temperatures when the thermodynamic stability of the oxonium cation is lower than or comparable with that of the C-protonated form has analogies. Thus, mesomeric anions formed by the heterolysis of α-C—H bonds of aliphatic ketones and nitro compounds are usually more readily protonated at O than at C though a thermodynamically less stable tautomer (enol, an acinitro form) is formed. These deviations from the Brönsted principle are explained [341,342] by the O-protonation requiring a smaller rearrangement of the electron system as compared with the C-protonation. This explanation, however, is scarcely acceptable for the nonconformity, between kinetics and thermodynamics in the formation of isomeric C-protonated forms (see also Sect. IV.1.A).

These examples demonstrate a complicated dependence of the ratio of O-protonated and isomeric C-protonated forms arising from aromatic hydroxyderivaties and their ethers on the nature of the acid medium and temperature. The part of C-protonation usually grows with the rising medium acidity (cf. [324,340,343]).

A similar conclusion was drawn from the C- and O-protonation of anisole and phenol in sulphuric acid of various concentration by UV spectroscopy [68, 136, 344]. In 30–90% H_2SO_4 at 0 °C anisole and phenol are only protonated at oxygen [345]. In 100% H_2SO_4 at about 25 °C anisole, according to [344] yields a methoxybenzenium ion. It was shown later that in 92—98% H_2SO_4 and in HSO_3F at room temperature anisole is sulphonated so rapidly that it is impossible to observe its protonation [147, 148] (cf., however, [346]). So the above conclusion should be considered to be wrong. However, for phenol in H_2SO_4 at —64 °C only C was protonated [68].

In the system HF—BF_3-sulfolane, as well as in 30–90% H_2SO_4, phenol behaves as an O-base [347]. When interacting with HBr and 2 mol of $AlBr_3$ in CS_2 at —70 °C, however, it is entirely converted into 4-hydroxybenzenium heptabromodialuminate [69].

The above regularity can be interpreted by the assumption that in the solvation of arenium ions an essential role is played by hydrogen bonds with the attached proton taking part. The hydrogen atom of the $—O^+(H)CH_3$ group is bound to take a more effective part in the formation of such bonds with nucleophilic particles of the medium than that of the ring CH_2 group of the benzenium ion[19], so with decreasing medium acidity (increasing anion nucleophilicity) the equilibrium shifts to the side of the oxonium cation while with increasing acidity to that of the benzenium one.

The accumulation of electron-releasing substituents in the aromatic ring increases its basicity without appreciably affecting that of the O—R group; the latter remains coplanar with the aromatic ring. Thus, 2,5- and 3,5-dimethylanisoles in HF at —20 to —80 °C are only protonated at the ring [70]:

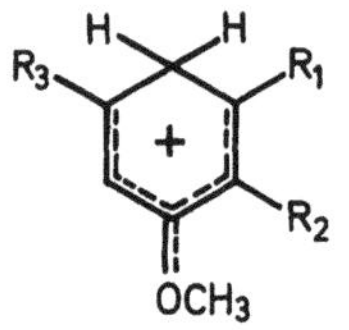

$R_1 = R_3 = CH_3$; $R_2 = H$
$R_2 = R_3 = CH_3$; $R_1 = H$

By contrast, 2,6-dimethylanisole in HF [70] and 2,4,6-trimethylanisole in HSO_3F [324] are protonated at the oxygen; this was explained by the distortion of the coplanar arrangement of the methoxy group and the benzenium ring. The distortion leads to an increasing basicity of the oxygen and a decreasing stability of the benzenium ion as a result of reducing the efficiency of the CH_3O group participation in delocalizing the positive charge over the conjugation. In the absence of steric inhibition an overlap of the p-orbitals of the oxygen and the ring carbon tends to fix the CH_3O group in the ring plane inhibiting its rotation relative to the ring. For some methoxybenzenium ions the hindered rotation of the CH_3O group has been experimentally proved by the NMR spectroscopy [71, 319].

19 Proton exchange with acid medium is effected, according to the PMR spectra, more readily for O-protonated forms than for C-protonated ones [70].

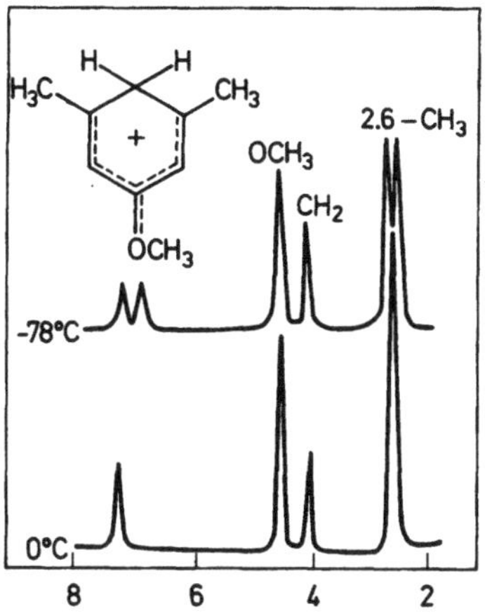

Fig. 4. PMR spectra of 3,5-dimethylanisole in HF—BF_3 [71]

Figure 4 presents the PMR spectra of 4-methoxy-2,6-dimethylbenzenium ion [71]. At —78 °C 2- and 6-CH_3 groups as well as H_3 and H_5 atoms give separate signals reflecting their nonequivalence when the CH_3O group rotation is inhibited. An increase of temperature contributes to overcoming the barrier hindering the rotation due to which the states of the methyl groups and those of the ring hydrogens are averaged and the corresponding signals coalesce (see Fig. 4).

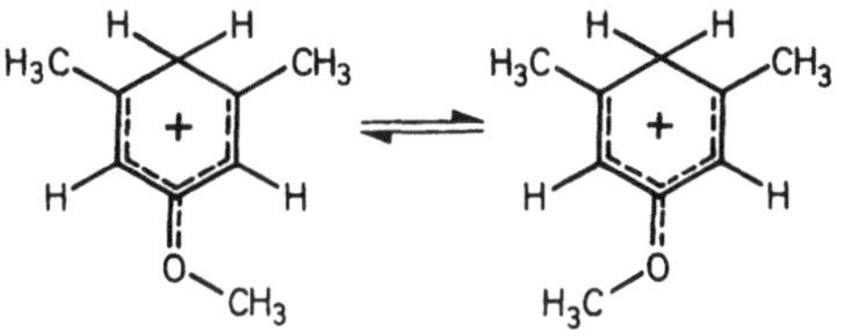

The barrier hindering the rotation in methoxybenzenium ions depends on the order of the $CH_3O—\overset{+}{C}$ bond; this is determined by the involvement of the group in the delocalization of the positive charge. The electron-withdrawing properties of the ring part of the ion increase as other electron-releasing substituents are removed from it. Therefore, the barrier inhibiting the rotation of the CH_3O group in the 4-methoxybenzenium ion turned out to be somewhat higher than that for the 4-methoxy-2,6-dimethylbenzenium ion [71].

For the rotation hindrance of the methoxy group in 2-halogen-4-methoxybenzenium ions see [340], in 2-methoxy-1-naphthalenium ion [300].

The hindrance of rotation in hydroxyarenium ions was observed in 4-hydroxybenzenium heptabromodialuminate [69]. In the PMR spectrum of a solution of this salt in CS_2 at —70 °C the H_3 and H_5 atoms give separate signals which coalesce when the temperature rises to —60 °C.

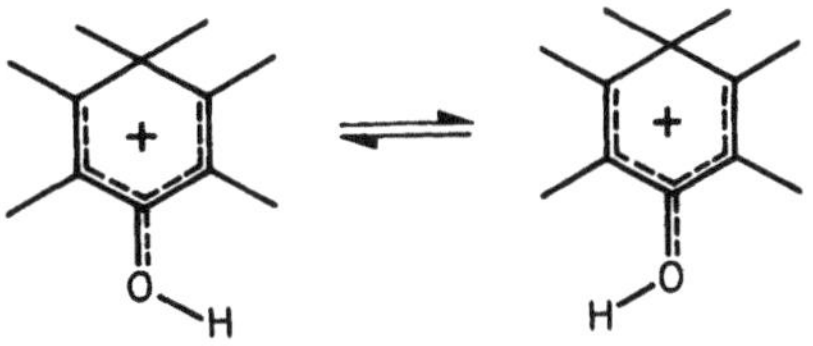

Table 11. PMR Data of 4-Hydroxybenzenium Ions and their O-Derivatives

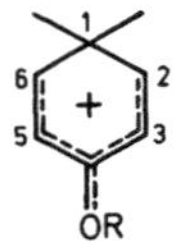

Substituents	Acid system	Temp. °C	Chemical shifts (ppm) of substituent and ring (in parentheses) protons				Standard	Ref.
			C_1	$C_{2,6}$	$C_{3,5}$	OR(OH)		
—; R=H	HSO_3F	−64	(4.34)	(8.87)	(7.50)		TMS, ext	68)
	HBr—2 $AlBr_3$—CH_2Br_2	−70	(4.46)	(8.70)	(7.54)	(10.40)	CH_2Br_2, int	69)
—; R=CH_3	HF—BF_3	−10	(4.28)	(8.62)	(7.52)	4.62	TMA, int	70)
	HF—BF_3	−83		(8.46, 8.78)	(7.40, 7.64)		TMA, int	34, 71)
	HSO_3F—SO_2		(4.5)	(8.6, 9.0)	(7.5, 7.8)			188)
	HSO_3F—SbF_5	−64	(4.46)	(8.79)	(7.66, 8.02)	4.79	TMS, ext	68, 271)
	HF—SbF_5—SO_2ClF			(8.85, 9.17)	(7.82)	4.95	TMS, ext	63)
1,1-$(CH_3)_2$; R=H	HSO_3F	−50	1.61	(8.48)	(7.20)		TMA, int	324, 325)
	H_2SO_4 (70%)	+36	1.45	(8.11)	(6.98)		TMA, int	349)
1-CH_3, 1-$CHCl_2$; R=H	H_2SO_4 (90%)	+35	1.65	(8.33)	(7.36)		TMA, int	224)
	H_2SO_4	+35	1.75	(8.42)	(7.39)		TMS, ext	326)
2-CH_3; R=H	HSO_3F	−67	(4.29)	2.76 (8.50)	(7.28, 7.35)			68)
	HBr—2 $AlBr_3$—CS_2	−70	(4.29)	2.66 (8.44)	(7.25, 7.32)	(9.99)	CH_2Br_2, int	69)
1,2-$(CH_3)_2$; R=H	HSO_3F	−50	1.37 (3.42)				TMS, ext	319)
	HSO_3F	−50	1.72 (3.92)	2.69 (8.31)			TMA, int	324)
	HSO_3F	−50	1.62 (3.88)	2.66 (8.26)			TMA, int	320)
	HSO_3F—SbF_5	−60	1.73 (3.90)	2.68 (8.34)	(7.10)		TMA, int	327)
1,1,2-$(CH_3)_3$; R=H	H_2SO_4 (70%)	+36	1.45	2.47 (8.20)	(7.01, 7.05)		TMA, int	349)
1,2-$(CH_3)_2$-1-$CHCl_2$; R=H	H_2SO_4	+35	1.70	2.62 (8.53)	(7.15, 7.38)		TMS, ext	326)
2,6-$(CH_3)_2$; R=H	HSO_3F	−50	(4.19)	2.62	(7.04)		TMA, int	324, 325)
	HSO_3F	−50	(3.72)	2.21	(6.71)		TMS, ext	319)
	HSO_3F	−50	(4.10)	2.58	(7.07)		TMA, int	320)
	HSO_3F—SbF_5	−60	(4.10)	2.60	(7.03)		TMA, int	327)
R=CH_3	HF	−20	(4.05)	2.57[a]	(7.12)	4.49	TMA, int	70)

1,2,6-$(CH_3)_3$; R=H	HSO_3F	−50	1.38 (3.37)	2.22	(6.65)		TMS, ext	319),
	HSO_3F	−50	1.70 (3.76)	2.63	(7.03)		TMA, int	324)
	HSO_3F	−50	1.69 (3.76)	2.60	(7.02)		TMA, int	321)
R=CH_3	HSO_3F	−40	1.32 (3.40)	2.27	(6.75)	3.98	TMS, ext	319)
	HSO_3F	−64	1.32 (3.40)	2.17, 2.30	(6.60, 6.93)	3.98	TMS, ext	319)
R=$COCH_3$	HSO_3F	−54	1.36 (3.43)	2.27	(6.78)	2.50	TMS, ext	319)
1,1,2,6-$(CH_3)_4$, R=H	H_2SO_4 (70%)	+36	1.49	2.51	(7.00)		TMA, int	349)
1,2,6-$(CH_3)_3$-1-$CHCl_2$; R=H	H_2SO_4 (90.5%)	+35	1.73	2.69	(7.14)		TMA, int	224)
3-CH_3; R=H	$HSO_3F—SbF_5$	−60	(4.28)	(8.40, 8.62)	2.38 (7.42)		TMA, int	327)
	HBr—2 $AlBr_3—CS_2$	−70	(4.39)	(8.43, 8.67)	2.29 (7.52)	(10.34)	CH_2Br_2, ext	69)
3-$\overset{+}{S}$(H)CH_3; R=H	$HSO_3F—SbF_5$	−60	(4.90)	(9.20, 9.32)	(7.96)	4.96	TMA, int	321)
1,3-$(CH_3)_2$; R=H	HSO_3F	−50	1.58 (4.10)		2.34		TMA, int	320)
1,1,3-$(CH_3)_3$; R=H	H_2SO_4 (70%)	+36	1.43	(7.92, 8.13)	2.17 (7.05)		TMA, int	349)
1,3-$(CH_3)_2$-1-$CHCl_2$; R=H	H_2SO_4	+35	1.73	(8.20, 8.40)	2.35 (7.43)		TMS, ext	326)
2,3-$(CH_3)_2$; R=H	HSO_3F	−50	(4.25)	2.64 (8.36)	2.32 (8.28)		TMA, int	324, 325)
	HSO_3F	−50	(4.25)	2.62 (8.40)	2.28 (8.32)		TMA, int	320)
	HSO_3F	−60	(4.28)	2.66 (8.34)	2.38 (8.28)		TMA, int	327)
2,5-$(CH_3)_2$; R=H	HSO_3F	−50	(4.17)	2.67 (8.16)	2.33 (7.13)		TMA, int	324)
	$HSO_3F—SbF_5$	−60	(4.18)	2.64 (8.16)	2.34 (7.18)		TMA. int	327)
R=CH_3	HF	−20	(4.15)	2.68 (7.93)	2.26 (7.39)	4.57	TMA, int	70)
2-$CH(CH_3)_2$-5-CH_3; R=H	HSO_3F	−45	(4.23)	1.45, 3.15- (8.20)	2.35 (7.25)		TMA, int	260)
3,5-$(CH_3)_2$; R=H	HSO_3F	−50	(4.27)	(8.45)	2.46		TMA, int	324)
	HSO_3F	−50	(4.25)	(8.46)	2.40		TMA, int	320)
	$HSO_3F—SbF_5$	−60	(4.26)	(8.39)	2.42		TMA, int	327)
	HBr—2 $AlBr_3—CS_2$	−70	(4.39)	(8.43)	2.38	(9.65)	CH_2Br_2, int	69)
1,2,3-$(CH_3)_3$; R=H	HSO_3F	−50	1.64 (3.92)	2.59 (8.34)	2.27		TMA, int	324)
	HSO_3F	−50	1.65 (3.92)	2.60 (8.66)	2.31 (7.25)		TMA, int	320)
1,2,5-$(CH_3)_3$; R=H	HSO_3F	−50	1.65 (3.83)	2.63 (8.13)	2.31 (7.13)		TMA, int	324)
	HSO_3F	−50	1.65 (3.84)	2.62 (8.15)	2.29 (7.15)		TMA, int	320)
1,3,5-$(CH_3)_3$; R=H	HSO_3F	−50	1.62 (3.98)	(8.33)	2.43		TMA, int	324)
2,3,5-$(CH_3)_3$; R=H	HSO_3F	−50	(4.21)	2.63 (8.14)	2.34, 2.36		TMA, int	324)
	$HSO_3F—SbF_5$	−60	(4.22)	2.62 (8.13)	2.34, 2.38		TMA, int	327)
1,2,3,5-$(CH_3)_4$; R=H	HSO_3F	−50	1.66 (3.83)	2.60 (8.12)	2.33, 2.37		TMA, int	324)
2,3,6-$(CH_3)_3$; R=H	HSO_3F	−50	(4.11)	2.56	2.25 (7.05)		TMA, int	324)
	HSO_3F	−50	(4.12)	2.52	2.24 (7.07)		TMA, int	320)

Table 11. (continued)

Substituents	Acid system	Temp. °C	Chemical shifts (ppm) of substituent and ring (in parentheses) protons				Standard	Ref.
			C_1	$C_{2,6}$	$C_{3,5}$	OR(OH)		
1,2,3,6-$(CH_3)_4$; R=H	HSO_3F	−50	1.66 (3.76)	2.56	2.23 (7.01)		TMA, int	320)
1,1,2,3,6-$(CH_3)_5$; R=CH_3	CF_3SO_3H		1.52	2.47, 2.60	2.18 (7.13)	4.48	TMA, int	265)
2,3,5,6-$(CH_3)_4$; R=H	HSO_3F	−50	(4.15)	2.56	2.30		TMA, int	324)
	HSO_3F	−70	(4.18)	2.53	2.27	(10.01)	TMA, int	202)
	HSO_3F—SbF_5	−60	(4.15)	2.51	2.31		TMA, int	327)
	HBr—2 $AlBr_3$—CS_2	−70	(4.18)	2.47	2.18	(8.90)	CH_2Br_2, int	69)
R=CH_3	HSO_3F	−70	(4.06)	2.50	2.34	4.74	TMA, int	
1,2,3,5,6-$(CH_3)_5$; R=H	HSO_3F	−50	1.62 (3.72)	2.52	2.26		TMA, int	324)
	HSO_3F	−40	1.57 (3.77)	2.50	2.22		TMA, int	265)
	HSO_3F—SbF_5	−60	1.66 (3.83)	2.66	2.36		TMA, int	327)
	HSO_3F—HNO_3	−30	1.50[b] (3.68)	2.43	2.15		CH_2Cl_2, int	169)
	CF_3SO_3H		1.56 (3.73)	2.47	2.20		TMA, int	265)
	HBr—2 $AlBr_3$—CS_2	−70	1.49 (3.74)	2.47	2.18	(9.04)	CH_2Br_2, int	69)
R=CH_3	HSO_3F	−60	1.61 (3.73)	2.50	2.30	4.75	TMA, int	265)
1,1,2,3,5,6-$(CH_3)_6$; R=H	HSO_3Cl		1.49	2.52	2.27	(10.08)	TMA, int	218, 329)
	HSO_3F		1.44	2.44	2.20		CH_2Cl_2, int	332)
	HSO_3F	−50	1.51	2.51	2.27		TMA, int	324)
	HCl—2 $AlCl_3$—CH_2Cl_2		1.33	2.50	2.28	(9.1)	CH_2Cl_2, int	219)
R=CH_3	HSO_3F	+30	1.52	2.47	2.30	4.72	TMA, int	243)
1—C_2H_5-1,2,3,5,6-$(CH_3)_5$; R=H	CF_3COOH—H_2SO_4		1.43, 0.32, 2.24	2.42	2.23		TMA, int	330)
1-C_6H_5-1,2,3,5,6-$(CH_3)_5$; R=H	CF_3COOH—H_2SO_4		1.76	2.06	2.23		TMA, int	330)
1-X-2,3,5,6-$(CH_3)_4$;								
R=H; X=Br	HSO_3F	+25	(5.43)	2.69	2.30	(10.66)	TMA, int	201, 202)
X=SO_2F	HSO_3F	+20	(5.76)	2.74	2.37	(11.54)	TMA, int	202)
X=SO_3H	CF_3SO_3H	−40	(5.76)	2.70	2.32	(11.80)	TMA, int	202)
R=CH_3; X=Br	HSO_3F	0	(5.44)	2.60	2.3	4.86	TMA, int	201)
X=SO_3H	CF_3SO_3H	−40	(5.76)	2.64	2.34	4.94	TMA, int	202)

$R=AlBr_3$; X=Br	CH_2Br_2	+20	(5.31)	2.49	2.19		CH_2Br_2, int	[336]
1-X-1,2,3,5,6-$(CH_3)_5$;								
R=H; $X=NO_2$	$HSO_3F—SO_2ClF$	—70	2.10	2.50	2.38	(11.54)	TMA, int	[223]
	HSO_3F	—30	2.00	2.37	2.25		CH_2Cl_2, int	[169]
	HSO_3F	—30	2.09	2.46	2.34		TMA, int	[169]
$X=SO_3H$	HSO_3F	—30	2.26	2.72	2.32		TMA, int	[175]
X=F	HSO_3F	—70	1.80	2.55	2.23	(10.97)	TMA, int	[220]
X=Cl	HSO_3F	—70	1.90	2.67	2.28	(10.83)	TMA, int	[220]
	HSO_3F	—80	1.84	2.24	2.60		CH_2Cl_2, int	[175]
X=Br	HSO_3F	—70	1.98	2.83	2.33	(11.03)	TMA, int	[220]
$R=CH_3$; $X=NO_2$	HSO_3F	—20	1.96	2.29	2.29	4.87	CH_2Cl_2, int	[175]
$X=SO_3H$	HSO_3F	—30	2.30	2.62	2.40	4.91	TMA, int	[175]
X=Cl	HSO_3F	—40	1.77	2.48	2.20	4.78	TMA, int	[175]
1,1,2,5,6-$(CH_3)_5$-3-NO_2; R=H	HSO_3F	+20	1.70	2.60, 2.90	2.33		CH_2Cl_2, int	[173]
1,1,3,5,6-$(CH_3)_5$-2-NO_2; R=H	HSO_3F	+20	1.67	2.60	2.17, 2.31		CH_2Cl_2, int	[173]

[a] At —78 °C two signals; [b] Data of [169] are erroneous.

Other things being equal, a rotation of the HO group seems to be easier than that of the CH_3O group due to a smaller difference in the energies of rotamers between the HO group located in the ring plane and that with the group in the plane normal to the ring.[20] This conclusion is based on the assumption that the H—O bond is more capable of σ,π-conjugation with the unsaturated ion system than the CH_3—O bond [350]. Taking the substitution of the hydroxy group hydrogen by a metal to enhance even more the ability of the group for hyperconjugation [351] it is easy to understand why the $O\overline{Al}X_3$ group in bipolar complexes of the type

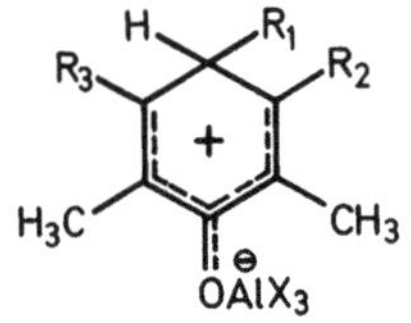

also proves to be rather a strong electron density donor [348] in spite of the steric inhibition to the location of the C—O—Al fragment in the ring plane.

The geometric isomers of the following type has been evidenced by PMR spectra for a large number of aliphatic hydroxycarbenium ions [136].

For asymmetric hydroxy- and methoxybenzenium ions such isomers could not be observed.

Data on proton chemical shifts of 4-hydroxy- and 2-hydroxybenzenium ions, as well as their O-derivatives are collected in Tables 10 and 11. The signals are assigned by comparison of PMR data for ions with various numbers and positions of substituents, while for hydroxyhexamethylbenzenium ions — additionally by selective deuteration of some of the CH_3 groups [218, 219].

The signal of the OH group of hydroxybenzenium ions due to the easy proton exchange with the acid medium can only be observed at low temperatures; its chemical shifts depends (cf. [219]) on the character of the acid medium.

The consideration of the data cited in Tables 10 and 11 reveal the following regularities:

1. The substitution of the hydroxy group hydrogen of hydroxybenzenium ions by the CH_3 and CH_3CO groups does not essentially affect the spectrum of the remaining part of the ion.

2. According to the greater electron-releasing ability of the RO group than the CH_3 group the signals of 4-hydroxybenzenium ions and their O-derivatives are located in somewhat higher fields than those of the respective methylbenzenium ions. In case the internal standard is used they are usually placed the following intervals (ppm):

---

20 These rotatory isomers can be regarded as n- and p-protonated cyclohexadienones [352,353] differing in the extent of transfer of the positive charge from the oxygen to the ring.

$C_1(R)$—CH_3	1.4–1.6	$C_{2(6)}$—CH_3	2.4–2.7	$C_{2(6)}$—H	8.2–8.8
$C_1(CH_3)$—H	3.4–3.7	$C_{3(5)}$—CH_3	2.2–2.4	$C_{3(5)}$—H	7.1–7.5
$C_1{<}^{H}_{H}$	4.1–4.4				

3. The signals of hydrogens and CH_3 groups located at the sp^2-hybridized carbons of 2-hydroxybenzenium ions and their O-derivatives are ordered according to the chemical shifts in the same sequence:

$$H_3 \leqq H_5 < H_6 < H_4 ,$$

$$3\text{-}CH_3 \leqq 5\text{-}CH_3 < 6\text{-}CH_3 < 4\text{-}CH_3 ,$$

reflecting an increased deficiency of π-electron density at the 4- and 6-positions.
4. The OH-group signals in the PMR spectra of hydroxybenzenium ions are usually located in a much higher field (by 4–6 ppm) than those in the spectra of the hydroxy-carbenium ions of the aliphatic series (cf. [136]); this is due to a more effective delocalization of the positive charge in the ions of the former type.

The spin-spin coupling constants of the ring protons for hydroxybenzenium ions and their O-derivatives are listed below (cps):

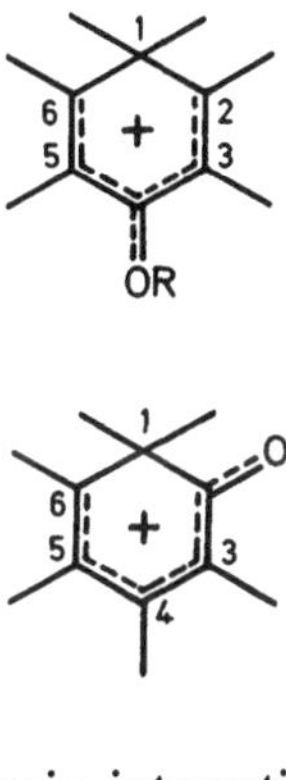

J_{12} and J_{16} = 2 - 3.5 [68,70,340]
J_{23} and J_{56} = 8 - 10 [69,70,324,326,340,349]
J_{26} and J_{35} = 1.0 - 2.5 [326,340,349]

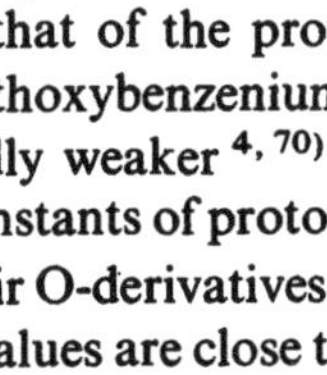

J_{16} = 0 [328], J_{34} = 7 - 9 [324,326]
J_{35} = 0 [326], J_{45} = 6 - 7 [326]
J_{46} = 1.5 - 2 [326], J_{56} = 8.5 - 9 [324,326]

The spin-spin interaction of the protons of the CH_2 group in 4-RO-benzenium ions[21] and that of the protons occupying the 2- and 6-positions (see the spectrum of the 4-methoxybenzenium ion in Fig. 3) is fairly large. For methylbenzenium ions it is markedly weaker [4,70] and does not display itself in the spectra. The spin-spin coupling constants of protons located at sp^2-hybridized carbons of hydroxybenzenium ions and their O-derivatives are also somewhat higher than those for methylbenzenium ions; their values are close to those recorded for the respective cyclohexadienones [326].

Summarized in Table 12 are the PMR data on a number of halogen-substituted hydroxy- and methoxybenzenium ions.

In Table 11 the chemical shifts of the CH_3-group protons in some 1-X-4-hydroxy-1,2,3,5,6-pentamethylbenzenium ions illustrate the effect of the X substi-

21 See also the data for polyhydroxybenzenium ions [354].

Table 12. PMR Data of Halogenated 4-Hydroxy- and 4-Methoxybenzenium Ions (TMS, ext. [340])

Substituents and their position	Acid system	Temp. °C	Chemical shifts (ppm)				
			C(H)(H)	H_2	H_3	H_5	H_6
4-OH, 3-F	$HF-SbF_5-SO_2ClF$	−20	4.80	9.00		7.83	8.50
$4-OCH_3$, 3-F	$HF-SbF_5-SO_2ClF$	−40	4.80	9.16		8.00	8.31
4-OH, 3-Cl	$HF-SbF_5-SO_2ClF$	−60	4.80	8.95		7.90	9.06
$4-OCH_3$, 3-Cl	HSO_3F-SbF_5	−40	4.63	8.74		7.90	9.04
4-OH, 3-Br	$HF-SbF_5-SO_2ClF$	−60	4.73	9.20		7.80	9.02
$4-OCH_3$, 3-Br	HSO_3F-SbF_5	−50	4.63	8.97		7.90	9.04
4-OH, 3-J	HSO_3F-SbF_5	−40	4.60	9.30		7.70	8.90
$4-OCH_3$, 3-J	$HSO_3F-SbF_5-SO_2ClF$	−80	4.50	8.60		7.80	9.00
4-OH, 2-F	$HF-SbF_5-SO_2ClF$	−16	4.50		7.48	7.31	8.40
$4-OCH_3$, 2-F	$HSO_3F-SbF_5-SO_2ClF$	−40	4.60		7.40	7.40	8.30
4-OH, 2-Cl	$HF-SbF_5-SO_2ClF$	−40	4.70		7.70	7.54	8.63
$4-OCH_3$, 2-Cl	$HF-SbF_5-SO_2ClF$	−30	4.60		7.60	7.70	8.80
4-OH, 2-Br	$HF-SbF_5-SO_2ClF$	−40	4.70		8.00	7.63	8.80
$4-OCH_3$, 2-Br	$HF-SbF_5-SO_2ClF$	−60	4.68		7.90	7.80	8.90
4-OH, 2-J	$HSO_3F-SbF_5-SO_2ClF$	−60	4.60		8.30	7.50	8.68
$4-OCH_3$, 2-J	$HSO_3F-SbF_5-SO_2ClF$	−60	4.62		8.22	7.88	8.90

tuent on the position of the $1-CH_3$-group signal. The authors [221, 222, 355] describe the generating of similar ions with X = OH, OCH_3 and $OCOCH_3$ by the protonation of the respective cyclohexadienones in HSO_3F at −70 °C. The analysis of the PMR spectra has shown that besides monoprotonation diprotonation results in the formation of dications (*48*). When SbF_5 is added to the system the equilibrium is, in fact, wholly shifted to the cations (*48*).

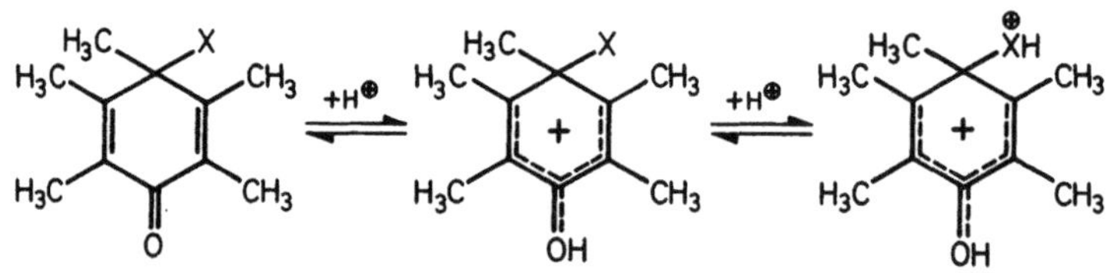

The PMR data of some 1-X-2-hydroxy-1,3,4,6-tetramethylbenzenium ions are given in the following publications:

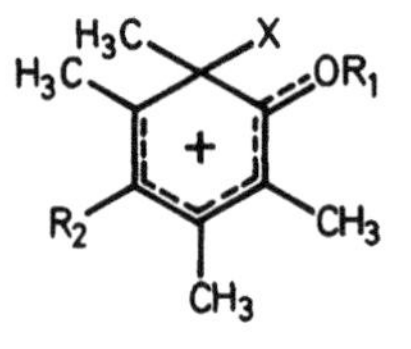

An interesting example of an ion with a meta-located oxygen function is the sulphonated 3-hydroxy-1,1,2,4,5,6-hexamethylbenzenium ion (*49*) formed by the action of SO_3 on 2- and 4-hydroxy-isomers. The chemical shifts of its CH_3-group protons are as follows:

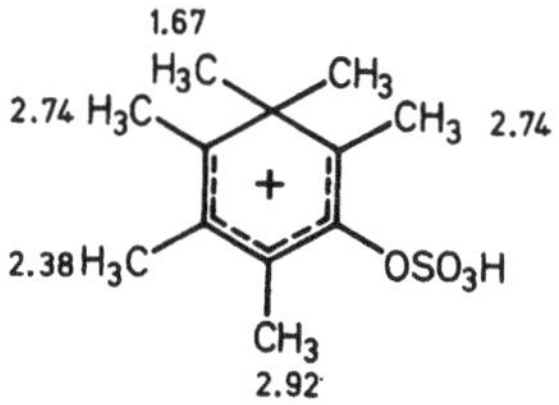

49

As indicated [324], a role similar to that of SO_3 is played by SbF_5 in the rearrangement of hexamethylcyclohexadienones. Similarly, the meta-orienting effect of the hydroxy group for the protonation of 2,4,6-trimethylphenol in $HSO_3F—SbF_5$ is attributed to the modification of the oxygen function by protonation, coordination with SbF_5 or formation of the $OSbF_5^-$ group [324].

The characteristics of the PMR spectra of 2-hydroxy-1-, 4-hydroxy-1- and 1-hydroxy-2-naphthalenium ions are listed in Tables 13–15. As shown with 2-naphthol-1-, 6- and 8-sulphonic acids, the PMR spectra of C-protonated forms can be effectively used to determine the position of the sulpho group in the ring [206].

The polycyclic hydroxyarenium ions studied by the PMR spectroscopy are: 9-H-10-(50b, c) [62]; hydroxyanthracenium ion (50a) [59, 60] and its 9,9dialkylderivatives also described are ions (*51*) and (*52*) generated by the protonation of the respective benzocyclohexadienones in the $CF_3COOH—H_2O—BF_3$ [60].

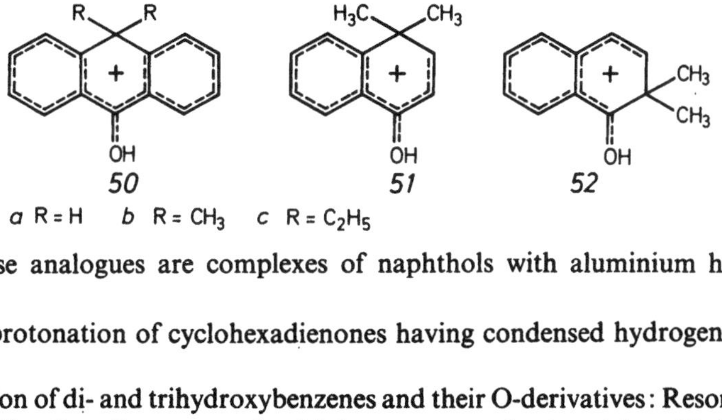

50 *51* *52*

a R = H *b* R = CH_3 *c* R = C_2H_5

Their close analogues are complexes of naphthols with aluminium halogenides (see p. 46).

For the protonation of cyclohexadienones having condensed hydrogenated cycles see [349].

Protonation of di- and trihydroxybenzenes and their O-derivatives: Resorcin and its ethers in acids of medium strength add the proton at the 4-position. Their chemical shifts are presented in Table 16.

1,2-Dialkoxybenzenes in $HF—BF_3$ at —20 °C are protonated at the 4-position:

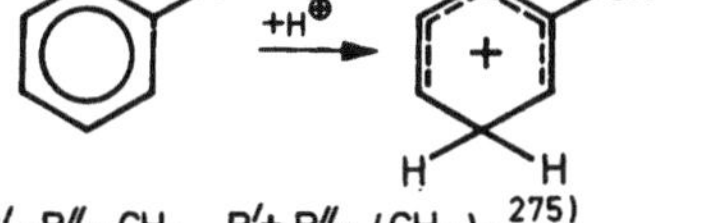

R' = R'' = CH_3, R' + R'' = $(CH_2)_2$ [275]

Table 13. PMR Data of 1-Hydroxy-2-Naphthalenium Ions and their O-Derivatives

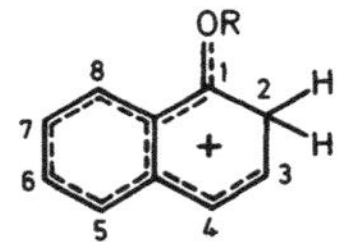

Substituents	Acid system	Temp. °C	Chemical shifts (ppm) of substituent and ring (in parentheses) protons				Standard	Ref.
			2	3	4	5–8		
R=H	$HSO_3F—SO_2F_2—SO_2ClF$	−70	(3.33)	(7.0)	(7.45)			203)
R=H; 4-CH_3	HSO_3F	−70	(4.26)	(6.74)	2.5			203)
R=$AlBr_3$; 4-CH_3	$CH_2Br_2—AlBr_3$	+25	(4.29)	(6.59)	2.42		CH, int	337)
R=H; 4-Br	HSO_3F	−70	(4.39)	(7.26)		7.95(H^7)[a], 8.71(H^8)[a], 8.2–8.6	TMA, int	203, 204)
R=$AlBr_3$; 4-Br	$CH_2Br_2—AlBr_3$	−50	(4.34)	(7.08)		7.76(H^7)[a]; 8.42(H^8)[a]	CH_2Br_2, int	336)
R=H; 4-Cl	HSO_3F	−70	(4.39)	(7.01)		7.95(H^7)[a]; 8.71(H^8)[a], 8.2–8.6	TMA, int	203, 204)
R=$AlBr_3$; 4-Cl	$CH_2Br_2—AlBr_3$	−50	(4.31)	(7.09)		7.79(H^7)[a]; 8.45(H^8)[a]	CH_2Br_2, int	336)
R=$AlBr_3$; 4-C_6H_5	$CH_2Br_2—AlBr_3$	+25	(4.47)	(6.62)		8.53(H^8); 7.0–8.2	CH, int	337)

[a] or H^6 and H^5.

Table 14. PMR Data of 2-Hydroxy-1-Naphthalenium Ions and their O-Derivatives

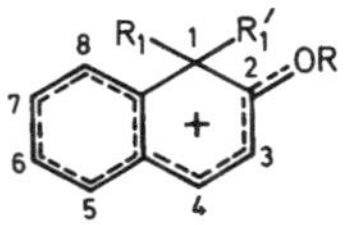

Substituents	Acid system	Temp. °C	Chemical shifts (ppm) of substituent and ring (in parentheses) protons					Standard	Ref.
			1	2	3	4	5–8		
$R_1=R_1'=R=H$	$HF—SbF_5—SO_2ClF$		(5.10)		(7.57)	(9.51)	(8.3)	TMS, ext	300)
$R_1=R_1'=H$; $R=AlCl_3$	$CH_2Cl_2—AlCl_3$	+25	(4.58)		(7.02)	(8.60)	(7.5–7.9)	CH, int	334)
$R_1=R_1'=H$; $R=AlBr_3$	$CH_2Br_2—AlBr_3$	+25	(4.73)		(7.18)	(8.74)	(7.6–8.2)	CH, int	334)
$R_1=R_1'=H$; $R=CH_3$	$HF—SbF_5—SO_2ClF$		(5.00)	4.92	(7.63)	(9.40)	(8.3)	TMS, ext	300)
	CF_3SO_3H	+20	(4.8)	4.66	(7.34)	(9.1)	(7.7–8.2)	TMA, int	206)
$R_1=CH_3$; $R_1'=H$; $R=AlCl_3$	$CH_2Cl_2—AlCl_3$	+25	1.77 (4.22)		(7.08)	(8.55)	(7.4–8.1)	CH, int	334)
$R_1=CH_3$; $R_1'=H$; $R=AlBr_3$	$CH_2Br_2—AlBr_3$	+25	1.85 (4.34)		(7.26)	(8.68)	(7.4–8.1)	CH, int	336)
$R_1=Br$; $R_1'=H$; $R=AlBr_3$	$CH_2Br_2—AlBr_3$	−50	(6.07)		(7.17)	(8.60)	(7.8)	CH_2Br_2, int	204, 205)
$R_1=Br$; $R_1'=R=H$	HSO_3F	−70	(6.30)		(7.23)	(9.05)	(7.8–8.2)	TMA, int	204, 205)
$R_1=Br$; $R_1'=H$; $R=CH_3$	HSO_3F	−70	(6.30)	4.76	(7.33)	(9.10)	(7.1–8.3)	TMA, int	204, 205)
$R_1=Cl$; $R_1'=H$; $R=CH_3$	HSO_3F	−70	(6.00)	4.76	(7.33)	(9.30)	(7.1–8.3)	TMA, int	204, 205)
$R_1=SO_3H$; $R_1'=R=H$	HSO_3F	−60	(6.30)		(7.52)	(9.33)	(8.0–8.5)	TMA, int	206)
	CF_3SO_3H	−30	(6.20)		(7.33)	(9.20)	(7.3–8.6)	TMA, int	206)
$R_1=SO_3H$; $R_1'=H$; $R=CH_3$	HSO_3F	−60	(6.23)	4.91	(7.44)	(9.60)	(8.0–8.5)	TMA, int	206)
	CF_3SO_3H	−30	(6.12)	4.84	(7.42)	(9.24)	(7.5–8.6)	TMA, int	206)
$R_1=R_1'=H$; $R=AlBr_3$; 4-C_6H_5	$CH_2Br_2—AlBr_3$	+25	(4.75)		(7.11)		(7.5–7.9)	CH, int	337)

Table 15. PMR Data of 4-Hydroxy-1-Naphthalenium Ions and their O-Derivatives

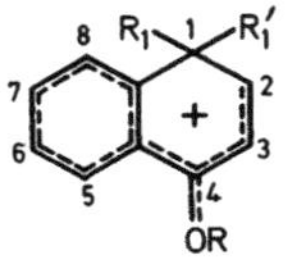

Substituents	Acid system	Temp. °C	Chemical shifts (ppm) of substituent and ring (in parentheses) protons						Standard	Ref.
			1	2	3	4	5–7	8		
$R_1=R_1'=R=H$	$HSO_3F—SO_2ClF$	−40	(4.7)	(8.9)	(7.70)		(8.2)	(8.9)	TMS, ext	[300]
	$HSO_3F—SO_2F_2—SO_2ClF$	−60	(4.64)		(7.56)					[203]
$R_1=R_1'=H$; $R=CH_3$	$HSO_3F—SO_2ClF$	−30	(4.90)	(9.20)	(8.00)	5.04	(8.2)	(8.9)	TMS, ext	[300]
$R_1=R_1'=H$; $R=AlBr_3$	$CH_2Br_2—AlBr_3$	+25	(4.41)	(8.45)	(7.63)		(7.6–8.1)	(8.5)	CH, int	[334]
$R_1=CH_3$; $R_1'=R=H$	HSO_3F	−50	1.81 (4.40)		(7.40)					[203]
$R_1=Br$; $R_1'=R=H$	HSO_3F	−20	(6.45)	(8.45)	(7.26)			(8.65)	TMA, int	[203, 204]
$R_1=Br$; $R_1'=H$; $R=AlBr_3$		−40	(6.26)		(7.38)				CH_2Br_2, int	[336]
$R_1=R_1'=H$; $R=AlBr_3$; 2-CH_3	$CH_2Br_2—AlBr_3$	+25	(4.43)	2.48	(8.19)		(7.7–8.1)	(8.61)	CH, int	[334]

Table 16. PMR Data of 2,4-Dihydroxybenzenium Ions and its O-Derivatives

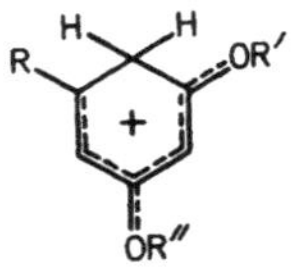

Substituents			Acid system	Temp. °C	Chemical shifts						Standard	Ref.
R	R'	R"			H_6	H_3 H_5	CH_2	R	R'	R"		
H	H	H	$HF—BF_3—(CH_2)_4SO_2$	+35	7.53	6.03–7.18	3.6–4.1				CH_2Cl_2, int	[347]
H	CH_3	CH_3	HF	−70	7.66	6.69; 6.95	4.01		4.33	4.45	TMA, int	[34, 70]
			H_2SO_4 (95%)	+38	7.74	6.53; 7.02	3.98		4.26	4.36	TMA, int	[346]
CH_3	H	H	$HF—BF_3—(CH_2)_4SO_2$	+35		6.16; 6.46	3.71	2.13			CH_2Cl_2, int	[347]
CH_3	C_2H_5	C_2H_5	$HF—BF_3—(CH_2)_4SO_2$	+35		6.30; 6.54	3.78	2.18	4.41 1.33	4.41 1.33	CH_2Cl_2, int	[347]

If R′ + R″ = $(CH_2)_3$ the conformation of a seven-membered ring distorts the coplanarity of the O—CH_2 groups and the benzene ring; as a result the ring basicity drops and no benzenium ion is observed [70].

For the protonation of the series of dihydroxynaphthalenes see [338].

G. Olah and Y. Mo [354] have studied the protonation of trihydroxybenzenes and their ethers in four acid systems of various strength:

A) $HSO_3F—SO_2ClF$;
B) $HSO_3F—SbF_5(4:1)—SO_2ClF$;
C) $HSO_3F—SbF_5(1:1)—SO_2ClF$;
D) $HF—SbF_5—SO_2ClF$.

In the first two systems phloroglucinol is protonated at —40 °C to form the 2,4,6-trihydroxybenzenium ion (*53*) generated earlier in other acids (in HSO_3F [68]; in a mixture of H_2SO_4 and CF_3COOH [346]; in 70% $HClO_4$ [68, 122, 346]; in sulpholane saturated by HF and BF_3 [347]). In the stronger system C phloroglucinol is protonated to yield the dication (*54*)[22]

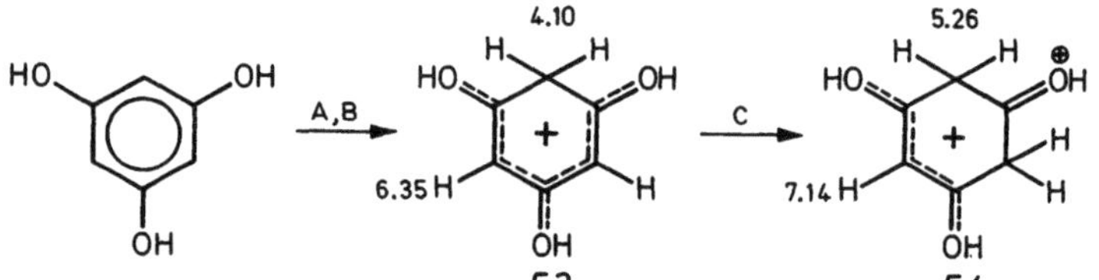

53 *54*

Similar is the behaviour of the phloroglucinol di- and trimethyl ethers (for their monoprotonation in different acid systems see also [70, 122, 346, 356, 357]), the dimethyl ether yielding two isomeric monocations and two dications.

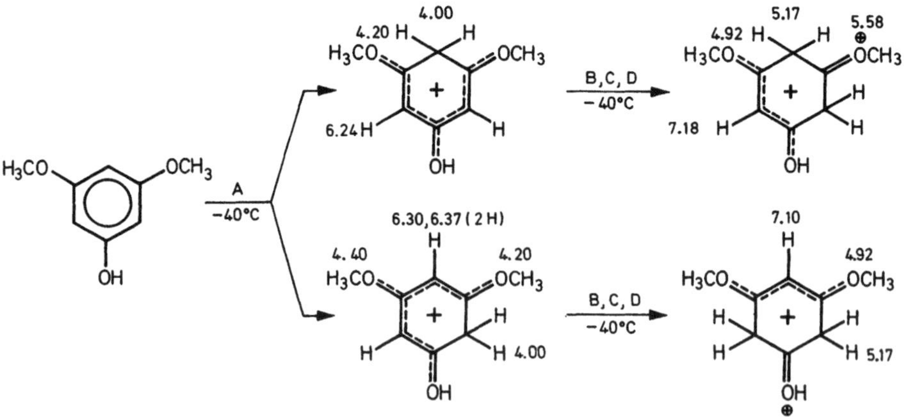

The C-monoprotonation of phloroglucinol monomethyl ether, as well as of mono-, di- and triethyl ethers in 70% $HClO_4$ and a mixture of H_2SO_4 and CF_3COOH

22 The chemical shifts shown in the following schemes have been measured relative to the TMS as the external standard.

is described [346]. Just as in the previous case, the PMR spectra demonstrated for mono- and diethers a mixture of isomeric ions of which the one containing the 4-hydroxy group (see Sect. IV.1.A) was predominant.

2,4,6-Trimethoxytoluene in system A is protonated both at an unsubstituted position and at a substituted one with a CH_3 group (80:20 at —70 °C); 2,4,6-trimethoxy-m-xylene, however, only at an unsubstituted carbon atom. In the systems B—D both compounds are subject to double protonation at the ring [358].

The PMR spectra of C-protonated polyphenol ethers have the signals of para alkoxy groups usually located in a weaker field than those of ortho groups [346, 354]. In several cases extra signals are due to decomposition products of alkoxybenzenium ions at the Alk—O bond [346].

Protonation of pyrogallol in the acid systems A and B at —20 °C yields the 2,3,4-trihydroxybenzenium ion (*55*) which in the system C is further protonated at the 3-hydroxy group [354].

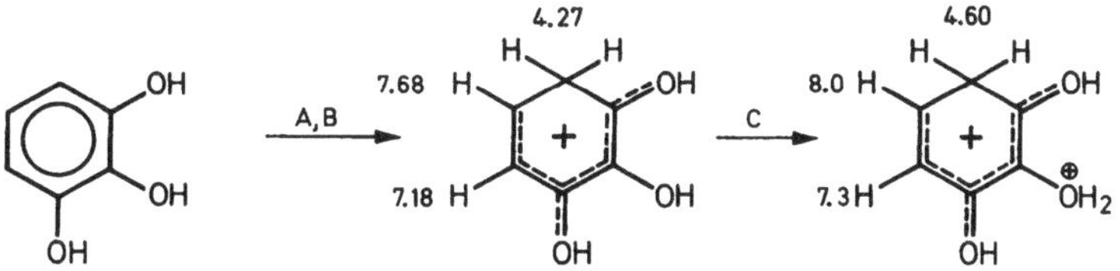

55

The PMR spectra of (*55*) reveal distinctly a strong spin-spin coupling (J_{16} = 3 cps, J_{56} = 10 cps) typical of hydroxybenzenium ions. Protonation of tri-, di- and monomethyl ethers of pyrogallol proceeds analogously but in the latter two cases mixtures of isomeric ions are formed. An example below illustrates the protonation of 2,3-dimethoxyphenol:

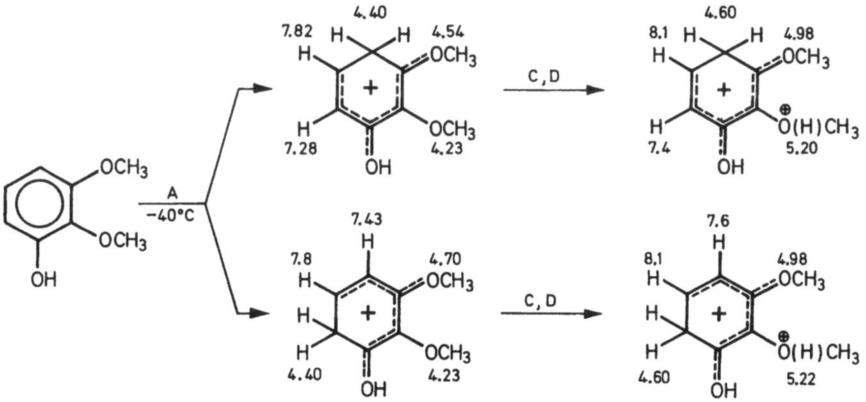

In all four acid systems 1,2,4-trihydroxybenzene yields a 2,4,5-trihydroxybenzenium ion

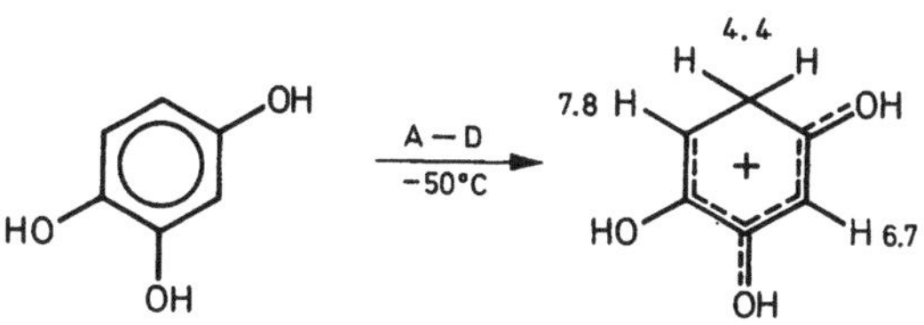

The spin coupling constant of the protons of the CH_2 group and the H_6 amounts to 4 cps.

Unlike 1,2,4-trihydroxybenzene, its trimethyl ether can be subject to both mono- and diprotonation.

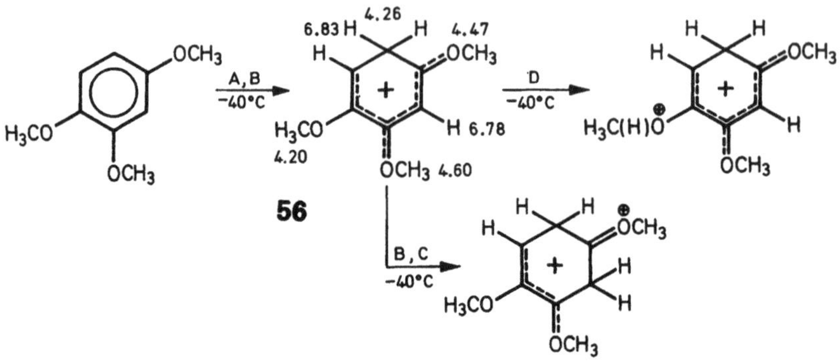

The direction of the second proton addition to 2,4,5-trimethoxybenzenium ion (*56*) proves to be different for systems C and D.

Among the C-substituted trihydroxybenzenes the protonation of 2,4,6-trihydroxybenzoic acid was studied in 70% $HClO_4$ and in HSO_3F at -60 °C [68].

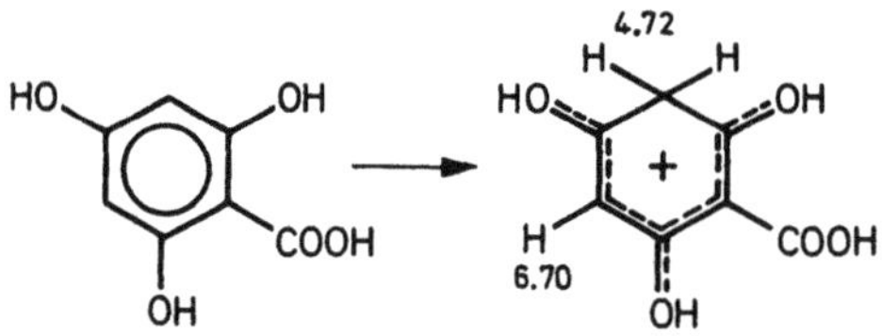

The strong stabilizing effect of hydroxy groups seems to make it possible to generate similar ions with other electronegative substituents as well.

C Aminobenzenium Ions

As noted in Sect. II, sym-tris(dialkylamino)benzenes interact with electrophilic reagents to generate triaminobenzenium ions (*29*). The ring hydrogens of these ions are listed in Table 17. The concurrence of the N- and C-protonation of triaminobenzenes is discussed in Sect. III.4.

Interesting results were obtained from the study [360] of hydrazobenzene in strongly acidic media by PMR. The dissolution of hydrazobenzene in HSO_3F-SO_2 or $HF-SO_2$ at -78 °C is accompanied by a rearrangement resulting in a benzenium dication (*57*) which, by adding SbF_5 to raise the acidity or by raising the temperature to about 25 °C, is converted into di-N-protonated benzidine (*58*).

The PMR spectrum of ion (*57*) shows a singlet at 4.38 ppm belonging to the proton of the $C_{sp3}-H$ fragment and doublets ($J = 10$ cps) at 7.00 and 7.60 ppm (TMS, ext) of the protons at sp^3-hybridized carbons.

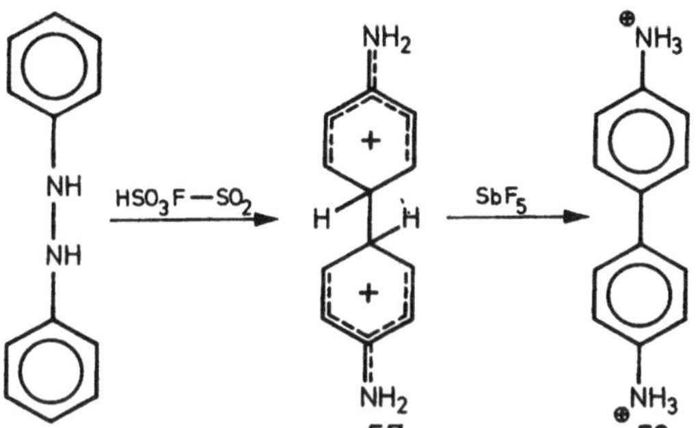

57 58

Table 17. PMR Data of 2,4,6-Triaminobenzenium Ions

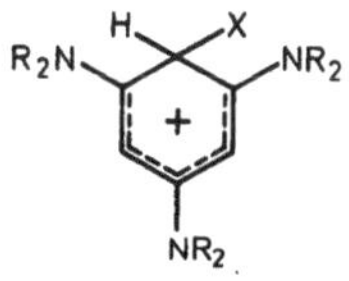

Substituents		Anion	Solvent	Chemical shifts (ppm) of ring protons		Ref.
X	NR_2			H_1	$H_{3,5}$	
	NH_2	Cl^-	$(CH_3)_2SO$	3.3	6.1	[124]
H	$N(CH_2)_4$	I^-	$CDCl_3$	3.80	4.83	[359]
Cl	$N(CH_2)_4$	Cl^-	$CDCl_3$	6.14[a]	4.71[a]	[197,199]
Cl	$N(CH_3)_2$	Cl^-	$CDCl_3$	5.88	5.01	[197]
Br	$N(CH_2)_4$	Br^-	$CDCl_3$	5.95[a]	4.80[a]	[197,199]
Br	$N(CH_2)_5$	Br^-	$CDCl_3$	5.89[a]	5.24[a]	[197,199]
Br	$N(CH_2CH_2)_2O$	Br^-	$CDCl_3$	6.18[a]	4.93[a]	[197,199]
Br	$N(CH_3)_2$	Br^-	$CDCl_3$	6.14[a]	4.88[a]	[197,199]
CH_3CO	$N(CH_2)_4$	Cl^-	$CDCl_3$	4.86	5.07	[198]
C_6H_5CO	$N(CH_2)_4$	Cl^-	$CDCl_3$	6.49	4.92	[198]
CH_3SO_2	$N(CH_2)_4$	Cl^-	$CDCl_3$	5.66	5.00	[198]
p-$C_7H_7SO_2$	$N(CH_2)_4$	Cl^-	$CDCl_3$	5.55	4.63	[198]
SCN	$N(CH_2)_4$	SCN^-	$CDCl_3$	5.08[a]	4.90[a]	[199]

[a] TMS as internal standard; in other cases — external

The authors [360] assume the dication (*59*) to be the immediate precursor of the ion (*57*).

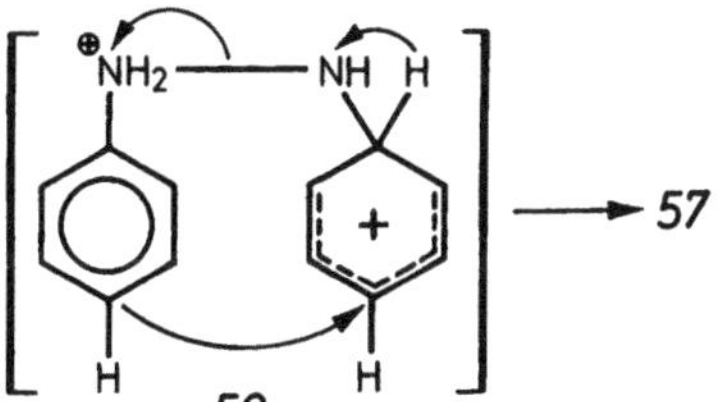

59

Considering, however, the reports [266, 267] on generating analogues of the dication (*57*) — the ions (*33*) — by oxidizing sym-tris(dialkylamino)benzenes the dication (*57*) may be obtained by the homolytical rupture of the N—N bond in the N,N'-diprotonated hydrazobenzene with the subsequent dimerization of radical cations of aniline.

Aniline, N,N-dimethylaniline, p-toluidine, diphenylamine and other aromatic amines in $HSO_3F—SbF_5—SO_2(SO_2ClF)$ at —60÷ 90 °C are only protonated at the amino group [360]. 2,4,6-Trimethylaniline at —78 °C, however, yields the dication (*60*):

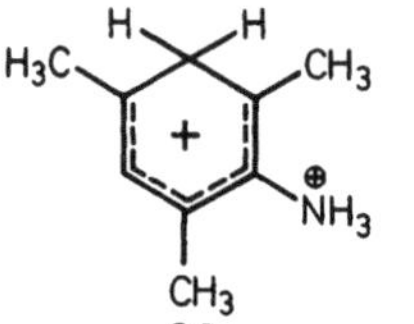

60

Possibly 2,4-dimethylaniline behaves likewise [4]; 2,4,6-trimethylphenol has been diprotonated on O- and meta-C [324], methyl(o-methoxyphenyl)sulphide on S- and C [321].

D Halogen-Substituted Arenium Ions

Halogen substituents are characterized by the duality of their electronic effect. Owing to their -I inductive effect they deactivate the benzene ring when it is attacked by electrophilic reagents. At the same time, due to unshared electron pairs, the halogen atoms delocalize the positive charge of benzenium ions by σ,π-conjugation; this fact accounts for their ortho-para-orienting effect in electrophilic substitutions.

To judge by the values of σ^+_{p-X} (+0.135; +0.150; +0.114 and —0.073 for I, Br, Cl and F, respectively [361]), only for fluorine the latter effect predominates over the former. Consequently, among the monohalobenzenium ions the most stable is the 4-fluorobenzenium ion.

Several halogen-containing ions have been generated by the protonation of halomethylbenzenes.

In all examples of generating halobenzenium ions by protonation of the halobenzene the sp^3-hybridized carbon[23] of ions is bound with two hydrogens or with a hydrogen and a methyl group. The arenium ions with two halogens at the sp^3-hybridized carbon can be obtained by removing the halogen anion from perhalogenated dihydroaromatic compounds. Evidence for these ions is found in Sect. III.3.

As pointed out earlier, substituents of the +M-type delocalize more effectively the positive charge of benzenium ions if they are para to the proton than in case of their ortho location. Accordingly the fluoro-, chloro- and bromobenzenes are protonated, to judge from the NMR [67, 363, 364], mostly at the para-position. Nevertheless for chlorine and bromine the difference between the electron effect from the para- and that from the ortho-position is not so pronounced as with the fluorine

23 Protonation of bromomesitylene, together with 3-bromo-2,4,6-trimethylbenzenium ion, yielded a small amount of an ion seeming to attach the proton to the carbon bonded with bromine [362].

atom[24]. This is evidenced, in particular, by a small amount of 2-halo-6-methylbenzenium ions by protonation of m-chloro- and m-bromotoluenes [363)]. For example, m-bromotoluene in HF—SbF_5 yields at —35 °C the following isomeric ions:

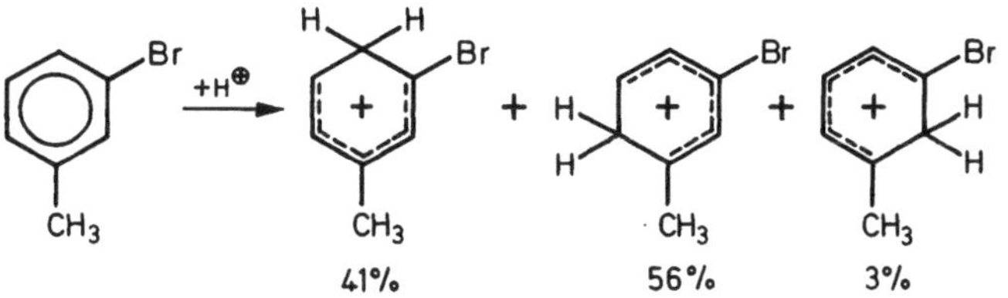

In the case of m-fluorotoluene no 2-fluoro-6-methylbenzenium ion has been detected [146, 363)].

The PMR spectra of fluorine-containing benzenium ions usually consist of rather complicated multiplets. This is due to the overlap of the spin-spin interaction of protons and the $^1H—^{19}F$ interaction. Fig. 5, e.g., presents the spectrum of the 2,4-

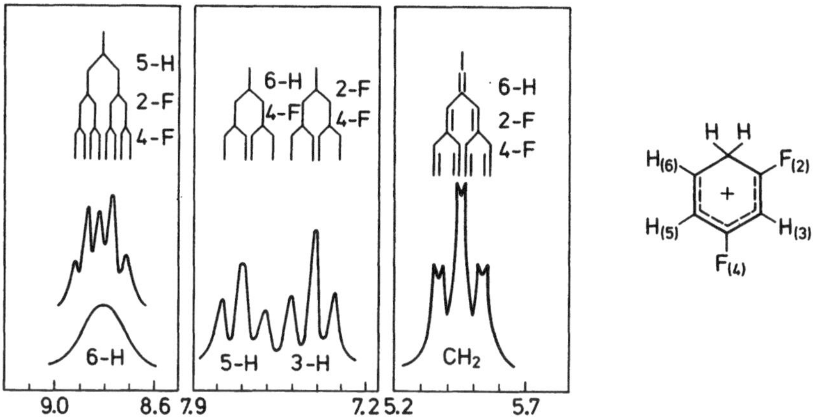

Fig. 5. PMR spectrum (100 MHz) of hexafluorantimonate of the 2,4-difluorobenzenium ion at —10 °C (the H_6 signal is recorded repeatedly with CH_2 irradiation) [363)]

Table 18. Coupling Constants $J_{^1H-^{19}F}$ (Hz) for Fluorobenzenium Ions [73,146,283,340,362,363,365)]

Position of fluorine atom	CH_2	ortho-H	meta-H
ortho-F	7.5–11[a]	—	8–10
meta-F	0	10	—
para-F	7.5–11[b]	5–5.5	7–10
meta′-F	0	0	4–6
ortho′-F	7.5–9.5	5–5.5	0–1.5

[a] 3.5 Hz for 2-fluoro-4-hydroxybenzenium ion [340)]; [b] 5 Hz in some cases [146)]

24 For the ratio of the para- and ortho-effects of halogen atoms see Sect. IV.1.A.

Table 19. ^1H and ^{19}F NMR Data of Halogenated Benzenium Ions

Substituents	Acid system	Temp. °C	Chemical shifts of protons (δ_{TMS}) at				Chemical shifts of fluorine atoms ($\delta_{C_6F_6}$)[a] at			Standard (PMR)	Ref.
			C_1	$C_{2,6}$	$C_{3,5}$	C_4	$C_{2,6}$	$C_{3,5}$	C_4		
4-F	HSO_3F—SbF_5	—40	5.50	10.00	8.40				—151.7	TMS, ext	[67]
	HSO_3F—SbF_5								—152.2		[4]
	HSO_3F—SbF_5—SO_2ClF	—80							—151.9		[366]
	HF—SbF_5—SO_2ClF	—84	5.43	9.93	8.33				—152.9	TMS, ext	[283]
2,4-F_2	HF—SbF_5	—10	4.95	8.80	7.42; 7.70		—153.8		—160.8	TBC, int	[4,363]
	HSO_3F—SbF_5	—40	5.32	8.94	7.64		—154.6		—159.2	TMS, ext	[67]
	HF—SbF_5—SO_2ClF	—30	5.40	9.3	7.90; 8.10		—155.6		—162.9	TMS, ext	[283]
2,4-Cl_2	HF—SbF_5	—10	4.93	8.72	7.98; 8.16					TBC, int	[363]
2,5-F_2	HF—SbF_5—SO_2ClF	—82	5.73	8.6	8.3	9.9	—162.4	56.9		TMS, ext	[283]
3,4-F_2	HF—SbF_5—SO_2ClF	—103	5.80	9.5	8.7			—38.5	—130.5		[367]
2,4,5-F_3	HF—SbF_5—SO_2ClF	—78	5.43	8.5	8.9		—144.1	—37.3	—161.9	TMS, ext	[283]
2,4,6-F_3	HF—SbF_5—SO_2ClF	—78	5.30		7.64		—141.4		—170.4	TMS, ext	[283]
	HSO_3F—SbF_5	—20	4.95							TBC, int	[67]
	HSO_3F—SbF_5	—40	5.35				—141.7		—168.3	TMS, ext	[283]
2,6-Cl_2-4-F	HSO_3F—SbF_5	—15	5.13		7.81				—156.0	TBC, int	[365]
2,4-Cl_2-6-F	HSO_3F—SbF_5	—15	4.93		7.81; 8		—146.0			TBC, int	[365]
2,4,6-Cl_3	HF—SbF_5	—30	4.98		8.00					TBC, int	[363]
	HSO_3F—SbF_5	—20	5.00		8.07					TBC, int	[365]
2,4,6-Br_3	HSO_3F—SbF_5	—20	4.36		8.35					TBC, int	[365]
2,6-Br_2-4-F	HSO_3F—SbF_5	—20	4.87		7.97				—151.5	TBC, int	[365]
2,4-Br_2-6-F	HSO_3F—SbF_5	—20	4.39		7.97, 8.07		—146.5			TBC, int	[365]
2,6-Br_2-4-Cl	HSO_3F—SbF_5	—20	4.80		8.28					TBC, int	[365]
2,4-Br_2-6-Cl	HSO_3F—SbF_5	—20	4.50		8.28; 8.53					TBC, int	[365]
4-Br-2,6-Cl_2	HSO_3F—SbF_5	—20	4.55		8.28					TBC, int	[365]

2-Br-4,6-Cl_2	$HSO_3F—SbF_5$	—20	4.90		8.05, 8.28					TBC, int	[365]
2-Br-4-Cl-6-F	$HSO_3F—SbF_5$	—20	4.85		7.79, 8.28					TBC, int	[365]
4-Br-2-Cl-6-F	$HSO_3F—SbF_5$	—20	4.44		8.03, 8.28					TBC, int	[365]
2-Br-6-Cl-4-F	$HSO_3F—SbF_5$	—20	5.00		7.79, 8.03				—153.6	TBC, int	[365]
2,4,5-F_3	$HSO_3F—SbF_5$		5.69	8.94	8.20		—145.2	—44.7	—163.6	TMS, ext	[67]
2,3,4,6-F_4	$HSO_3F—SbF_5$		5.0		7.2		—112.9	—19.9	—148.7	TMS, ext	[67]
	$HF—SbF_5—SO_2ClF$	—78	5.62		7.8		—109.7	—10.9	...	TMS, ext	[283]
2,3,4,5-F_4	$HF—SbF_5—SO_2ClF$	—83	5.8	8.8			—121.8	—19.6–25.4	—134.9	TMS, ext	[283]
2,3,5,6-F_4	$HF—SbF_5—SO_2ClF$	—80	5.8			8.7	—117.2	—41.5		TMS, ext	[283]
2,3,4,5,6-F_5	$HF—SbF_5—SO_2ClF$	—78	5.8				—115.3	—16.3	—129.4	TMS, ext	[283]

[a] $\delta_{C_6F_6(ext)} = -162.9 + \delta_{CCl_3F(ext)}$ [368]

Table 20. ^{1}H and ^{19}F NMR Data of Halogenated Methylbenzenium Ions

Substituents	Acid system	Temp. °C	Chemical shifts of substituent and ring (in parentheses) protons (δ_{TMS}) at				fluorine atoms ($\delta_{C_6F_6}$)[a] at			Standard (PMR)	Ref.
			C_1	$C_{2,6}$	$C_{3,5}$	C_4	$C_{2,6}$	$C_{3,5}$	C_4		
3-CH_3-4-F	HSO_3F—SbF_5	—60	(5.44)	(9.76)	2.84 (8.46)				—147.2	TMS, ext	146)
2-CH_3-4-F	HF—SbF_5	—30	(4.87)	3.00 (9.06)	(7.60, 7.70)				—142.0	TBC, int	4,363)
	HSO_3F—SbF_5	—70	(5.36)	3.53 (9.84)	(8.30)				—143.6	TMS, ext	146)
2-CH_3-4-Cl	HF—SbF_5	—40	(4.80)	2.94 (8.78)	(7.95, 8.07)					TBC, int	363)
2-CH_3-4-Br	HF—SbF_5	—35	(4.41)	2.85 (8.63)	(8.06, 8.23)					TBC, int	363)
4-CH_3-2-F	HF—SbF_5	—30	(4.6)	...	(7.87)	3.13	—139.9			TBC, int	363)
4-CH_3-2-Cl	HF—SbF_5	—40	(4.89)	(8.72)	(7.95, 8.07)	3.08				TBC, int	363)
4-CH_3-2-Br	HF—SbF_5	—35	(4.80)	(8.85)	(7.94, 8.23)	2.99				TBC, int	363)
2-CH_3-6-Cl	HF—SbF_5	—40	(5.06)	...	...					TBC, int	363)
2-CH_3-6-Br	HF—SbF_5	—35	(4.97)	...	...	...				TBC, int	363)
3-CH_3-6-F	HSO_3F—SbF_5	—75	(5.40)	(8.96)	2.84 (8.24)	(9.94)	—148.7			TMS, ext	146)
2,4-$(CH_3)_2$-6-F	HF—SbF_5	—20	(4.59)	2.78	(7.44, 7.67)	3.01	—129.8			TBC, int	4,362)
2,4-$(CH_3)_2$-6-Cl	HF—SbF_5	—20	(4.73)	2.80	(7.73, 7.77)	2.96				TBC, int	362)
2,4-$(CH_3)_2$-6-Br	HF—SbF_5	—20	(4.68)	2.78	(7.75, 8.02)	2.89				TBC, int	362)
2,6-$(CH_3)_2$-4-F	HF—SbF_5	—20	(4.69)	2.85	(7.44)				—136.3	TBC, int	4,362)
2,6-$(CH_3)_2$-4-Cl	HF—SbF_5	—20	(4.62)	2.80	(7.80)					TBC, int	362)
2,6-$(CH_3)_2$-4-Br	HF—SbF_5	—20	(4.62)	2.73	(7.80)					TBC, int	362)
3-CH_3-4,6-F_2	HSO_3F—SbF_5	—40	(5.40)	(9.03)	3.03 (7.97)		—150.7		—160	TMS, ext	362)
4-CH_3-2,5-F_2	HSO_3F—SbF_5	—40	(5.54)	(8.3)	(8.3)	3.53	—147.8	—59.1		TMS, ext	146)
2,4-$(CH_3)_2$-5-Cl	HSO_3F—SbF_5	—40	(4.95)	2.92 (8.63)	(8.03)	3.04				TBC, int	362)
2,5-$(CH_3)_2$-4-Cl	HF—SbF_5	—40	(4.78)	2.87 (8.67)	2.57 (8.00)					TBC, int	362)
2,4,6-$(CH_3)_3$-3-F	HF—SbF_5	—20	(4.78)	2.69, 2.81	(7.71)	2.90		—43.7		TBC, int	4,362)
	HSO_3F—SbF_5	—50	(4.99)	...	(7.85)	...		—51.7		TMS, ext	146)
2,4,6-$(CH_3)_3$-3-Cl	HF—SbF_5	—20	(4.80)	2.81, 2.83	(7.77)	3.03				TBC, int	362)
2,4,6-$(CH_3)_3$-3-Br	HF—SbF_5	—20	(4.83)	2.78, 2.86	(7.77)	3.06				TBC, int	362)
2,4,6-$(CH_3)_3$-3,5-F_2	HSO_3F—SbF_5	—40	(5.38)	3.20		3.33		—59.2		TMS, ext	146)

4-CH_3-2,3,5,6-F_4	$HSO_3F—SbF_5$	−50	(5.60)			3.50	−108.3	−40.7		TMS, ext	146)
1,3,5-$(CH_3)_3$-2,4,6-F_3	$HSO_3F—SbF_5$	−40	2.42 (4.88)		2.70		−127.1		−161.1	TMS, ext	146)
1,3,5-$(CH_3)_3$-1-Cl--2,4,6-F_3	$HSO_3F—SbF_5$-SO_2ClF	−70	2.60		2.65		−119.8		−182.1		166)
1,3,5-$(CH_3)_3$-1--NO_2-2,4,6-F_3	$HSO_3F—SO_2$	−70	2.26		2.33		−119.9		−179.1		166)
1,2,3,5,6-$(CH_3)_5$--1-NO_2-4-Cl	$HSO_3F—SO_2$	−80	2.07	2.62	2.48					TMS, ext	174)
	HSO_3F	−70	..	2.65	2.22					TMA, int	331)
1,2,3,5,6-$(CH_3)_5$-4-F	$HSO_3F—SbF_5$	−70	1.75 (4.35)	2.70	2.25					TMA, int	263)
	$HF—SbF_5—SO_2ClF$	−80	1.77 (4.30)	2.74	2.34				−131.6	TMS, ext	174, 366)
1,2,3,5,6-$(CH_3)_5$--1-NO_2-4-F	$HSO_3F—SO_2$	−80	2.02	2.57	2.24					TMS, ext	174)
	HSO_3F	−70	2.23	2.75	2.44					TMA, int	331)
1,1,3,4,5,6-$(CH_3)_6$--2-Cl	HSO_3F	+15	1.77	2.74	2.45, 2.59	2.93				TMA, int	177)

[a] See footnote in table 19

difluorobenzenium ion with a schematic description of signal splitting. The observed coupling constants of ^1H and ^{19}F nuclei depending on their position in the benzenium ion are given in Table 18. For the respective data on the monofluoronaphthalenium ions see [300].

The spin-spin interaction constants for the protons of the CH_2 and p-CH_3 groups in methylhalobenzenium ions range from 3 to 4.5 cps, while the protons of H_2 and H_3 range from 8 to 10 cps [362,363].

The proton chemical shifts of halobenzenium ions (Tables 19 and 20) obtained using internal standards show some regularities in the influence of halogen atoms (cf. [365]).

Thus, for 2,4-di- and 2,4,6-trihalobenzenium ions the CH_2-group signal is determined in the first place by the halogen at the 4-position and to a small extent by the halogens at the 2- and 6-positions. If the 4-position is occupied by fluorine or chlorine, the signal of the CH_2 fragment ranges from 4.80 to 5.10 ppm while that of 4-bromo-substituted benzenium ions ranges from 4.35 to 4.55 ppm. For methylhalobenzenium ions this simple relationship is not fulfilled.

The chemical shifts of H_3 and H_5 in 2,4-di- and 2,4,6-trihalobenzenium ions can be roughly estimated in an additive scheme by describing the influence of substituents on the signals of aromatic hydrogen atom [369]:

$$\delta_H = \delta_0 - \Sigma S_{i-x}$$

where δ_0 = const. and S_{i-x} is the contribution made to the chemical shift by the substituent X located at the i-position to the hydrogen. If S_{ortho} and S_{para} of the F, Cl and Br atoms[25] are taken to be, respectively, +0.25 and +0.17[25a]; −0.05 and +0.13; −0.22 and +0.06 ppm [369], then the chemical shifts of the H_3 and H_5 atoms of 2,4-di- and 2,4,6-trihalobenzenium ions are described by the above equation with an accuracy of about 0.1 ppm at δ_0 = 8.10. This equation is also applicable to the 2,4-di- and 2,4,6-trihalomethylbenzenium ions if S_{ortho} and S_{para} for the CH_3 group are +0.13 and +0.23, respectively.

The value of δ_0 must correspond to the chemical shift of the H_3 and H_5 atoms of the unsubstituted benzenium ion. Unfortunately, its chemical shifts have been measured relative to the external standard, so the direct comparison is impossible.

The PMR spectra of 1-H-4-, 1-H-2-, 2-H-1- and 2-H-3-halonaphthalenium ions are described [300].

E Benzenium Ions with Other Types of Substituents at Ring sp^2-Hybridized Carbon Atoms

For the study of the σ,π-conjugation in the CH_2X-type substituents the benzenium ions *(61)*–*(63)* containing the $CH_2Si(CH_3)_3$ were generated [66]. The chemical shifts in the PMR spectra of the solution of heptachlorodialuminates of these ions in CH_2Cl_2 are:

25 Relative to the hydrogen atom under consideration.

25a This value of S_{para} for fluorine yields better results than the earlier suggested one [365] of +0.22.

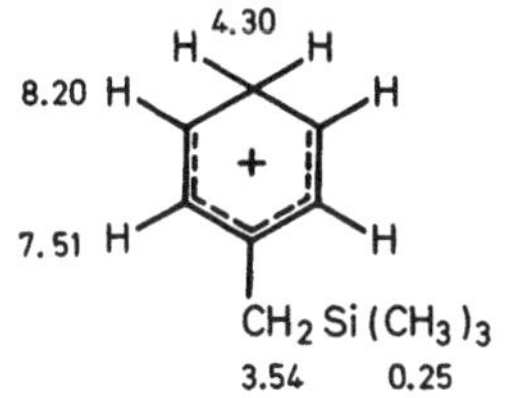

61

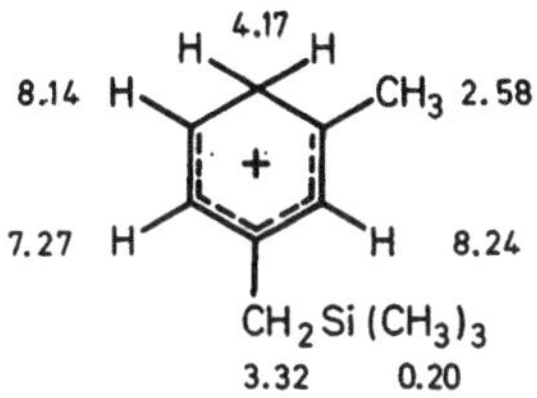

62

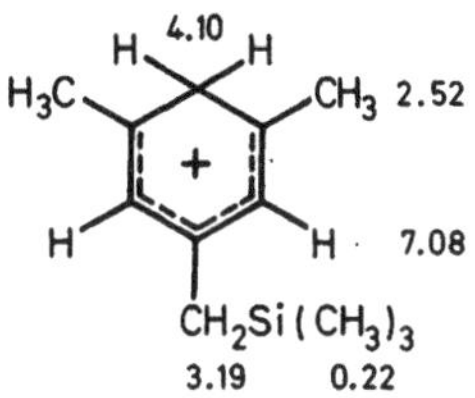

63

Comparison with the data in Table 5 shows the substitution of the CH_3 group in methylbenzenium ions by the $(CH_3)_3SiCH_2$ group to cause an appreciable shift of the other proton-containing groups to a high field. This seems to be due to a significant transfer of the charge in the ions *(61)*–*(63)* to the $(CH_3)_3SiCH_2$ group via the σ,π-conjugation (cf. [370]). On protonation of 1,3-dimethyl-5-ethylbenzene and of 1,3-dimethyl-5-trimethylsilylmethylbenzene the down-field shift of the CH_2 fragment of the CH_3CH_2 group does not exceed 0.6 ppm; in the $(CH_3)_3SiCH_2$ group it amounts to 1.26 ppm [66].

The spin-spin interaction of the ring CH_2 group protons with the ortho-positioned protons and the protons of the CH_2 fragment of the $(CH_3)_3SiCH_2$ group amount to 2.5 and 2 cps, while those of the ring vicinal protons, to 9 cps.

The attempts to generate 2,6-dimethyl-4-trimethylgermylmethyl- and 2,6-dimethyl-4-trimethylstannyl-methylbenzenium ions by interaction between the derivatives of benzene and $HCl/AlCl_3$ at -80 °C in CH_2Cl_2 have resulted in splitting the CH_2-metal bond with the formation of the methylbenzenium ion [66].

The PMR spectra of methylbenzenium ions containing CH_2Cl, $CH_2C(=O)H$, $CH_2C(=\overset{+}{O}H)H$ and $CH_2C_6H_5$ at the ring sp^2 hybridized carbon atom are given in [155].

F 1-X-1,2,3,4,5,6-Hexamethylbenzenium Ions

The methods of generating methylbenzenium ions containing the CH_3 group and an electronegative substituent X at the ring sp^3-hybridized carbon atom are discussed in Sect. II. The influence of X on the chemical shifts of the CH_3-group protons (above all those of the 1-CH_3 group) can be estimated from Table 21 (cf. also Table 11).

2 Carbon-13 Magnetic Resonance Spectroscopy

Recently carbon-13 NMR spectroscopy has become an important tool in the field of carbenium ions. From the chemical shifts of the ^{13}C nuclei and the spin-spin interaction constants $J_{^{13}C-^{1}H}$ [26] one can estimate the distribution of the positive charge in an ion and the valence characteristics of some of the carbon atoms.

---

26 $J_{^{13}C^{1}H}$ is often measured by PMR, but it is discussed in this Section.

Table 21. PMR Data of 1-X-1,2,3,4,5,6-Hexamethylbenzenium Ions

X	Acid system	Temp. °C	Chemical shifts					Standard	Ref.
			X	1-CH_3	2(6)-CH_3	3(5)-CH_3	4-CH_3		
CH_3	HSO_3F	+20	1.54	1.54	2.64	2.35	2.84	TMA, int	Table 5
$CH_2C_6H_5$	$HSO_3F—SO_2ClF$	—84	3.75 6.3–7.5	1.79	2.73	2.14	2.26		[244]
CH_2Cl	$HCl—2\ AlCl_3—CH_2Cl_2$	+40	4.18	1.55	2.68	2.40	2.89	CH_2Cl_2, int	[155]
	HSO_3F	—60	4.23	1.49	2.67	2.36	2.85	CH_2Cl_2, int	[254]
	HSO_3F	—20	4.25	1.54	2.70	2.41	2.89	CH_2Cl_2, int	[155]
	HSO_3Cl	+40	4.22	1.55	2.71	2.42	2.91	CH_2Cl_2, int	[155]
	CF_3COOH	—15	4.26	1.60	2.75	2.48	2.97	CH_2Cl_2, int	[152, 153, 155]
	CF_3COOH	—15	4.18	1.50	2.64	2.37	2.85	TMA, int	[155]
$CHCl_2$	$HCl—2\ AlCl_3—CH_2Cl_2$	+40	6.25	1.69	2.72	2.35	2.87	$CHCl_3$, int	[155]
CH_2CHO[a]	HSO_3F	—40	4.35, 9.5	1.5	2.6	2.4	2.9	CH_2Cl_2, int	[155]
C_6H_5	CF_3SO_3H	0	. . .	1.85	2.31	2.4	2.95	CH_2Cl_2, int	[242]
OCH_3 [a]	HSO_3F	—75	3.52	1.82	2.78	2.47	3.03	TMA, int	[243]
$\overset{+}{O}(H)CH_3$	$HSO_3F—SbF_5—SO_2ClF$	—90	4.20	2.16	2.85	2.56	3.20	TMA, int	[243]
Cl	$HSO_3F—SbF_5—SO_2ClF$	—70		2.03	2.96	2.54	3.11		[166]
	$HCl—AlCl_3—CH_2Cl_2$	—30		1.83	2.75	2.39	2.93	TMA, int	[177]
	CF_3SO_3H	—30		1.90	2.83	2.46	2.99	TMA, int	[177]
	$AlCl_3—CH_2Cl_2$	—90		1.80	2.72	2.33	2.91	CH_2Cl_2, int	[172, 259]
Br	$AlBr_3—CH_2Cl_2$	—100		1.75	2.80	2.30	2.80	CH_2Cl_2, int	[172, 259, 371]
NO_2	HSO_3F	—70		2.01	2.53	2.37	3.00	CH_2Cl_2, int	[169]
	HSO_3F	—44		1.97	2.49	2.31	2.95		[372]
	$HSO_3F—SO_2$	—70		2.00	2.60	2.40	3.00		[166]
	$HSO_3F—SO_2$	—70		2.30	2.81	2.43	3.01	CH_2Cl_2, int	[171]
	$HSO_3F—SO_2$	—80		1.95	2.50	2.32	2.97	TMS, ext	[174]
SO_3H	SO_2	—60		2.20	2.75	2.36	2.95	CH_2Cl_2, int	[171]
	HSO_3F	—60		2.26	2.79	2.42	2.97		[372]

[a] Partly protonated oxygen

The $J_{^{13}C-^{1}H}$ constants are rather sensitive to the changes in the s-character (α_C^2) of the hybridized orbital used by the carbon atom to form the C—H bond: the interrelation of these parameters in neutral compounds is well described by the semi-empirical equation $J_{^{13}C-^{1}H} \approx 500\,\alpha_C^2$ [373–375]. Thus, in hydrocarbons the values of $J_{^{13}C-^{1}H}$ for the fragments C_{sp^3}—H ($\alpha_C^2 = 0.25$) and C_{sp^2}—H ($\alpha_C^2 = 0.33$) are close to 125 and 166 cps, respectively.

The $J_{^{13}C-^{1}H}$ constants depend not only on the type of carbon hybridization, but also on other factors (see [376]), e.g. the changes of electron density under the influence of the polar substituents [373,377–384]. However, the influence of this factor is not too great to exceed the effects induced by the change of carbon hybridization and to make the relationship between $J_{^{13}C-^{1}H}$ and α_C^2 inapplicable to charged particles. Neutral $(CH_3)_nX$ molecules and the respective onium cations show the following values:

$(CH_3)_3P$	127 cps [385]	$(CH_3)_4P^+$	134 cps [385]
$(CH_3)_3N$	131 cps [385]	$(CH_3)_4N^+$	145 cps [385]
C_6H_6	159 cps [375]	$C_7H_7^+$	171 cps [386]
$H_3C—CH_3$	126 cps [375]	$H_2C=CH_2$	159 cps [375]
$(CH_3)_2CH^+$	130 cps (CH_3) [387,388]		168 cps (CH) [387,388]

A more complete collection of $J_{^{13}C-^{1}H}$ values for organic cations is presented in [376].

On the described basis $J_{^{13}C-^{1}H}$ were used to indicate the valent state of the carbon atom in arenium ions and their analogues. Thus, a comparison of $J_{^{13}C-^{1}H}$ for anthracene, 9-H-anthracenium ion and diphenylmethane [389] has supported the assumption that on protonation of anthracene the carbon atom capturing the proton passes from the sp^2 state into the sp^3 type.

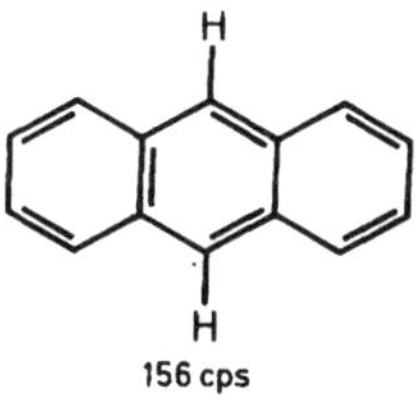

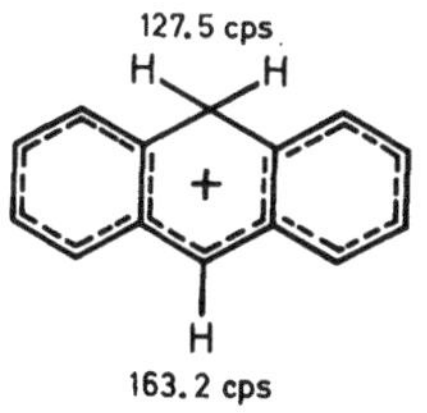

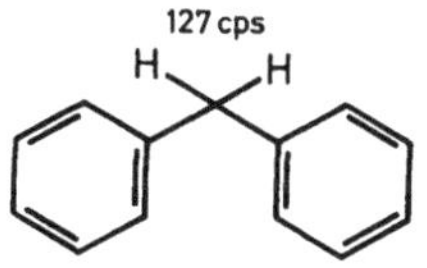

Likewise, the constant of the $C(CH_3)$—H in the 9-H-9,10-dimethylanthracenium ion is 132 cps [278]. Similar values have been found for the ring CH_2 and $C(CH_3)H$ groups of benzenium ion [157,278,299,390,391].

The change in $J_{^{13}C-^{1}H}$ of the $C(CH_3)$—H fragment from 127 to 165 cps on photochemical conversion of 1-H-1,2,3,4,5,6-hexamethylbenzenium ion (*15*) into the bicyclic ion (*64*) was used as an argument in favour of a cyclopropane ring (the values of $J_{^{13}C-^{1}H}$ for cyclopropane and other strained systems are abvormally high, see [376]).

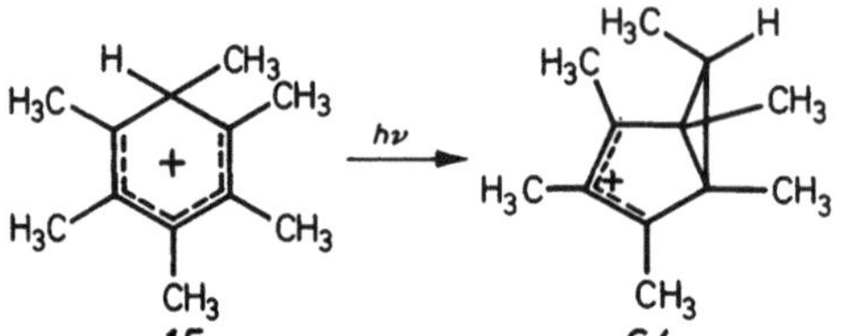

15 *64*

The $J_{^{13}C-^{1}H}$ constants for arenium and other carbenium ions are usually somewhat higher than for the respective neutral precursors. For carbanions, the values, by contrast, decrease as compared with the neutral precursors. The variations of $J_{^{13}C-^{1}H}$ in ions with the same sign and charge magnitude are comparatively small. Thus, in the anthracene dication they vary for different carbons from 178.2 to 175.2 cps, in anthracene from 162.4 to 157.5 cps and in its dianion from 147.5 to 145.0 cps [392]. For the 9-H-anthracenium ion the variation is 163.5–168.5 cps. The values of $J_{^{13}C(sp^2)-^{1}H}$ prove to be determined first all by the electron deficiency of a hydrogen atom ($q_H = 1 - \varrho_H$, where ϱ_H is the electron density on the hydrogen atom) and in a much lesser degree by that of the carbon atom. The above ions of anthracen reveal the following correlation [288,392]:

$$J_{^{13}C(sp^2)-^{1}H} = 161 + 187.5\, q_H \qquad (r = 0.973,\ s = 2.5,\ n = 19)$$

Important evidence on the electronic structure of arenium ions is provided by the ^{13}C chemical shift values[27]. For the first time this was demonstrated by comparing the NMR-^{13}C spectra of mesitylene and 2,4,6-trimethylbenzenium heptachlorodialuminate [390,394]:

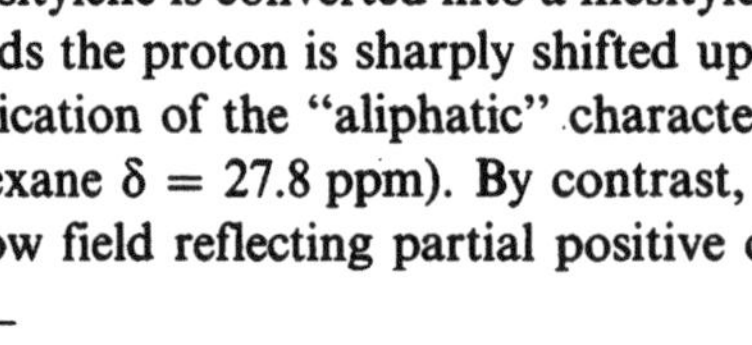

When mesitylene is converted into a mesitylenium ion the signal of the ring carbon which bounds the proton is sharply shifted up-field from 127.6 to 54.5 ppm. This is another indication of the "aliphatic" character of the ring CH_2 group in such ions (for cyclohexane $\delta = 27.8$ ppm). By contrast, the signals of the other ring carbons shift to a low field reflecting partial positive charges on these atoms, the strongest

27 Carbon chemical shifts are cited in ppm of tetramethylsilane (TMS). Literature data were recalculated using the relationships $\delta_{TMS(ext)} = 193.7 - \delta_{CS_2(ext)} = 192.8 - \delta_{CS_2(int)}$ [376]; $\delta_{CS_2(ext)} = \delta_{K_2CO_3(ext)} + 23.2$ [348]. The chemical shift of CH_2Cl_2 relative to tetramethylsilane is taken to be 53.3 ppm [393].

deshielding being observed for the atoms C_2, C_4 and C_6. By neglecting the transfer of the positive charge to the CH_2 and CH_3 groups and by assuming that the changes in the chemical shifts of the carbons retaining the sp^2 hybridization are proportional to the positive charge fractions (π-electron deficiency) [386] one arrives at the following positive charge distribution [390,394].

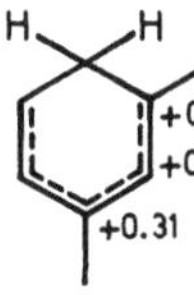

Here the coefficient of proportionality between the change of the chemical shift and the deficiency of π-electron density on the sp^2-hybridized carbon turns out to equal 187 ppm/electron.

NMR-^{13}C spectra of arenium ions are summarised in Table 22. The changes of the chemical shifts of the five sp^2-hybridized carbons in the pentadienyl fragment of methylbenzenium ions relative to the shifts of the respective carbons of methylbenzenes are nearly constant and equal to 180 ± 2 ppm [299]. This value is close to the slope of the linear dependence of carbon-13 chemical shifts of charged monocyclic nonbenzenoid aromatic systems on the value of the π-electron density on these atoms (160 [376,386,399], 168 [400] ppm/charge unit). From these one can assume that the substitution of hydrogens at the sp^2-hybridized carbons of the benzenium ions by the CH_3 groups does not change the π-electron deficiency in the dienyl system; it only causes a redistribution of the π-electron density in this system. Introduction of the CH_3 group to the electron-deficient positions results in "pushing" the π-electron density from the substituted position to other positions increasing the π-electron deficiency at the substituted position [282,299]. On the contrary, electronegative substituents of NH_3^+ or NO_2 type are bound "to pull" the π-electron density to the site of their attachment which is illustrated by the examples [173,360]:

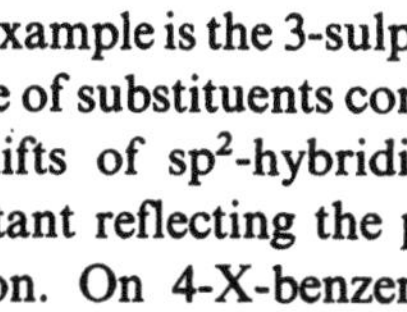

C_1	51.5
C_2	205.5
C_3	107.6
C_4	179.1
C_5	136.0
C_6	181.9

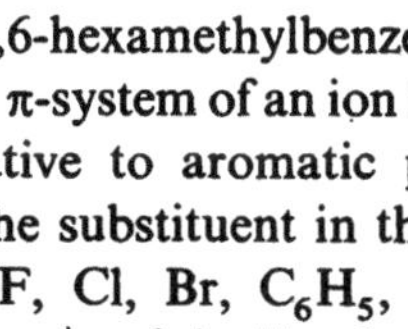

C_1	99.6
C_2	193.2 (174.5)
C_3	152.1
C_4	174.5 (193.2)
C_5	145.2
C_6	196.1

Another example is the 3-sulphonated 1,1,2,4,5,6-hexamethylbenzenium ion (*49*)[201].

In the case of substituents conjugating with the π-system of an ion the changes of the chemical shifts of sp^2-hybridized carbons relative to aromatic precursor are no longer constant reflecting the participation of the substituent in the positive charge delocalization. On 4-X-benzenium ions (X = F, Cl, Br, C_6H_5, OH, OCH_3) the values of $\Sigma\,\delta(C_i)$ are shown [288] to correlate with σ_p^+- of the X-substituents:

$$\Sigma\,\delta(C_i) = 42.5\,\sigma_p^+ + 169.8 \qquad (r = 0.999;\ s = 0.9)$$

Table 22. ^{13}C NMR Data of Alkyl- and Aryl-substituted Benzenium Ions

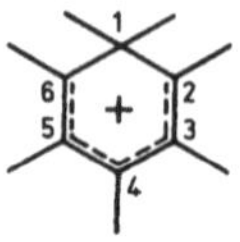

Position of substituents	Acid system	Temp. °C	Chemical shifts (from	
			C_1	C_2
—	$HSO_3F-SbF_5-SO_2ClF-SO_2F_2$	−135	52.2	186.6
4-CH_3	$HF-SbF_5-SO_2ClF$	−90	48.6	180.3
	HSO_3F-SbF_5	−80	46.7	178.6
	$HF-SbF_5$	−70	46.4	178.0
	HSO_3F-SO_2			181
4-C_2H_5	$HF-SbF_5-SO_2ClF$	−90	48.3	180.4
	$HF-SbF_5$	−100	47.3	179.4
	$HF-TaF_5$	−75	47.0	179.1
4-n-C_3H_7	$HF-SbF_5-SO_2ClF$	−90	48.9	180.6
4-$(CH_3)_2CH$	$HF-SbF_5-SO_2ClF$	−90	47.2	179.0
4-$C(CH_3)_3$	$HF-TaF_5$	−60	44.8	176.7
4-CH_2CH_2Cl	$HF-SbF_5$	−90	48.9	181.2
4-C_6H_5	$HSO_3F-SbF_5-SO_2ClF$	−60	45.1	174.8
2,4-$(CH_3)_2$	HSO_3F-SbF_5	−70	48.9	201.1 (25.5)
2-C_2H_5	$HF-SbF_5$	−100	52.1	215.0 (34.4, 11.7)
2,4-$(C_2H_5)_2$	$HF-SbF_5-SO_2$	−65	48.5	207.0[a]
	$HF-TaF_5$	−60	48.4	207.1
2,4,5-$(CH_3)_3$	HSO_3F-SbF_5	−70	49.5	198.9 (24.9)
2,4,6-$(CH_3)_3$	HSO_3F-SbF_5	−70	51.2	193.8 (24.2)
	$HCl-2\ AlCl_3$	+25	54.5	194.2 (27.5)
	$HCl-2\ AlBr_3$	+25	55.5	194.2 (28.5)
	$HF-SbF_5-SO_2ClF$	−90	53.3	196.0
	$HF-SbF_5-SO_2$	−65	51.9	194.0 (25.2)
2,4,6-$(C_2H_5)_3$	$HF-SbF_5-SO_2$	−65	49.8	199.2
	$HF-TaF_5$	−60	49.3	199.9
2,4,6-$[C(CH_3)_3]_3$	$HF-TaF$.	−60	44.7	206.0
2,3,4,6-$(CH_3)_4$	HSO_3F-SbF_5	−70	53.3	190.2[a] (23.0)[b]
	$HF-SbF_5-SO_2$	−65	53.9	190.1[a] (23.8)[b]
2,3,4,6-$(C_2H_5)_4$	SO_2	−65	50.4	196.5
2,3,4,5-$(CH_3)_4$	HSO_3F-SbF_5	−80	50.5	194.4 (24.1)[a]
2,3,5,6-$(CH_3)_4$	HSO_3F-SbF_5	−80	57.6	193.4 (22.5)
	HSO_3F	−78	57.90	194.25 (23.08)
2,3,4,5,6-$(CH_3)_5$	HSO_3F	−70	54.8	185.8 (22.8)
	$HCl-AlCl_3-CH_2Cl_2$	0	61.6	188.0 (26.2)
	HSO_3F	−78	55.08	186.39 (23.11)
1,2,3,4,5,6-$(CH_3)_6$	see table 23			
1,2,3,4,5,6-$(C_2H_5)_6$	$HSO_3F-SbF_5-SO_2$	−65	58.2	198.6
$(CH_3)_7$	see table 23			
$(C_2H_5)_7$	SO_2	−70	70.1	203.8

[a], [b] Possibly these values should be exchanged.

TMS) of ring and substituents (in parentheses) carbon atoms				Standard	Ref.
C_3	C_4	C_5	C_6		
136.9	178.1	136.9	186.6	TMS, ext	282) cf. 65, 139)
138.5	201.0	138.5	180.3		63, 65)
136.8	200.0 (28.5)	136.8	178.6	CH_2Cl_2, int	299)
136.5	200.9 (27.9)	136.5	178.0	TMS, ext	72)
139	201	139	181		188)
137.5	209.7	137.5	180.4		65)
136.0	204.6 (36.0, 11.7)	136.0	179.4	TMS, ext	72)
135.6	204.2	135.6	179.1	TMS, ext	398)
138.0	204.2	138.0	180.6		65)
134.1	211.7	134.1	179.0		65)
129	210.7 (41.0)	129	176.7	TMS, ext	398)
136.8	197.7 (43.2, 40.7)	136.8	181.2	TMS, ext	391)
133.0	181.5	133.0	174.8	TMS, ext	288)
136.1[a]	195.8 (27.5)	134.3[a]	172.1	CH_2Cl_2, int	299)
134.3	173.9	132.2	177.9	TMS, ext	72)
133.9[b]	200.5[a]	133.5[b]	172.8	TMS, ext	157)
133.7	199.6	133.7	172.8	TMS, ext	398)
136.6	196.1 (26.1)	143.1 (17.5)	169.6	CH_2Cl_2, int	299)
133.7	194.6 (26.7)	133.7	193.8 (24.2)	CH_2Cl_2, int	299)
135.4	194.2 (27.5)	135.4	194.2 (27.5)		390, 394)
136.2	194.2 (28.5)	136.2	194.2 (28.5)		390, 394)
135.0	196.0	135.0	196.0		65, 139)
133.9	194.3 (27.6)	133.9	194.0 (25.2)	TMS, ext	157)
130.8	198.3	130.8	199.2	TMS, ext	157)
130.8	198.8	130.8	199.9	TMS, ext	398)
126.6	202.0	126.6	206.0	TMS, ext	398)
140.8 (12.8)	192.5[a] (26.1)	134.7	189.7[a] (23.7)[b]	CH_2Cl_2, int	299)
141.1 (13.6)	190.4[a] (26.8)	135.1	192.7[a] (24.4)[b]	TMS, ext	157)
144.6	195.9	131.8	195.9	TMS, ext	157)
143.4 (14.1)	194.4 (22.7)[a]	143.4 (19.2)	166.7	CH_2Cl_2, int	299)
141.5 (16.6)	174.9	141.5 (16.6)	193.4 (22.5)	CH_2Cl_2, int	299)
142.36 (17.15)	175.32	142.36 (17.15)	194.25 (23.08)	TMS, ext	395)
140.9 (14.2)	191.6 (22.8)	140.9 (14.2)	185.8 (22.8)	CH_2Cl_2, int	299)
141.7 (17.9)	193.7 (26.2)	141.7 (17.9)	188.0 (26.2)		396)
141.35 (14.50)	192.65 (23.11)	141.35 (14.50)	186.39 (23.11)	TMS, ext	395)
145.2	195.8	145.2	198.6	TMS, ext	570)
148.6	198.2	148.6	203.8	TMS, ext	570)

Introducting an electronegative substituent (Cl, Br, NO_2, SO_3H) at the 1-position of benzenium ions polarizes the π-system with the shift of π-electrons from C_4 to C_2 and C_6 atom [174,299)28]. The chemical shift of the sp^3-hybridized C_1 atom grows with the increasing electronegativity of the substituent [299,393)] (see Table 23) similar to the observed changes for the compounds of the RCH_2X type. The chemical shifts of the C_1 atoms in 1-X-1,2,3,4,5,6-hexamethylbenzenium and 1-X-4-hydroxy-1,2,3,5,6-pentamethylbenzenium ions are linearly interdependent [299)]. For the 4-X-benzenium ions (X = F, Cl, Br, C_6H_5, OH, OCH_3) $\delta(C_1)$ depends on the σ^+_{para} constants of the substituents [288)]:

$$\delta(C_1) = 10.6\,\sigma_p^+ + 47.2 \qquad (r = 0.989,\ s = 0.8)$$

For methyl-substituted ions with two hydrogens at the ring sp^3-hybridized carbon the chemical shift is described by the relationship:

$$\delta(C_1) = 50 + 2.5\,n_o + n_m - 3\,n_p$$

where n_o, n_m and n_p are the numbers of CH_3 groups at the ortho, meta and para positions of the ion, respectively [299)].
The chemical shifts of the sp^2-hybridized carbons for methylbenzenium ions were compared with the π-electron (q_π) and the total ($q = q_\pi + q_\sigma$) electron deficiency calculated by the Hückel LCAO MO and CNDO/2 methods. To judge by the values of root-mean-square deviations (s) and of correlation coefficients (r) the chemical shifts of the sp^2-hybridized carbons of the ions under consideration correlate better with the π-electron deficiency calculated by the simple Hückel MO method [299)]:

$$\delta(C_i) = 138\,q_\pi(C_i) + 141 \qquad (r = 0.990,\ s = 3.6,\ n = 32)$$

If the changes of $\delta(C_i)$ and $q_\pi(C_i)$ are considered relative to the respective values for methylbenzenes, the correlation takes the form

$$\Delta\delta(C_i) = 157\,\Delta q_\pi(C_i) + 6.3 \qquad (r = 0.987,\ s = 3.8,\ n = 32)$$

in which the coefficient at Δq_π is close to the above ones obtained for charged monocyclic aromatic systems. Taking, for five sp^2-hybridized carbons of methylbenzenium ions, $\Sigma\,\Delta q_\pi = 1$ we find the sum of chemical shift variations of the sp^2-hybridized carbons in the ion relative to the respective methyl-substituted benzene (188 ppm) close to the experimental value (180 ppm).

The chemical shifts of sp^2-hybridized carbons in 4-methoxy-, 4-hydroxy-, 4-methyl- and unsubstituted benzenium ions [65,333)] are connected with the total atom charges

28 The correlation of the electronic effects of para and ortho substituents to the transition state structurally similar to the σ-complex must depend on the character of the added electrophil [174)].

Table 23. ^{13}C NMR Data of 1-X-4-R-1,2,3,5,6-pentamethylbenzenium Ions

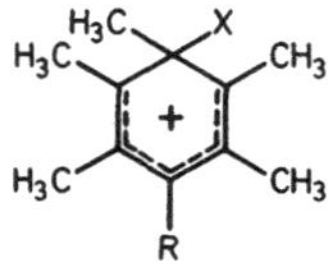

Substituents R	Substituents X	Acid system	Temp. °C	C_1	$C_{2(6)}$	$C_{3(5)}$	C_4	1-CH_3	2(6)-CH_3	3(5)-CH_3	4-CH_3	Standard	Ref.
CH_3	H	HF—SbF_5—SO_2ClF	−80	57.5	193.6	139.5	191.8	20.6	23.5	14.8	23.5	TMS, ext	[174]
		HSO_3F	−85	57.1	193.2	138.9	191.2	20.0	22.8	14.0	22.8	CH_2Cl_2, int	[393]
		HSO_3F	−78	57.85	193.85	139.52	191.85	20.51	23.26	14.50	23.26	TMS, ext	[395]
CH_3	CH_3	HSO_3F	+25	56.6	197.7	139.2	191.1	21.5	23.9	14.9	23.7	CH_2Cl_2, int	[393]
		HSO_3F		58.5	199.2	140.7	193.7	22.1	25.5	16.1	25.5	K_2CO_3, ext	[396]
		H_2SO_4 (98%)		58.3	198.5	140.3	192.3	23.7	26.1	17.1	26.1	K_2CO_3, ext	[396]
CH_3	$CH_2C_6H_5$	HSO_3F—SO_2ClF	−110	63.3	192.4	140.7	189.4	19.3	21.2	14.2	22.6	CH_2Cl_2, int	[244]
CH_3	Br	HSO_3F—SO_2ClF	−95	58.7	187.1	137.9	197.3	24.9	21.5	14.6	20.5	CH_2Cl_2, int	[393]
CH_3	Cl	HSO_3F—SO_2ClF	−70	68.1	190.3	139.2	199.6					CH_2Cl_2, int	[177]
		HCl—$AlCl_3$—CH_2Cl_2	−75	67.0	187.3	137.4	196.8	26.1	22.5	15.2	24.6	CH_2Cl_2, int	[393]
CH_3	SO_3H	HSO_3F	−60	78.7	180.8	145.5	198.2	16.8	22.3	14.9	24.8	CH_2Cl_2, int	[393]
CH_3	NO_2	HSO_3F	−60	98.4	181.5	141.8	204.7	23.6	19.7	14.6	26.2	CH_2Cl_2, int	[393]
		HSO_3F—SO_2	−80	98.7	181.4	142.5	205.8	23.7	20.4	15.2	27.0	TMS, ext	[174]
OH	H	HSO_3F	+25	51.6	185.4	126.4	184.1	18.4	21.2	9.8		CH_2Cl_2, int	[393]
		HBr—$AlBr_3$—CH_2Br_2	−30	52.7	187.7	127.7	187.7	—	—	13.7		K_2CO_3, ext	[348]
OH	CH_3	HSO_3F	+25	51.8	190.9	126.5	184.2	19.7	23.3	10.3		CH_2Cl_2, int	[393]
		HBr—$AlBr_3$—CH_2Br_2	−30	54.2	195.7	131.7	187.7	25.2	25.2	15.7		K_2CO_3, ext	[348]
OH	Br	HSO_3F	+25	57.9	181.5	125.9	185.7	23.4	21.1	10.5		CH_2Cl_2, int	[393]
OH	Cl	HSO_3F	+25	66.5	183.5	126.9	186.9	25.4	19.7	10.2		CH_2Cl_2, int	[393]
OH	SO_3H	HSO_3F	−60	74.8	173.9	134.0	186.4	16.6	20.9	11.2		CH_2Cl_2, int	[393]
OH	NO_2	HSO_3F	−60	94.9	172.2	130.1	188.6	22.1	17.7	10.3		CH_2Cl_2, int	[393]
OH	OH	HSO_3F	−80	83.8	178.5, 179.5	128.0, 130.1	187.1	23.9	17.4	10.1		CH_2Cl_2, int	[288]
F	H	HSO_3F	−70	56.8	201.6	129.5	186.7					TMS, ext	[397]
		HF—SbF_5	−80	57.1	202.3	129.7	187.0	20.2	23.2	9.4		TMS, ext	[174]
F	NO_2	HSO_3F—SO_2	−80	98.4	191.9	133.1	192.2	24.2	21.0	10.1		TMS, ext	[174]
Cl	H	HF—SbF_5—SO_2ClF	−80	58.7	196.8	139.7	185.9	21.1	24.1	15.6		TMS, ext	[174]
Cl	NO_2	HSO_3F—SO_2	−80	99.4	184.3	142.7	197.7	24.3	21.6	16.3		TMS, ext	[174]
Br	NO_2	HSO_3F—SO_2	−80	99.9	181.5	145.5	200.5	24.5	21.7	19.5		TMS, ext	[174]
C_6H_5	CH_3	CF_3SO_3H—SO_2ClF	−110	57.6	202.4	138.0	184.3					CH_2Cl_2, int	[242]

(calculated by the CNDO/2 method) by a linear relation [401)] which is displaced relative to the similar dependence for unprotonated compounds by 20 ppm. This circumstance, as well as the free term in the last relationship allows one to assume that the variation of the chemical shifts of sp^2 hybridized carbons in benzenium ions relative to the respective benzenes reflects not only the electronic charges on these atoms, but also another factor. This is the variation of anisotropic contributions to the shielding of carbon nuclei [392)] which can be analyzed in the ring π-electron current model. This factor is explained by comparing the NMR-^{13}C spectra of anthracene dication (A), anthracene (B) and its dianion (C), as well as 9-H-anthracenium ion (D), 9,10-dihydroanthracene (E) and the respective anion (F)

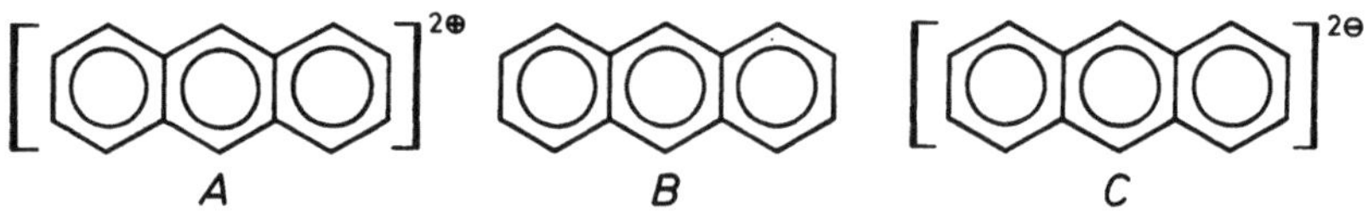

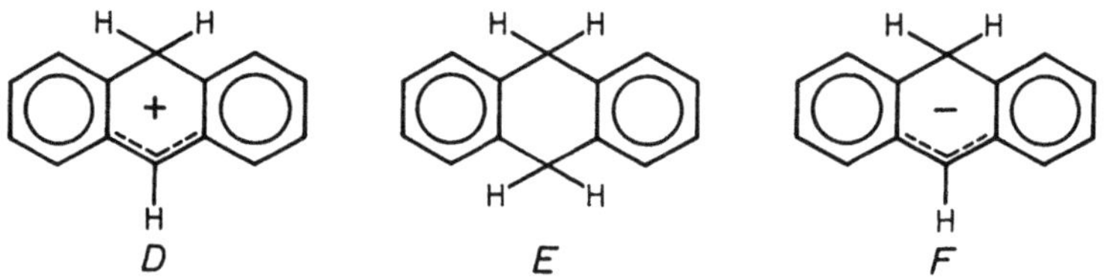

The changes of the chemical shifts of carbons for dication (A)[30] and dianion (C) compared to anthracene are sharply different in their absolute value (+416.4 and —159.6) though the change in the total electronic charges equals 2 units. Calculations show the intensities of the ring π-electron currents in dication (A) and dianion (C) to be the same and to differ in magnitude and direction from the ring currents in the anthracene molecule. The change in the chemical shifts of the dication (A) relative to dianion (C) amounts to $(416.4 + 159.6)/4 = 144$ ppm per unit of π-electron charge corresponding to the proportionality coefficients in the above mentioned relationships. Therefore, the difference of the changes in the carbon chemical shifts of dication (A) and dianion (C) relative to anthracene and their difference from the expected value (288 ppm) are due to the difference in the anisotropy of magnetic susceptibility of the A(C) and B structures.

Likewise the changes in the chemical shifts of the sp^2-hybridized carbons in passing from anthracene (B), to cation (D) and anion (F) are different in their absolute value and equal to +188.1 and —103.9 ppm. As to the change of the shifts of the cation relative to those of the anion (292 ppm) it is in good conformity with the expected one on changing the π-charge by 2 units. Also, the average chemical shifts in the lateral rings of the cation (D) and anion (F) are close to the respective shifts in 9,10-dihydroanthracene (E). This means that in passing from anthracene (B) to cation (D) and anion (E) the change in the chemical shifts of the sp^2-hybridized

29 For the NMR-^{13}C spectra of other polycyclic hydrocarbon dications see [402–405)].

carbons is determined by that in the electronic charges and by the "switch-off" of the anisotropic influence of the central ring.

The above data lead to the conclusion that the change in the chemical shift of sp^2-hybridized carbons per unit of π-electron charge should be about 145 ppm.[30] The mentioned change in the chemical shifts of the five sp^2-hybridized carbons of methylbenzenium ions relative to those of the respective methylbenzenes is equal to 180 ppm. The difference between these values (35 ppm) is comparable with the calculated correction for the anisotropy of the benzene ring magnetic susceptibility [392)]. This correction is necessary if the anisotropy of magnetic susceptibility is assumed to produce no effect on the chemical ring carbon shifts in methylbenzenium ions. However, the PMR spectra of benzenium ions (see Sect. III.1) led to a conclusion that a considerable ring current is likely to be retained due to the hyperconjugation of the CHR fragment with the charged pentadienyl ion. This discrepancy can be removed if the ring carbons of benzenium ions are assumed to be located in the zero ring current zone.

If calculated corrections for the anisotropic contribution $\delta^a(C_i)$ are introduced into the experimentally measured values of $\delta^{exp}(C_i)$, then the corrected values of the chemical shifts reveal their common linear dependence on the values of π-electron deficiency calculated by the Hückel MO method [288)]:

$$\delta^{corr}(C_i) = \delta^{exp}(C_i) - \delta^a(C_i) = 152\, q_\pi(C_i) + 136$$
$$(r = 0.984,\ s = 6.5,\ n = 79)$$

This dependence covers the methylbenzenium ions, the above cations and anions of the anthracene series, the monocyclic nonbenzenoid π-systems and the π-conjugated anions of the benzyl and allyl types (q_π varies in the range of +0.5 to −0.6 units while the chemical shifts in that of 50 to 200 ppm). Using the total electronic charges calculated by the CNDO/2 method does not make it possible to obtain the single correlation through in structurally similar ions the correlations are good [392)].

The data on the NMR-^{13}C spectra of various benzenium ions are collected in Tables 22–26.[31]

The ions of the phenonium type [63,391,406)] are close in structure to the benzenium ones. The shifts of the 2,4,6-trimethylphenonium ion [63)] are presented below. A marked shielding of the $C_{2,6}$ and C_4 atoms as compared to the 2,4,6-trimethylbenzenium ones points to the participation of the cyclopropane ring in the charge delocalization.

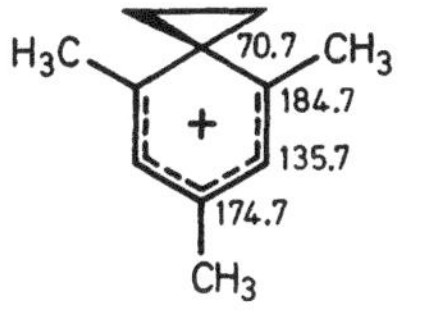

30 Or 220–230 ppm per unit of the complete charge of the atom [392)].

31 The dependence of chemical shifts on the character of acid media is weak (see [288)]). For heptamethylbenzenium tetrachloroaluminate the carbon shifts change little when passing from the solution to solid salt as well [269)].

Table 24. ^{13}C NMR Data of 2-Hydroxybenzenium Ions and their O-Derivatives

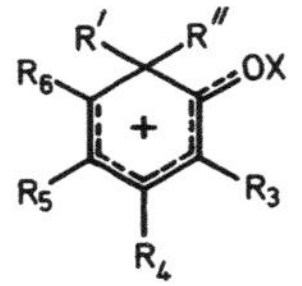

Substituents							Acid system	Temp. °C	Chemical shifts (from TMS) of ring carbon atoms						Standard	Ref.
R′	R″	R_3	R_4	R_5	R_6	OX			C_1	C_2	C_3	C_4	C_5	C_6		
CH_3	CH_3	CH_3	CH_3	CH_3	CH_3	OH	HSO_3F	−50	49.7	201.7	122.4[a]	188.7	132.7[a]	172.2	CH_2Cl_2, int	175)
H	CH_3	CH_3	CH_3	CH_3	CH_3	OCH_3	HSO_3F	−50	45.4	204.9	125.9[a]	188.4	131.7[a]	163.3	CH_2Cl_2, int	175)
C_2H_5	CH_3	CH_3	CH_3	CH_3	CH_3	OH	$CF_3COOH—H_2SO_4$	−5	56.1	203.8	125.2	188.3	134.6	170.9		330)
C_6H_5	CH_3	CH_3	CH_3	CH_3	CH_3	OH	HSO_3F	−20	58.1	201.1	123.1	189.9	131.6	170.7		330)
NO_2	CH_3	CH_3	CH_3	H	CH_3	OCH_3	HSO_3F	−70	95.5	196.7[a]	125.8	194.9[a]	130.6	160.8	CH_2Cl_2, ext	202)

[a] Possibly these values should be exchanged.

Table 25. ^{13}C NMR Data of 4-Hydroxybenzenium Ions and their O-Derivatives

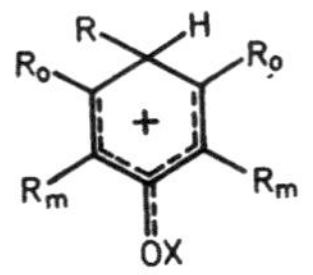

Substituents				Acid system	Temp. °C	Chemical shifts (from TMS)							Standard	Ref.
R	R_o	R_m	OX			C_1	$C_{2(6)}$	$C_{3(5)}$	C_4	R	R_o	R_m		
H	H	H	OH	$HSO_3F—SbF_5—SO_2ClF$	−80	38.0	171.7, 171.1	123.0	188.8					288)
						42.1	176.4	126.8	181.6					354)
H	H	H	OCH_3			41.7	168.7, 175.7	122.7, 128.7	192.7					63)
				$HSO_3F—SO_2$			169, 176	123, 129	193					188)
				$HSO_3F—SbF_5—SO_2ClF$	−80	38.2	166.3, 173.0	116.7, 124.5	189.7				TMS, ext	288)
				$HSO_3F—SO_2ClF$	−40	39.9	168.4, 175.1	120.7, 126.7	191.9				TMS, ext	391)
H	CH_3	H	OH	$HBr—AlBr_3—CS_2$	−30	46.8	187.8	122.8	187.8		22.8		CS_2, int	348)
H	CH_3	H	$OAlBr_3^-$	$HBr—AlBr_3—CS_2$	+25	51.3	181.8	125.8	190.3		23.3		CS_2, int	348)
H	CH_3	CH_3	OH	$HBr—AlBr_3—CH_2Br_2$	−30	51.7	183.7	131.7	188.7		25.7	13.7	K_2CO_3, ext	348)
H	CH_3	CH_3	$OAlBr_3^-$	$HBr—AlBr_3—CH_2Br_2$	+25	47.7	173.7	132.2	188.7			13.2	K_2CO_3, ext	348)
CH_3	CH_3	CH_3	OH	$HBr—AlBr_3—CH_2Br_2$	−30	52.7	187.7	127.7	187.7			13.7	K_2CO_3, ext	348)
CH_3	CH_3	CH_3	$OAlBr_3^-$	$HBr—AlBr_3—CH_2Br_2$	+25	50.7	183.2	132.7	190.2	...	...	14.2	K_2CO_3, ext	348)
CH_3	CH_3	CH_3	$OAl_2Br_6^-$	$HBr—2\ AlBr_3—CS_2$	+25	50.8	181.3	132.8	187.3	21.8	21.8	14.8	CS_2, int	348)
CH_3	CH_3	CH_3	$OAlCl_3^-$	$HCl—AlCl_3—CH_2Cl_2$	+25	50.2	180.2	132.2	189.7	21.7	21.7	13.7	K_2CO_3, ext	348)
H	OH	H	OH	HSO_3F	−40	...	184.0	127.7	189.3					354)
Br	CH_3	CH_3	OH	HSO_3F	+25	43.9	177.4	127.6	186.8		20.8	9.5	CH_2Cl_2, int	201)
Br	CH_3	CH_3	$OAlBr_3^-$	CH_2Br_2	+20	47.1	173.5	133.3	189.2					336)

Table 26. ^{13}C NMR Data of Halogenated Benzenium Ions

Position of substituents	Acid system	Temp. °C	Chemical shifts (from TMS)				Standard	Ref.
			C_1	C_2	C_3	C_4		
4-F	HSO_3F—SbF_5—SO_2ClF	−80	47.7	188.1	126.2	194.3	TMS, ext.	[366]
	HSO_3F—SbF_5—SO_2ClF	−60	47.6	187.5	125.9	193.8	TMS, ext.	[288]
4-Cl	HSO_3F—SbF_5—SO_2ClF	−70	48.2	181.1	137.5	192.0	TMS, ext.	[364]
	HSO_3F—SbF_5—SO_2ClF		48.0	180.5	136.9	191.4	TMS, ext.	[288]
4-Br	HSO_3F—SbF_5—SO_2ClF	−70	48.6	179.1	141.2	188.5	TMS, ext.	[364]
	HSO_3F—SbF_5—SO_2ClF	−80	48.5	178.4	140.6	187.9	TMS, ext.	[288]
2,4,6-F_3	HSO_3F—SbF_5	−60	41.0	193.3	109.1	204.1	TMS, ext.	[397]
F_7	SbF_5	−10	103.5	166.5	140.5	198.8	TMS, ext.	[397]

The ring carbon chemical shifts of the 1-naphthalenium ion and its methylated and halogenated derivatives are summarized in Tables 27 and 28. Substitution of a hydrogen by a CH_3 group causes a shift of the respective ring sp^2-hybridized carbon signal to the low field, the shift being 20–30 ppm for C_2, 15–20 ppm for C_4, 10–15 ppm for C_7 and 7–10 ppm for C_3, C_5, C_6 and C_8. The changes of the chemical shifts ($\Sigma\,\Delta\delta_i$) of the C_2–C_{10} atoms of the unsubstituted, as well as 2- and 4-methyl-1-H-naphthalenium ions relative to the respective neutral compounds amounts, just as in the benzene series, to 175–180 ppm [282,407]; for polymethylnaphthalenium ions this sum increases by 25–30 ppm, and on diprotonation of polymethylnaphthalenes [80,81,303] it increases twofold. Estimation of positive charge distribution in the unsubstituted naphthalenium ion by means of the $\Delta\delta_i/180$ ratio yields the values of Δq_π which are in fair agreement with those predicted by the Hückel MO calculations [407]. The study of dimethylnaphthalenium ions indicated that between the values of $\Delta\delta_i$ and those of Δq_π calculated by the Hückel MO (or CNDO/2) method there exist separate linear correlations — for unsubstituted and for substituted carbon atoms [301].

Among the 2-naphthalenium ions NMR-^{13}C data are available for the 1,2,2,3,4-pentamethyl-substituted cation [210] and for a few hydroxynaphthalenium ions [204].

The NMR-^{13}C spectra of anthracenium ions are presented in Table 29. The spectra of 9-R-9,10-dimethylphenanthrenium ions (R=CH_3, C_6H_5, p-$CH_3C_6H_4$, p-ClC_6H_4, p-FC_6H_4 and p-$CF_3C_6H_4$) are also analysed [303]. Cited below are the data for R=CH_3[32] (HSO_3F, −70 °C).

32 The signals of this ion are assigned differently [408].

Table 27. ^{13}C NMR Data of Methyl substituted 1-Naphthalenium Ions 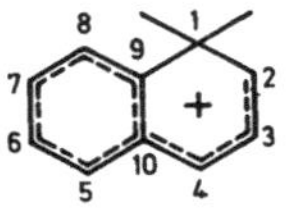

Substituents	Acid system	Temp. °C	C_1	C_2	C_3	C_4	C_5	C_6	C_7	C_8	C_9	C_{10}	Standard	Ref.
—	$HF-SbF_5-SO_2ClF$	−90	44.7	185.0	132.8	180.7	139.9	131.4	144.2	129.9	157.3	134.3	CH_2Cl_2, int	407)
	$HSO_3F-SbF_5-SO_2ClF$	−80	44.9	185.6	132.9	180.9	140.1	130.1	144.4	131.6	157.4	134.3	TMS, ext	282)
4-CH_3	HSO_3F	−70	43.2	179.7	133.2	200.1	136.1	130.8	142.1	129.9	154.1	133.0	CH_2Cl_2, int	407)
	$HSO_3F-SbF_5-SO_2ClF$	−80	43.2	180.1	133.4	200.4	136.3	130.1	142.2	131.0	154.2	133.5	TMS, ext	282)
	$HSO_3F-SbF_5-SO_2ClF$	−80	...	176.4	136.4	198.9	132.7	131.6	142.1	129.5	152.9	132.8	TMS, ext	301)
2-CH_3	HSO_3F	−70	47.6	210.2	132.6	177.1	138.1	130.4	141.3	129.0	153.2	132.2	CH_2Cl_2, int	407)
	$HSO_3F-SbF_5-SO_2ClF$	−80	47.9	210.9	133.0	177.5	138.4	129.4[a]	141.8	130.8[a]	153.4	132.4	TMS, ext	282)
	$HSO_3F-SbF_5-SO_2ClF$	−80	...	206.5	132.3	176.5	137.6	...	...	...	...	...	TMS, ext	301)
3,4-$(CH_3)_2$	$HSO_3F-SbF_5-SO_2ClF$	−80	42.8	176.9	144.2	200.8	133.0	130.8	141.2	129.9	153.4	133.9	TMS, ext	301)
2,4-$(CH_3)_2$	$HSO_3F-SbF_5-SO_2ClF$	−80	46.4	203.0	135.8	195.9	132.5	130.4	140.2	129.4	151.0	132.1	TMS, ext	301)
4,8-$(CH_3)_2$	$HSO_3F-SbF_5-SO_2ClF$	−80	41.3	177.9	136.1	199.7	132.0	130.7	144.0	138.9	152.1	133.6	TMS, ext	301)
	$HSO_3F-SbF_5-SO_2ClF$	...	42.9	178.8	144.9[a]	201.1	137.2[a]	131.7	133.1[a]	140	153.1	134.5	CS_2, ext	300)
4,7-$(CH_3)_2$	$HSO_3F-SbF_5-SO_2ClF$	−80	42.0	174.9	135.3	197.0	133.4	132.6	158.0	130.5	154.6	131.8	TMS, ext	301)
2,8-$(CH_3)_2$	$HSO_3F-SbF_5-SO_2ClF$	−80	46.1	208.4	132.6	177.3	137.3	130.7	143.8	138.6	151.1	132.6	TMS, ext	301)
4,5-$(CH_3)_2$	$HSO_3F-SbF_5-SO_2ClF$	−80	43.8	175.4	138.4	201.2	149.4	134.9	141.5	128.8	156.3	134.3	TMS, ext	301)
2,3-$(CH_3)_2$	$HSO_3F-SbF_5-SO_2ClF$	−80	49.5	208.3	142.4	177.6	137.2	130.7	140.3	128.7	152.4	133.0	TMS, ext	301)
2,6-$(CH_3)_2$	$HSO_3F-SbF_5-SO_2ClF$	−80	47.7	208.6	132.8	176.7	137.5	142.0	143.7	129.2	151.5	132.8	TMS, ext	301)
2,7-$(CH_3)_2$	$HSO_3F-SbF_5-SO_2ClF$	−80	46.8	204.0	131.5	175.3	138.5	132.4	157.7	130.1	154.0	130.9	TMS, ext	301)
1,2,3,4-$(CH_3)_4$	HSO_3F	−70	50.8	204.6	140.9	194.5	132.5	129.4	140.9	128.5	156.0	131.4	CH_2Cl_2, int	407)
1,1,2,3,4-$(CH_3)_5$	HSO_3F	+25	51.4	207.3	141.0	194.6	133.6	129.6[a]	140.5	128.1[a]	160.8	131.3	CH_2Cl_2, int	210)
$(CH_3)_8$	HSO_3F	+25	48.5	196.7	140.6	189.9	143.5	140.1	152.9	133.1	153.5	134.9	CH_2Cl_2, int	407)

[a] Possibly these values should be exchanged.

Table 28. ^{13}C NMR Data of Halogenated 1-Naphthalenium Ions

Substituents	Acid system	Temp. °C	Chemical shifts (ppm from TMS)										Standard	Ref.
			C_1	C_2	C_3	C_4	C_5	C_6	C_7	C_8	C_9	C_{10}		
4-F	$HSO_3F—SbF_5—SO_2ClF$	−80	42.6	187.9	123.5	193.0	130.6	131.6	143.9	130.3	159.2	124.0	TMS, ext	[366]
F_7	SbF_5	+25	108.0	168.2	141.9	191.7	161.6	153.4[a]	161.6	146.6[a]	119.4	108.3	TMS, ext	[397]

[a] Possibly these values should be exchanged.

Table 29. ^{13}C NMR Data of 9-Anthracenium Ions

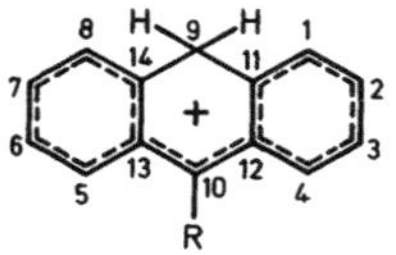

R	Acid system	Temp. °C	Chemical shifts (ppm from TMS)									Standard	Ref.
			$C_{1,8}$	$C_{2,7}$	$C_{3,6}$	$C_{4,5}$	C_9	C_{10}	$C_{11,14}$	$C_{12,13}$	R		
H	$HCl—2\,AlCl_3—C_6H_6$	+25	...	145	...	141	38	181	154	...		CS_2, ext	[390]
	$HCl—2\,AlCl_3—CH_2Cl_2$	+25	130.6	146.1	130.8	141.6	39.2	182.3	154.6	132.7		CH_2Cl_2, int	[392]
	$HSO_3F—SO_2ClF$	−80	130.6	146.1	130.6	141.6	38.5	183.4	155.2	132.8		TMS, ext	[282]
CH_3	$HSO_3F—SbF_5—SO_2ClF$	−80	130.5	144.2	130.5	135.3	38.5	200.5	153.0	133.4	20.7	TMS, ext	[282]
CH_2CH_3	$HSO_3F—SbF_5—SO_2ClF$	−75	130.7	144.2	130.7	134.8	38.6	205.1	153.6	132.2	18.9	TMS, ext	[282]
											27.6		
Cl	$HSO_3F—SO_2ClF$	−80	130.2[a]	145.1	131.1[a]	135.4	38.2	191.8	152.8	132.9		TMS, ext	[282]
Br	$HSO_3F—SO_2ClF$	−70	130.3[a]	145.2	131.6[a]	139.1	38.8	194.2	152.7	134.9		TMS, ext	[282]

[a] Possibly these values should be exchanged.

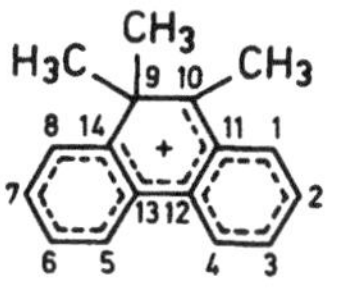

C_1 – 136.8 C_6 – 128.1 C_{11} – 125.9[c]
C_2 – 125.1[a] C_7 – 127.0[a] C_{12} – 149.1
C_3 – 152.8 C_8 – 126.2 C_{13} – 132.5[c]
C_4 – 130.2[b] C_9 – 52.9 C_{14} – 149.1
C_5 – 133.3[b] C_{10} – 232.4
(the values with the same letters may have to be exchanged).

The estimation of the partial positive charge of the C_{10} atom obtained from $\delta(C_{10})$ yields 0.56 units; this agrees well with the value 0.52 units predicted by the Hückel MO method for the o-phenylbenzylic cation.

The NMR-^{13}C spectra of phenanthrenium ions are also discussed in other publications [252,307,409,410].

Listed below are the carbon shifts of the 2-H-biphenylenium ion [314] and of the 2-H-acenaphthylenium ion [315]:

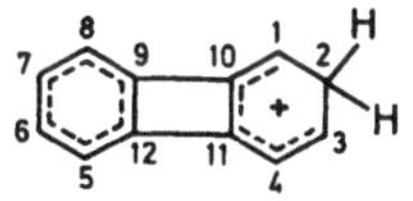

C_1 – 131.4; C_2 – 44.8; C_3 – 195.1;
C_4 – C_8 – 124.1, 126.9, 132.3, 135.0, 146.6;
C_9 – 162.5; C_{10} – 149.0; C_{11} – 204.2;
C_{12} – 146.6

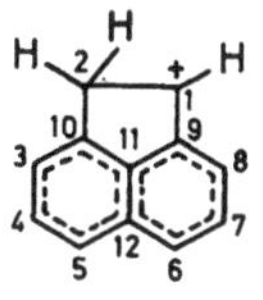

C_1 – 200.1; C_2 – 45.0; C_3, C_4, C_5, C_7 – 128.9, 127.6, 122.7, 120.9; C_6 – 156.0;
C_8 – 131.9; C_9, C_{10}, C_{11}, C_{12} – 147.6, 139.7, 130.9, 124.4

For 1- and 2-substituted acenaphthylenium ions see [315].

An interesting regularity has been revealed when comparing the NMR-^{19}F and the -^{13}C spectra of fluorinated arenium ions. The fluoro-, 2,4,6-trifluoro- and heptafluorobenzenium, as well as perfluorinated 1-naphthalenium and 9-anthracenium ions have shown the changes of the carbon ($\Delta\delta_C$) and fluorine ($\Delta\delta_F$) chemical shifts of the C_{sp^2}—F fragment with respect to its aromatic precursor to be bound by the linear relation [397]:

$$\Delta\delta_C = 0.32 \cdot \Delta\delta_F \qquad (\pm 3.5, r = 0.978).$$

This means: while the positive charge on the ring C_{sp^2}—F fragment changes, the ratio of the partial charges inside the fragment remains roughly the same (a proportional response of the F atom to an electronic request).

The fluorine atom involvement in the conjugation with the electron-deficient system of the arenium ion is also reflected by the spin-spin interaction constant $J_{^{13}C-^{19}F}$ of C_{sp^2}—F. The absolute values of the constants for arenium ions are higher than those for the respective aromatic compounds; the data on the above set of ions show [288] the changes of the $J_{^{13}C-^{19}F}$ and fluorine chemical shifts to be bound by the following relationship:

$$\Delta J_{^{13}C_i-^{19}F_i} = 0.49 \cdot \Delta\delta_{F_i} + 24.7 \qquad (r = 0.975,\ s = 6.5,\ n = 12)$$

For p-, m- and α-fluorine-substituted ions of benzyl type, 4-fluorine-substituted benzenium, 1,2,3,5,6-pentamethylbenzenium and 1-naphthalenium ions a similar dependence has been established [366]:

$$\Delta J_{^{13}C_i-^{19}F_i} = -0{,}892 \cdot \delta_{F_i} \quad (\text{from } CCl_3F) + 347.6$$

$$(r = 0.994,\ s = 4.3)$$

The long-range spin-spin interaction in fluorine-containing carbenium (among them arenium) ions is described in [174, 366, 397]. Examples are the $J_{i,j}$ constants of the C_i and F_j atoms spin-spin interaction in mono- and perfluorinated benzenium and naphthalenium ions [366, 397]:

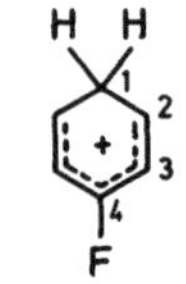

$J_{2,4} = 26.7$
$J_{3,4} = 19.7$
$J_{4,4} = 324.8$

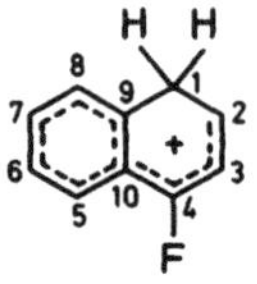

$J_{2,4} = 28.6$ $J_{4,9} = 18.2$
$J_{3,4} = 21.0$ $J_{4,10} = 8.5$
$J_{4,4} = 329.0$ $J_{4,5} = 15.4$

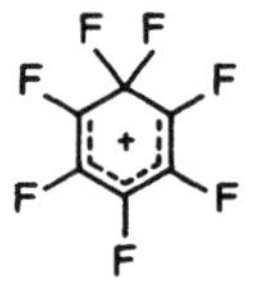

$J_{2,2} = 319$ $J_{2,4} = 27$
$J_{4,4} = 322$ $J_{4,2} = 27$
$J_{1,2} = 14$ $J_{2,6} = 15$

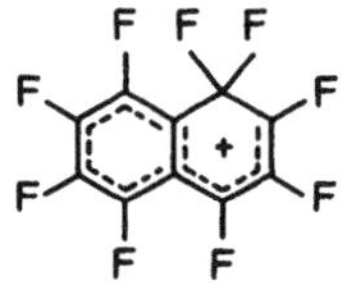

$J_{1,1} = 260$ $J_{5,5} = 304$
$J_{2,2} = 350$ $J_{6,6} = 270$
$J_{3,3} = 280$ $J_{7,7} = 304$
$J_{4,4} = 358$ $J_{8,8} = 270$

The highly characteristic chemical shifts of the ring carbon atom in arenium ions and the spin-spin interaction of carbons and other elements with magnetic nuclei are widely used in establishing the structure of new ions. An example is the corroboration of the bipolar structure in the complexes formed from phenols and aluminium halides [348] and proving the structure of the ions generated by the interaction of polymethylated aromatic compounds with heteroatomic electrophilic reagents (see [393]).

An interesting example of determining the nature of an added particle is reported in [170]. To prove that the interaction of hexamethylbenzene with HNO_3 in HSO_3F yields the 1-nitro-1,2,3,4,5,6-hexamethylbenzenium ion (*22a*) nitric acid was enriched with the ^{15}N isotope. The accompanying doublet splitting of the C_1 atom signal (8.3 cps) has corroborated the C_1—N bond in the ion (see also [202]).

3 Fluorine-19 Magnetic Resonance Spectroscopy

The fluorine chemical shifts of mono-, di-, tri- and tetrahalogenbenzenium ions and their analogues are presented in Tables 19 and 20. There is also evidence available for 2- and 3-fluoro-4-hydroxy- and 4-methoxybenzenium ions [340], as well

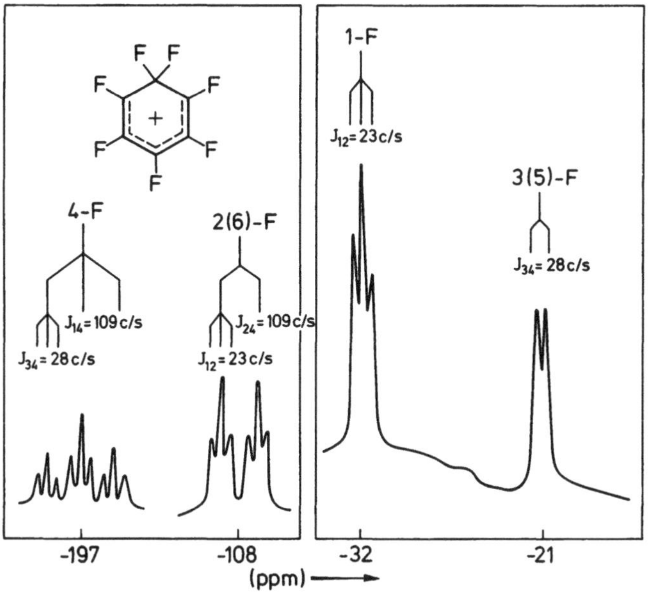

Fig. 6. NMR ^{19}F spectrum of hexafluorantimonate of the heptafluorobenzenium ion in SbF_5 at 20 °C [238)]

as for monofluoronaphthalenium ions [300)]. Also, as mentioned in Sect. II, by the action of SbF_5 on polyfluorinated cyclohexa-1,4-dienes and by other procedures a large number of polyfluorinated arenium ions have been generated.

The NMR-^{19}F spectrum of the heptafluorobenzenium ion is given in Fig. 6; the data on the chemical shifts of the ^{19}F nuclei of this ion and its analogues are presented in Table 30[33]. The fluorine signals of perhalobenzenium ions are markedly shifted to the low field as compared with those of the ions containing, in addition to the halogen atoms, hydrogen and a CH_3 group. From all the NMR-^{19}F spectra of fluorinated benzenium ions the chemical shifts of fluorine atoms, depending on their position in the ion, vary in the following ranges (ppm):

4-F	from	—140 to —230
2(6)-F	from	— 90 to —155
3(5)-F	from	— 20 to — 60
CF_2	from	— 30 to — 60

The down-field shift of ^{19}F signals when protonation of benzene derivatives leads to benzenium ions are usually as follows (ppm) [67,146,283)]:

4-F	90–110
2(6)-F	80–110
3(5)-F	8– 15

33 With respect to C_6F_6. To pass over to the ∅ scale one should use the relationship $\delta_{CCl_3F} = \delta_{C_6F_6} + 162.9$ [368)].

Table 30. ^{19}F NMR Data of Polyhalogenated Benzenium Ions

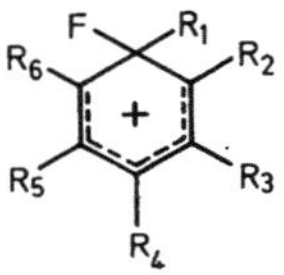

Substituents						Chemical shifts (from C_6F_6)						Standard	Ref.
R_1	R_2	R_3	R_4	R_5	R_6	1-F	2-F	3-F	4-F	5-F	6-F		
F	F	F	F	F	F	—32	—108	—21	—197	—21	—108	C_6F_6, ext	237, 238)
F	F	F	F	F	Cl	—44	—144	—22	—176	—45		C_6F_6, ext	237, 238)
F	F	F	F	Cl	F	—33	—104	—23	—212		—135	C_6F_6, ext	237, 238)
F	F	F	Cl	F	F	—31	—92	—37		—37	—92	C_6F_6, ext	237, 238)
F	Cl	F	F	F	Cl	—53		—44	—157	—44		C_6F_6, ext	237, 238)
F	F	Cl	F	Cl	F	—33	—130		—229		—130	C_6F_6, ext	237, 238)
F	F	F	Cl	F	Cl	—42	—98	—38		—60		C_6F_6, ext	237, 238)
F	Cl	Cl	F	Cl	Cl	—58			—200				237, 238)
F	F	H	F	F	F	—27.0	—138.5		—209.2	—16.7	—92.8	C_6F_6, ext	411)
F	Br	F	F	F	F	—43.2		—46.8	—164.8	—17.0	—111.5	C_6F_6, ext	411)
F	F	Br	F	F	F	—28.6	—140.1		—215.2	—19.0	—96.0	C_6F_6, ext	411)
F	F	CH_3	F	F	F	—28.3	—133.2		—206.5	—17.3	—85.5	C_6F_6, int	239)
F	F	C_6F_5	F	F	F	—31.8	—143.3		—213.1	—23.1	—101.9	C_6F_6, int	413)
F	F	C_6F_5	F	C_6F_5	F	—29.7	—134.5		—228.5		—134.5	C_6F_6, int	412)
Cl	F	F	F	F	F	—6.1	—106.3	—16.5	—174.6	—16.5	—106.3	C_6F_6, int	178)
Cl	F	CH_3	F	F	F	—6.5	—131.6		—188.7	—14.2	—88.7	C_6F_6, int	178)
Cl	F	C_6F_5	F	F	F	—7.3	—139.8		—192.3	—17.6	—100.8	C_6F_6, int	178)
Cl	F	F	Cl	F	F	—7.4	—91.7	—33.0		—33.0	—91.7	C_6F_6, int	179)
Cl	F	Cl	F	F	F	—7.7	—128.2		—190.2	—16	—100.5	C_6F_6, int	179)
Cl	Cl	F	F	F	F	—16		—37.6	—155.4	—16	—114.4	C_6F_6, int	179)
Cl	F	Cl	Cl	F	F	—9.0	—119.7			—36.3	—87.5	C_6F_6, int	179)
Cl	Cl	F	Cl	F	F	—18.6		—55.1			—100.0	C_6F_6, int	179)
Cl	Cl	Cl	F	F	F	—21.5			—177.4	—17.0	—109.9	C_6F_6, int	179)
Cl	F	Cl	F	Cl	F	—9.0	—125.4		—207.8		—125.4	C_6F_6, int	179)

As to the differences between perhalogenated benzenium ions and perhalobenzenes corresponding to the formal removal of the F^+ cation from the C_1 atoms of these ions, their fluorine chemical shifts vary within the ranges (ppm):

4-F	from	−150 to −200
2(6)-F	from	− 90 to −120
3(5)-F	from	− 10 to − 20

Interestingly, the CF_2 signals of the ions in question are shifted relative to those of the parent perhalocyclohexa-1,4-dienes by 12–19 ppm to the high field. The authors [238] believe the unexpected up-field shift to be due to a specific change in the van der Waals component of the chemical shift for CF_2 upon formation of the ion.

Table 31. Coupling Constants $J_{^{19}F-^{19}F}$ for Fluorobenzenium Ions (cps) [67,146,166,178,237–239,283,411–413]

Position	1-F	2-F	3-F	4-F
2-F	22–27	—	4–6	70–115
3-F	4–7	4–6	—	20–30
4-F	0–2	70–115	20–30	—
5-F	4–7	7–10	0–2	20–30
6-F	22–27	10–15	7–10	70–115

The ^1H-^{19}F spin-spin coupling constants of fluorine-substituted ions are presented in Sect. III.1.D, those of the ^{19}F-^{19}F spin interaction in Table 31. The spin-spin coupling constants of the fluorine atoms located at the ortho and para positions is abnormally high ($J_{24} = J_{46} = 70$–115 cps). They are tens of times higher than the average value of the meta-positioned fluorine atoms. This fact, added to the chemical shifts of ortho- and para-positioned fluorine atoms, indicates the electron density distribution in fluorinated benzenium ions to correspond with the predominant contribution of the following resonance structures:

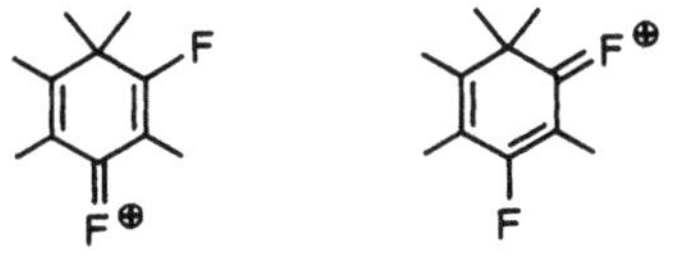

The relation between $J(F_i, F_j)$ and the chemical shifts of the F_i and F_j atoms depend on the extent of the electron deficiency on the C_i and C_j atoms (see Sect. III.2) [19].

The spectra of polyfluorinated 4-hydroxy- and 2-hydroxybenzenium ions are listed in [233,235].

The NMR-^{19}F spectroscopy has been used to investigate the polyfluorinated 1-naphthalenium ions [178,180,238,239,413,414]. Examples are the chemical shifts of the heptafluoro-1-naphthalenium ion (the shifts of α- and β-fluorine in octafluoronaphthalene amount to −17.1 and −8.4 ppm, respectively) [414]:

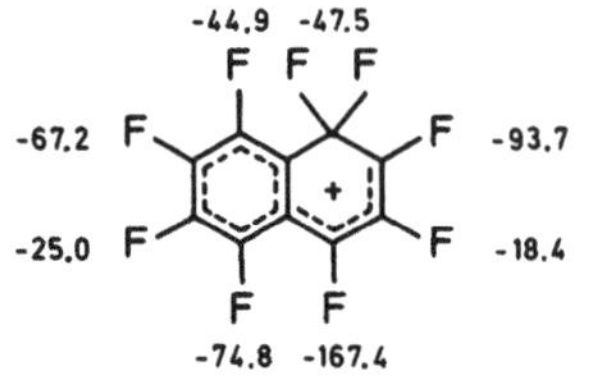

$J_{24}=100$, $J_{45}=150$, $J_{57}=50$ cps

Special note should be made of the high value of the J_{45} constant. For aromatic molecules the spin-spin coupling constant of para-positioned fluorine atoms is as low as 60–70 cps.

The NMR-^{19}F spectra of polyfluorinated 1- and 2-hydroxynaphthalenium ions are reported in [233–235]. Those of fluorinated anthracenium ions, among them the perfluoroanthracenium ion, are available in [240, 241].

4 Electronic Absorption Spectra

A Unsubstituted and Alkylated Arenium Ions

Spectroscopy in the ultraviolet and the visible region was one of the first physical methods applied to the studies of aromatic compounds in acid systems. By this method the first indications were obtained for arenium ions. In particular, V. Gold and F. Tye [58] showed the solutions of anthracene and 1,1-diphenylethylene in concentrated sulphuric acid to have absorption maxima very close in position and intensity (λ_{max} 422 and 431 nm, respectively). This similarity meant that the hydrocarbons are protonated to form, respectively, the 9-H-anthracenium ion and the diphenylmethyl cation having similar π-electron systems:

Electronic spectroscopy, however, should be handled with care since at low concentrations absorbing substances may be transformed in secondary photochemical and oxidizing reactions. For instance, complexes of toluene, 1- and 2-methylnaphthalenes with HCl and $AlCl_3$ [32, 46] and those of hexamethylbenzene with DF and BF_3 [415] have changed colour rapidly. For polycyclic hydrocarbons of the type of perylene, pyrene and naphthacene time-stable spectra can only be observed after a complete removal of the oxygen from the system [416] (cf. [417, 418]). In the presence of oxygen radical cations are formed (see Sect. IV.6). When recording the spectra in sulphuric acid the sulphonation reaction is also possible.

Cells used in recording the spectra of solutions in liquid hydrogen fluoride and chloride at room temperature and under high pressure are described in [416,418], cells for spectra recording at low temperatures in [417].

Table 32 lists the positions and intensities of absorption maxima of methylated benzenium ions; as an example, Fig. 7 shows the spectra of the mesitylene and the mesitylenium ion. Table 32 reveals rather big discrepancies for the first three ions by different authors. The reasons are not clear. One should therefore carefully consider the conclusion [4] drawn from the data [362] and from the molecular-orbital calculation [92] stating that introduction of CH_3 groups at ortho and para positions of the benzenium ion causes a hypsochromic shift of the long-wave absorption maximum.

Curiously, the electronic spectra of the mesitylenium ion in the HF—BF_3 system [416] and in the gas phase (λ_{max} = 355 and 250 nm [419]) practically do not differ in the position of the absorption maxima.

B Hydroxyarenium Ions

For sulphuric acid [68,147,344,345] the position of absorption maxima of unsubstituted 4-hydroxy- and 4-methoxybenzenium ions was the subject of discussions. No reliable experimental measurements are yet available.

Table 32. Electronic Absorption Spectra of Alkylbenzenium Ions

Ion	Acid system	Temp. °C	λ_{max}, nm (log ε)	Ref.
Benzene · H^+	HF—BF_3	<−20	~420 (4.2)	417)
	HBr—n$AlBr_3$—C_6H_6	25	332 (3.58)	33,48)
Toluene · H^+	HF—BF_3	−20	~400 (4.15)	417)
	HBr—n$AlBr_3$—C_6H_6	25	326	16,48)
Mesitylene · H^+	HF—BF_3	25	355 (4.05); 256 (3.95)	416)
	HF	...	354 (4.0); 257 (3.9)	420)
	HF—BF_3	−80	390 (4.43)	417)
Bimesityl · H^+ [a]	HF	25	362 (3.76); 258 (3.59)	416)
Isodurene · H^+	HF—BF_3	25	365 (3.88); 261 (3.73)	416)
Pentamethylbenzene · H^+	HF	25	377 (3.99); 272 (3.63)	416)
Hexamethylbenzene · H^+	HF		395 (3.87); 281 (3.63)	120,420)
	HF	...	390 (3.81); 290 (3.69)	421)
	HCl—n$AlCl_3$	25	390	418,421)
$C_6(CH_3)_7^+$	HCl-aq	25	397 (3.93); 287 (3.83)	3)
Hexaethylbenzene · H^+	HF—BF_3	−80	~400 (4.4); ~290 (4.3)	417)
	H_2SO_4 (96%)	25	400 (3.93); 285 (3.85)	422)
	H_2SO_4	25	413 (4.03); 291 (3.93); 216 (4.17)	156)
Hexa-n-propylbenzene · H^+ [b]	H_2SO_4 (100%)	25	405 (3.88); 298 (3.88)	422)

[a] Ref. [416,423] contain data for biphenyl in HF—BF_3 and HSO_3F—SbF_5

[b] For ions with polymethylene cycles as substituents see [422]

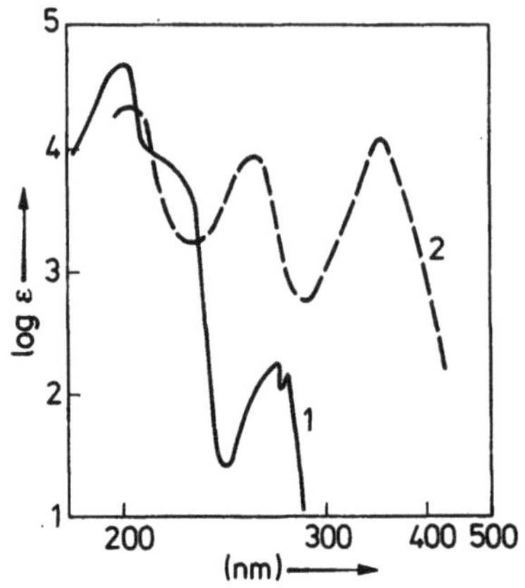

Fig. 7. Electronic absorption spectra of mesitylene in n-hexane 1) and in the HF—BF_3 2) [16,416]

However, the analysis of the positions of the long-wave absorption maxima of phloroglucinol, resorcinol and their derivatives, as well as of methyl-substituted phenols in perchloric acid in terms of the additive substituent effects has made it possible to conclude [346] that the long-wave absorption maximum of the 4-hydroxybenzenium ion is likely to be near 285 nm. The shift of this maximum by the 1-, 2(6)- and 3(5)-CH_3 groups amounts to about +5, +15 and +11, while for the 2-HO group it is +28 nm. Substitution of the hydroxy group hydrogens by alkyl leaves the absorption maximum almost unchanged. The electronic spectra of substituted 4-hydroxybenzenium ions listed in Table 33 corroborate the possibility of using this approach to predict the position of the long-wave absorption maximum.

A similar character to the spectra of the above ions is observed with the complexes of tautomeric phenols with AlX_3 [424].

The long-wave absorption maximum of the 2-hydroxybenzenium ion seems to be located near 360 nm. Introduction of the 4-HO group results in its hypsochromic shift by about 50 nm, and that of the 6-HO group, in a bathochromic shift by about 30 nm. The effect of 1-, 3-, 4-, 5- and 6-CH_3 groups is about +5, +15, —10, +25 and +20 nm, respectively (Table 34).

From these data one could expect the long-wave absorption maximum of the unsubstituted ion to be located near 330–335 nm (cf. [33,48]) rather than at 420 nm as found in [417]. Molecular orbital calculations in different approximations yield values for the long-wave absorption maximum of the unsubstituted benzenium ion varying from 375–400 nm [92,416,429].

In addition to the data in Tables 33 and 34 the absorption maxima for the 2,4,6-trimethoxybenzenium ion in 63.5% $HClO_4$ are located at 347 and 247 nm (lg ε is equal to 4.09 and 4.31) [356,357]; for the spectra of other ions containing alkoxy groups see [346].

Assuming, for the unsubstituted benzenium ion $\lambda_{max} = 335$ nm, the position of the long-wave absorption maximum of substituted ions is approximately described by the relationship

$$\lambda_{max} \approx 335 + \Sigma \Delta\lambda_i \text{ nm},$$

Table 33. Electronic Absorption Spectra of Substituted 4-Hydroxybenzenium Ions

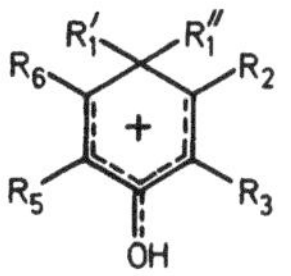

Substituents						Acid system	λ_{max}, nm (log ε)	Ref.
R_1'	R_1''	R_2	R_6	R_3	R_5			
CH_3	CH_3	H	H	H	H	$HClO_4$ (71%)	295 (3.57); 260 (4.12)	225)
						H_2SO_4 (95%)	295 (3.57); 260 (4.12)	425)
						H_2SO_4 (75%)	290 (4.08); 260 (4.14)	227)
						HSO_3F	298	325)
CH_3	$CHCl_2$	H	H	H	H	H_2SO_4 (90%)	294 (3.56); 262 (4.12)	224)
						H_2SO_4	297 (3.53); 263 (4.11)[a]	326, 427)
							295 (3.56); 267 (4.12)	425)
CH_3	CH_3	CH_3	H	H	H	H_2SO_4 (78%)	305 (3.78); 260 (4.17)	227)
CH_3	$CHCl_2$	CH_3	H	H	H	H_2SO_4	310 (3.73); 260 (4.11)	326)
CH_3	$CHCl_2$	H	H	CH_3	H	H_2SO_4	315 (3.57); 267 (4.07)	326)
CH_3	CH_3	CH_3	CH_3	H	H	H_2SO_4 (75%)	314 (3.88); 261 (4.24)	227)
CH_3	$CHCl_2$	CH_3	CH_3	H	H	H_2SO_4 (90%)	320 (3.84); 263 (4.22)	224)
						H_2SO_4 (80%)	318 (3.81); 264 (4.21)	227)
CH_3	C_2H_5	CH_3	CH_3	CH_3	CH_3	$CF_3COOH-H_2SO_4$	340 (3.92); 275 (4.16)	330)
CH_3	CH_3	CH_3	CH_3	CH_3	CH_3	H_2SO_4 (95%)	340 (3.88); 273 (4.08)	219)
H	H	CH_3	H	H	H	$HClO_4$	298	346)
H	H	CH_3	H	CH_3	H	HSO_3F	334 (3.26)	428)
H	H	CH_3	CH_3	H	H	$HF-BF_3-(CH_2)_4SO_2$	308 (3.80); 256 (4.09)	347)
						HSO_3F, −50 °C	309 (4.08); 254 (4.30); 250,5 (4.25)	319)
						$HClO_4$ (70%)	313 (3.93)	224)
						$HClO_4$	313	346)
H	H	CH_3	CH_3	CH_3	CH_3	$HClO_4$	330	346)
H	H	OH	H	H	H	$HF-BF_3-(CH_2)_4SO_2$	303 (3.94); 231 (4.01)	347)
						$HClO_4$ (70%)	311	346)
H	H	OH	CH_3	H	H	$HF-BF_3-(CH_2)_4SO_2$	324 (3.94); 244 (4.10)	347)
						$HClO_4$ (70%)	326	346)
H	H	OH	OH	H	H	$HF-BF_3-(CH_2)_4SO_2$	343 (3.90); 242 (4.09)	347)
						$HClO_4$ (70%)	340	346)
						$HClO_4$ (70%)	365 (~4); 250 (~4)	122)

[a] Data of ref. 426) are probably erroneous.

Table 34. Electronic Absorption Spectra of Substituted 2-Hydroxybenzenium Ions

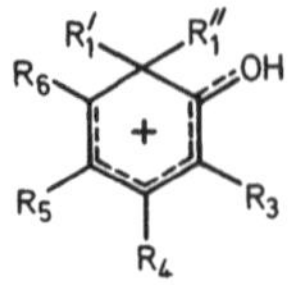

Substituents						Acid System	λ_{max}, nm (log ε)	Ref.
R_1'	R_1''	R_3	R_4	R_5	R_6			
CH_3	CH_3	H	H	H	H	HSO_3F	363	[325]
CH_3	$CHCl_2$	H	H	H	H	H_2SO_4	370 (3.77); 251 (3.43)	[326, a]
						H_2SO_4	369 (3.68)	[426]
CH_3	$CHCl_2$	CH_3	H	H	H	H_2SO_4	385 (3.78); 252 (3.43)	[326]
CH_3	$CHCl_2$	H	CH_3	H	H	H_2SO_4	360 (3.80); 252 (3.84)	[326]
CH_3	$CHCl_2$	H	H	CH_3	H	H_2SO_4	393 (3.78); 253 (3.46)	[326]
CH_3	$CHCl_2$	H	H	H	CH_3	H_2SO_4	397 (3.91); 275 (3.10); 237 (3.54)	[326]
CH_3	CH_3	CH_3	CH_3	CH_3	CH_3	H_2SO_4 (95%)	398 (3.95)	[219]
H	H	H	CH_3	CH_3	CH_3	HSO_3F, −50 °C	380 (3.60)	[319]
H	H	H	OH	H	H	$HF{-}BF_3{-}(CH_2)_4SO_2$	303 (3.94); 231 (4.01)	[347]
						$HClO_4$ (70%)	311	[346]
H	H	H	OH	H	CH_3	$HF{-}BF_3{-}(CH_2)_4SO_2$	324 (3.94); 244 (4.10)	[347]
H	H	H	OH	H	OH	$HF{-}BF_3{-}(CH_2)_4SO_2$	343 (3.90); 242 (4.09)	[347]
						$HClO_4$ (70%)	340	[346]

[a] cf. [427]

where $\Delta\lambda_i$ reflects the influence of the substituents: +30 and −50 for 2(6)- and 4-HO; 5–10, 15–20, 10–15 and −10 for 1-, 2(6)-, 3(5)- and 4-CH_3 groups, respectively.

C Aminoarenium Ions

An important part has been played by electronic spectra in studying the protonation of di- and triaminobenzenes [123,124]. Addition of the proton to the amino group is known to prevent its conjugation with the aromatic ring, so the electronic spectra of $ArNH_3^+$ cations coincide with those of the ArH compounds. Accordingly the UV spectrum of m-phenylenediamine with increasing medium acidity is first converted into that of aniline (formation of m-$NH_2C_6H_4NH_3^+$) and then into a spectrum close to that of benzene (formation of m-$NH_3^+C_6H_4NH_3^+$).

The UV spectrum of a solution of 1,3,5-triaminobenzene in concentrated sulphuric acid (Fig. 8) is similar to that of benzene and seems to correspond with the three times charged cation (*65*). At pH 1.2 the spectrum takes the form of aniline, i.e., the dication (*66*) is predominant. With further decrease of medium acidity down to

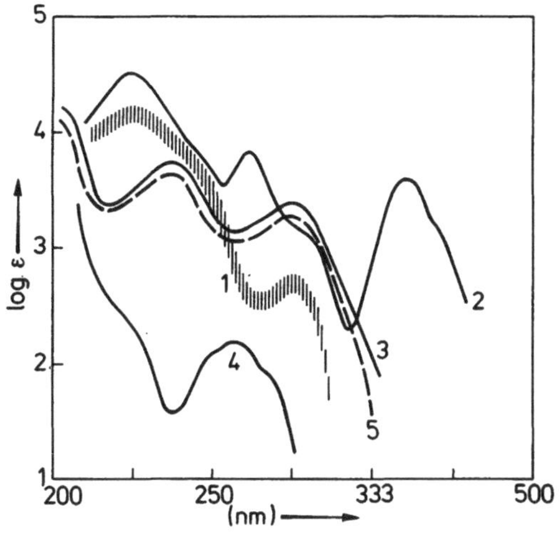

Fig. 8. Electronic absorption spectra of 1,2,5-triaminobenzene in a water solution (1), in buffered solutions at pH of 4.3 (2) and 1.2 (3) and in conc. H_2SO_4 (4), as well as aniline spectrum (5) [123,124]

pH 4.3[34] the spectrum suddenly shows a long-wave absorption maximum at 360 nm (see Fig. 8) which does not belong to the spectrum of m-phenylenediamine and hence cannot correspond to the N-monoprotonated form (*67*). This led to the suggestion [123] that monoprotonation of the 1,3,5-triaminobenzene does not yields the ammonium cation (*67*), but the 2,4,6-triaminobenzenium ion (*68*) which in stronger acids is converted into di- and triammonium cations (*66*) and (*65*).

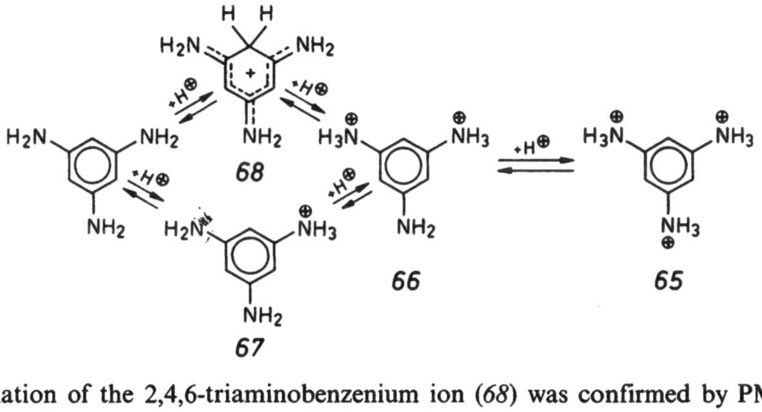

The formation of the 2,4,6-triaminobenzenium ion (*68*) was confirmed by PMR for a solution of 1,3,5-triaminobenzene in dimethyl sulphoxide acidified with hydrochloric acid [124].

Temperature-dependent changes in the UV spectra of 1,3,5-triaminobenzene in an acidified mixture of alcohol and ether at a pH corresponding to monoprotonation made it possible [124] to elucidate the monoprotonation. It both affects the C and the N atom, the benzenium ion (*68*) predominating in an equilibrium mixture of ions at room temperature and the ammonium cation (*67*) at about —60 °C (cf. C- and O-protonation of phenols and their ethers).

34 For 1,3,5-triaminobenzene $pK_a^1 = 5.5$ [123,124].

The competition between the N- and the C-protonation in weakly acid media is likely to occur for N,N-dimethyl-m-phenylenediamine as well [430]. There is an indication [373] that the protonation of 1,3,5-tris(dialkylamino)benzenes depends on the alkyl substituents. Thus, 1,3,5-tris(pyrrolidino)benzene in $CDCl_3$ attaches the proton to the unsubstituted ring position, whereas 1,3,5-tris(piperidino)benzene under the same conditions is protonated at the amino group.

C-monoprotonation in aqueous acids at room temperature also predominates for 2,4,6-triaminoanisole ($pK_a^1 = 5.2$), 2,4,6-triamino-m-xylene and 2,4,6-triamino-mesitylene [123]. The trihydrochlorides of the latter two compounds, when dissolved in water, turn into salts of benzenium ions.

The position and intensity of the absorption bands for the 2,4,6-triaminobenzenium ion and its analogues are summarized in Table 35; the molecular-orbital calculations for the energies and probabilities of electron transitions for ions of this type can be found in [124,431,432].

The ease with which benzenium ions are formed from sym-triaminobenzene and its derivatives seems to be due to the effectivity of amino groups in delocalizing the positive charge. According to the molecular-orbital calculations [124], about 40% of the charge in the 2,4,6-triaminobenzenium ion is transferred to the nitrogen atoms,

Table 35. Electronic Absorption Spectra of Aminobenzenium Ions

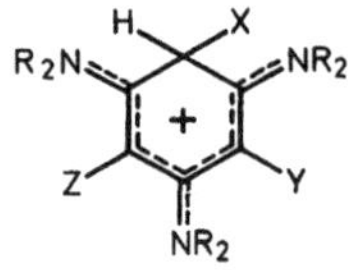

Substituents				Anion	Solvent	λ_{max}, nm (log ε)	Ref.
X	Y	Z	R_2				
H	H	H	H_2	HSO_4^-	H_2O	365 (3.6); 271 (3.8); 220 (4.5)	[124]
						360 (3.47); 272 (3.87); 220 (4.25)	[123]
H	CH_3	CH_3	H_2	HSO_4^-	H_2O	390 (3.98); 285 (4.35); 235 (4.25)	[123]
CH_3	CH_3	CH_3	H_2	HSO_4^-	H_2O	400 (3.90); 287 (4.35); 232 (4.25)	[123]
H	H	OCH_3	H_2	HSO_4^-	H_2O	375 (2.95); 277 (3.45); 220 (4.26)	[123]
Cl	H	H	$(CH_2)_4$	Cl^-	$CHCl_3$	466 (3.79); 348 (4.14); 250 (4.12)	[197]
Br	H	H	$(CH_2)_4$	Br^-	$CHCl_3$	490 (3.66); 360 (4.15); 250 (4.21)	[197]
Br	H	H	$(CH_2)_5$	BF_4^-	$CHCl_3$	488 (3.48); 368 (3.97); 256 (4.14)	[197]
Br	H	H	$(CH_2CH_2)_2O$	BF_4^-	$CHCl_3$	488 (3.74); 365 (4.15); 250 (4.27)	[197]
Br	H	H	$(CH_3)_2$	Br^-	$CHCl_3$	484 (3.56); 355 (4.02); 250 (4.11)	[197]

the general picture of the expected charge distribution being:

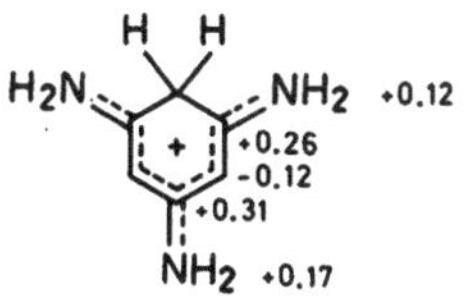

As a result of later calculations [432] the stabilizing effect of the amino groups in aminobenzenium ions is concluded to be additive and to increase in the series: meta-NH_2 < para-NH_2 < ortho-NH_2.

From 1,3,5-tris(dialkylamino)benzenes ions are formed by addition of electrophiles different from the proton, for example, halogen-cations. The evidence of these ions is given in the electronic spectra (Table 35), too. Of particular interest is a strong bathochromic shift of these ions in respect to C-protonated 1,3,5-triaminobenzenes. This shift seems to result from a stronger + M-effect of the dialkylamino groups than that of the unsubstituted amino groups.

D Polycyclic Arenium Ions

The electronic absorption spectra of the ions from protonation of naphthalene and its derivatives are listed in Table 36. The molecular-orbital calculations according to the SCF method have led to the conclusion that the spectra correspond to the ions resulting from the addition of the proton at the α-position of the naphthalene ring. Protonation at the β-position should have resulted in a long-wave absorption

Table 36. Absorption Spectra of Naphthalenium and Anthracenium Ions (25 °C)

Cation	Acid System	λ_{max}, nm log (ε)	Ref.
Naphthalene · H^+	$HF—BF_3$	410 (sh) ; 390 (4.04); 280 (4.17); 254 (4.11)	[416]
2-Methyl-	$HF—BF_3$	390	[421]
naphthalene · H^+	$HCl—AlCl_3$	460 (sh) ; 395	[418,421], cf. [48]
2,3-Dimethyl-naphthalene · H^+	$HF—BF_3$	381 (4.05); 270 (3.82); 253	[416]
1,4-Dimethyl-naphthalene · H^+	$HF—BF_3$	381 (4.27); 269 (3.95); 252 (3.88)	[416], cf. [300,301]
1,1,2,3,4-Penta-methylnaphthalenium ion	H_2SO_4 (75%)	409 (4.25) 281 (3.85); 253 (3.59); 227 (4.15); 203 (4.25)	[210]
Acenaphtene · H^+	$HF—BF_3$	420 (3.45); 354 (4.09); 264 (3.7) ; 255 (3.7)	[416]
Anthracene · H^+	HF	408 (4.57) 285 (3.27); 255 (3.93); 230 (4.13)	[416]
2-Methyl-anthracene · H^+	$HF—BF_3$	423 (4.63) 285 (3.57); 255 (3.93); 230 (4.24)	[416]
9-Methyl-anthracene · H^+	HF	397 (4.62)	[16,433]
9,9,10-Trimethyl-anthracenium ion	CF_3COOH	404 (4.41) 294 (3.69)	[153]

maximum at 550–580 nm [92,416]. The spectra of acenaphthene and 1,4-dimethylnaphthalene in HF have revealed in this region [416] very weak absorption maxima which are likely to belong to β-H-ions present in very small amounts (1–2%) in equilibrium with α-H-ions.

Similar calculations has been carried out for the addition of the proton at the meso-, α, and β-positions of anthracene [16,92,416,429,434]. Comparison of the results with the experimental spectrum of an anthracene solution in HF has confirmed the mentioned conclusion of V. Gold and F. Tye [58] that anthracene is mostly protonated at the meso position. This is also proved by PMR[35]. Among the anthracenium ions listed in Table 36, only protonated 2-methylanthracene has no PMR data available. The similarity of its electronic spectrum to that of the 9-H-anthracenium ion, however, testifies that too, the proton is attached at the meso position (for α-H- and β-H-ions long-wave absorption bands have been calculated near 550 and 720 nm, respectively [416]).

The electronic absorption spectra of acid phenanthrene solutions are more complicated. According to molecular-orbital calculations, the most basic positions must be 1(8) and 9(10) and close to them positions 4(5). The protonation of phenanthrene, therefore, could be expected to yield a mixture of isomeric ions.

Molecular-orbital calculations [416,423] predicted the 9-H-phenanthrenium ion to have in the long-wavelength region a stronger absorption band at about 530 nm, the 4-H-isomer near 400 nm (and a weaker one at about 600 nm) and the 1-H-isomer near 490 nm. In fact the spectrum of the 9,9,10-trimethylphenanthrenium ion (Table 37) shows only the first of these bands. Similar is the spectrum of a solution of 9-methylphenanthrene in the $HF—BF_3$ system, so this hydrocarbon must be mostly protonated at the 10-position.

By contrast, the spectrum of phenanthrene in the $HF—BF_3$ at —70 °C (Fig. 9, 1), in addition to the absorption maximum at 510 nm (lg ε = 4.11), shows a maximum at 410 nm (lg ε = 4.23) [417] which testifies [416] the mixtu ` of the 9-H- and the 4-H-isomers. The absorption of phenanthrene in HSO_3ı -SbF_5 at —70 °C [423] (see Fig. 9.2) differs from the previous one in the redistribution of the intensities near 500 and 400 nm. The observed change of the spectrum in passing from the

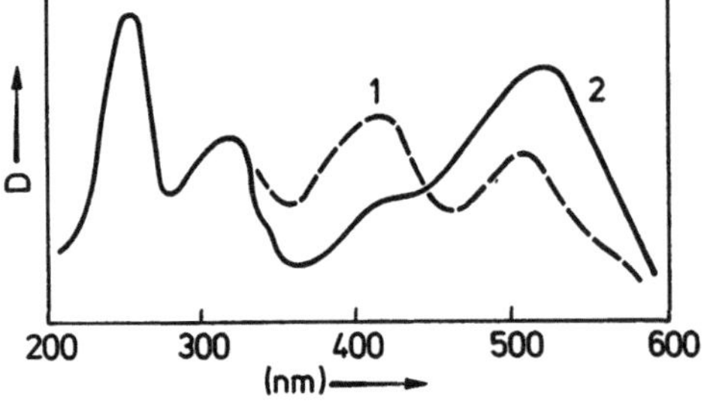

Fig. 9. Electronic absorption spectra of phenanthrene at —70 °C in $HF—BF_3$ [417] 1) and in $HSO_3F—SbF_5$ [423] 2)

35 The reactivity of the three positions in anthracene can be estimated from the isotopic hydrogen exchange rate: k(meso): k(α): k(β) = 7250:5:1 [436].

Table 37. Electronic Absorption Spectra of Phenanthrenium Ions (25 °C)

Substituents			Acid system	λ_{max}, nm (log ε)	Ref.
9-R_1	9-R_2	10-R_3			
CH_3	H	H	HF	510 (3.80) 320 (4.01); 259 (4.43)	416)
CH_3	CH_3	CH_3	H_2SO_4 (75%)	530 (3.59) 338 (3.92); 266 (4.28)	211, 313)
			$HClO_4$ (70%)	538 (3.64) 339 (3.89); 267 (4.29)	309, 313)
			H_2SO_4	537 (3.85) 337 (4.03); 266 (4.56)	310)
C_2H_5	CH_3	CH_3	H_2SO_4	528 (3.76) 340 (3.98); 266 (4.47)	212)
C_6H_5	CH_3	CH_3	$HClO_4$	549 (3.78) 333 (3.96); 268 (4.34)	435)
			$HClO_4$ (70%)	558 (3.79); 373 (sh) ; 335 (3.99); 269 (4.50)	309, 313)
			H_2SO_4 (96%)	555 (3.77); 390 (3.52); 334 (3.96); 270 (4.42)	309, 313)
p-$CH_3OC_6H_4$	CH_3	CH_3	$HClO_4$ (70%)	562 (3.66); 397 (3.35); 337 (3.84); 269 (4.32)	309)
			H_2SO_4 (96%)	556 (3.73); 406 (3.42); 338 (3.98); 271 (4.46)	309)
C_6H_5	CH_3	C_6H_5	H_2SO_4—CF_3COOH	594 (3.95); 402 (3.92); 316 (3.84); 270 (4.33)	217)
p-$CH_3OC_6H_4$	CH_3	p-$CH_3OC_6H_4$	H_2SO_4—CF_3COOH	577 (4.26); 448 (4.08); 318 ; 260 (4.31)	217)

solution of phenanthrene in $HF{-}BF_3$ to that in $HSO_3F{-}SbF_5$ seemingly [423] reflects the change in the ratio of isomeric ions. This conclusion may be significant in discussing the effect of the acid medium on the direction of preferable protonation (see Sect. IV.1.A); so it is desirable to verify it by NMR.

Table 37 presents two 9-aryl-9,10-dimethylphenanthrenium ions (for other examples see [213,216,217,309,312,313]). The weak band near 400 nm seems to be due to the donor-acceptor interaction between the π-system of the aryl fragment and the carbenium centre at C_{10}. This band for 9-p-X-phenyl-9,10-dimethylphenanthrenium ions correlates well with the ionization potentials of the respective X-substituted benzenes [309,313].

The electronic spectroscopy combined with molecular-orbital calculations and isotopic hydrogen exchange has also been applied in the protonation of other polycyclic hydrocarbons: pyrene [171,416,418], 1,2- and 3,4-benzpyrenes [171,416], perylene [171,416], 1,2-benzanthracene and its monomethylated derivatives [171,423,433,434] and tetracene [423,434]. The electronic spectra of polycyclic arenium ions ArH_2^+ are compared with their negatively charged analogues ArH_2^- [437].

5 Vibrational Spectra of Arenium Ion Salts

A Some Introductory Data

The infrared (IR) absorption spectra and Raman spectra have been used until recently to a far lesser extent than the electronic and the NMR spectroscopy. For IR this is due to the fact that the standard cell windows made of alkaline metal halides fail to stand the attack of the acids in which the carbenium ions are generated. These complications can be overcome by using cells made of special materials. Thus, the IR spectra of alkylcarbenium hexafluoroantimoniates in SbF_5 have been successfully recorded in cells made of "IRTRAN" [438,439]. Other materials resistant to acids and suitable for work in the IR region are [440]: quartz, calcium fluoride, silver chloride, diamond etc. However, they have a limited range of transparency (2600–3600 cm^{-1} for quartz, 1000–3600 cm^{-1} for CaF_2) or are not easily available (diamond). Therefore, the IR spectra of arenium ion salts and their solutions were recorded [50,53,441] using special cells with conventional salt windows (KBr, NaCl) protected by polymer films. It is proposed to prepare solid samples of $AH^+ \cdot MY_{n+1}^-$ for recording the spectra at low temperatures (when there is practically no interaction between the salt and the window material) by the condensation of metal halide MY_n, hydrocarbon and hydrogen halide HY from the gas phase right onto the salt windows of the evacuated cell [194,442–444]. The drawbacks of this technique are the complicated procedure of component proportioning, the lack of confidence that the reaction is completed in the component condensation (cf. [50,441]) and the low quality of the spectrum above 2000 cm^{-1} due to the strong scattering of radiation by the solid sample. However, unstable salts of the unsubstituted benzenium ion, and arenium ions, which readily isomerize (protonated o- and p-xylenes, asymmetric trimethylbenzenes etc.) have only been measured with this low-temperature technique [442,443].

Recording the Raman spectra implies no difficulties due to agressiveness of acid media since most acids do not affect the glass cells for this method. For $HF—BF_3$ and $HF—SbF_5$ systems, quartz cells can be used. For a long time the Raman spectra could not be used for carbenium ions since the solutions of their salts are coloured. This may be due to both the absorption of carbenium ions (e.g. arenium ions) and the formation of the coloured side products. Also, some types of carbenium ions when irradiated by the mercury lamp (until recently the main source of exitation) undergo photochemical transformations. The possibilities of the Raman spectra were drastically enlarged with the spectrometers using a helium-neon ($\lambda_{excit} = 6328$ Å) and other lasers to excite the spectra [441,445–447].

It is reasonable to discuss firstly the main results obtained by the vibrational spectroscopy in studying some other types of carbenium ions[36].

G. Olah and co-workers [439,447] have recorded the IR and Raman spectra of alkylcarbenium hexafluoroantimoniates of the $(CH_3)_2\overset{+}{C}R$ type (R=H, CH_3, C_2H_5 etc.) in SbF_5. In these ions the C—H stretching vibrations manifest themselves at exceptionally low frequencies (2730, 2830 and 2815 cm^{-1} for R=H, CH_3 and C_2H_5), while the symmetric stretching vibrations of the carbon skeleton, at exceptionally high ones (1260, 1290 and 1295 cm^{-1}). These peculiarities reflect a decrease of the force constants of the C—H bonds and their increase for the C—C bonds as a result of involving the alkyl groups in the hyperconjugation with the carbenium centre:

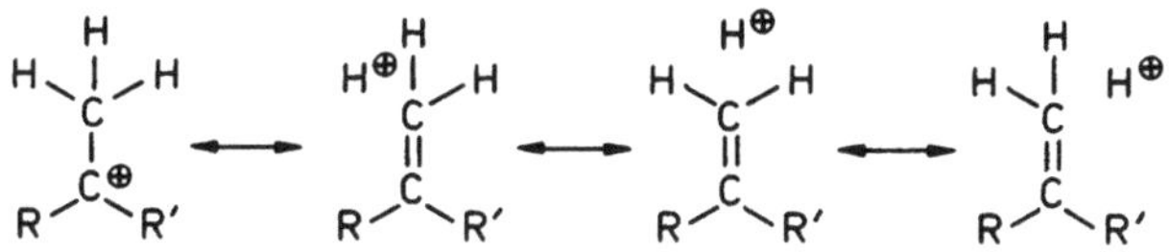

Comparison of the IR and Raman spectra of the trimethylcarbenium ion with those of its isoelectric analogue — trimethylboron — has confirmed the planar structure of alkylcarbenium ions. The planar structure of the carbenium fragment $—\overset{+}{C}<$ was also concluded from the IR spectra of salts of the triphenylmethyl cation $(C_6H_5)_3C^+$ [449–451] and its analogues having different substituents at the para positions of the benzene rings [451,452]. A detailed analysis of the IR spectrum of this cation has shown [449–452] the cation to have the form of a propeller; this conclusion is supported by the X-ray analysis [453]. The asymmetric stretching vibration of the C—C bonds of the $C-\overset{+}{C}\langle{}^{C}_{C}$ groups of this cation is characterized by a much higher frequency (1359 cm^{-1}) than is that of the $C-CH\langle{}^{C}_{C}$ group of triphenylmethane (1280 cm^{-1}) [451] indicating an increase in the force constants of the bonds between the phenyl group and the carbenium centre by conjugation.

36 see also the survey [448].

For the tricyclopropylcarbenium ion the antisymmetric stretching vibration of the 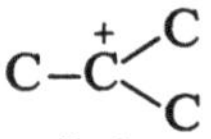 group is at 1279 cm^{-1} [448,454].

Information has been obtained on the C—C stretching vibrations of various enyl carbenium ions. The IR spectrum of $[CCl_2 \equiv CCl \equiv CCl_2]^+$ shows this pentachloroallyl cation to have a planar structure, the antisymmetric stretching vibration of its C—C bonds displaying itself at 1350 cm^{-1} and the symmetric, at 1040 cm^{-1} [455]. To judge by the frequencies of this cation, the C—C bond order is close to the expected value 1.5. The stretching vibrations of the bonds between the sp^2-hybridized carbon atoms of 4-X-1,2,3,4-tetramethylcyclobutenyl cations (*69*) are responsible for the very strong bands in the range of 1465–1475 cm^{-1} [456] (for polymethyl-

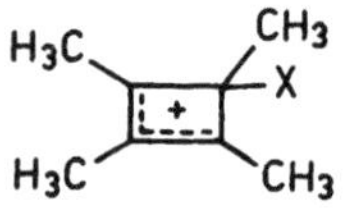

69
X = H, CH3, Cl, Br

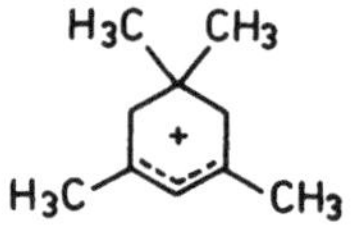

70

cyclobutenes the band corresponding to the C=C bond is observed at 1690 cm^{-1}); likewise for the 1,3,5,5-tetramethylcyclohexenyl cation (*70*) a very intensive absorption band at 1533 cm^{-1} is observed [457].

Vibrational spectroscopy was used to study positively charged nonbenzenoid aromatic systems: the substituted cyclopropenyl cations [458–462] and the tropylium cation [463–465]. The C—C stretching vibrations in the tropylium cation are correlated with the intensive band at 1477 cm^{-1} in the IR spectrum [463–465] and the line at 1594 cm^{-1} in the Raman spectrum [463,464]. They correspond to the absorption band at 1485 cm^{-1} in the IR and with a doublet at 1594 and 1605 cm^{-1} in the Raman spectrum of benzene.

These data on the C—C stretching vibration of various enyl carbenium ions are summarized in Table 38. For comparison, the frequency of the C^+—C stretching vibrations of alkylcarbenium ions is included.

The Raman spectra were also used [445,446] to choose between the nonclassic and classic structures of the norbornyl cation and its 2-alkyl derivatives.

B Unsubstituted and Alkyl-Substituted Arenium Ions

IR and Raman spectra of arenium ions were obtained for $AH^+ \cdot M_nY^-_{3n+1}$ formed by interaction of the aromatic hydrocarbon A, the hydrogen halide HY and the metal

Table 38. Carbon—Carbon Stretching Bands in IR Spectra of Various Carbenium Ions [448]

Cation	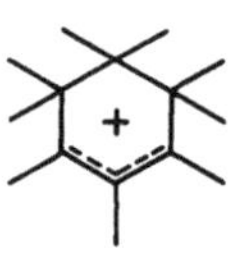	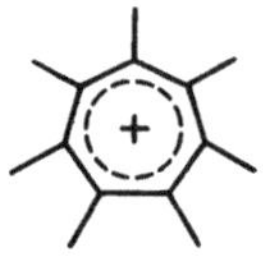		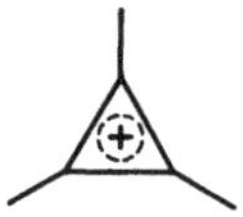	$C^{\oplus}—C_{sp3}$
ν (cm^{-1})	~1530	~1480	1465...1475	1400 ... 1430	1250...1300

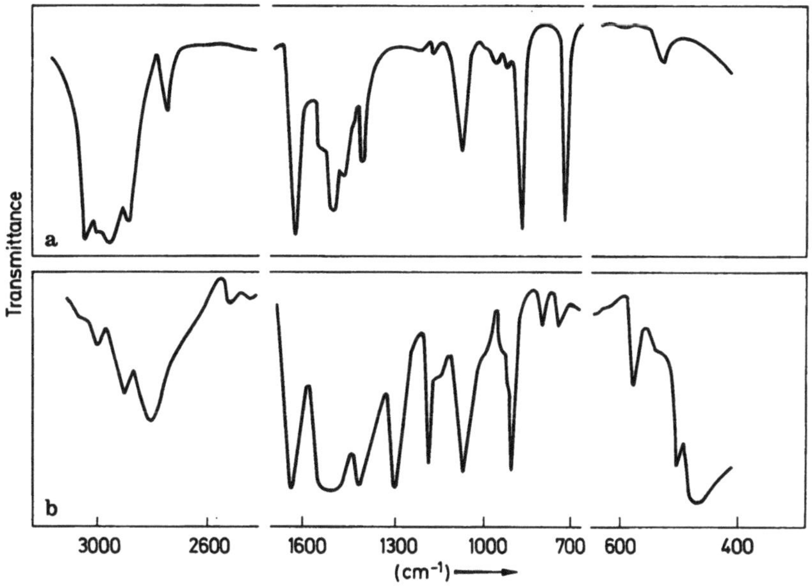

Fig. 10. Infrared spectra of mesitylene **a** and mesitylenium ion heptabromodialuminate **b** recorded at 25 °C [50)]

halide MY_3. Fig. 10b depicts as an example the IR spectrum of mesitylenium heptabromodialuminate.

The formation of the arenium ions from the hydrocarbons greatly changes the IR spectra in the region of the C—H stretching, in particular, the wide absorption bands at about 2800 cm^{-1} is characteristic [50,444)]. Figure 11 depicts the IR spectra of methyl-

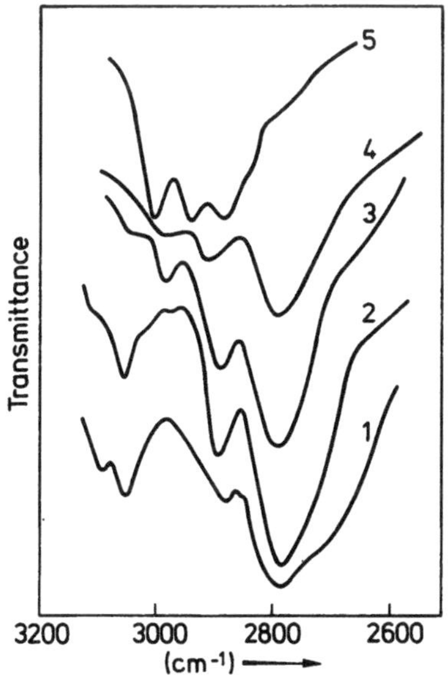

Fig. 11. Infrared spectra of the salts $AH^+ \cdot Al_2Br_7^-$ prepared from toluene 1), m-xylene 2), mesitylene 3), pentamethylbenzene 4) and hexamethylbenzene 5) at 2000–3200 cm^{-1} at 25 °C [50)]

benzenium heptabromodialuminates at 2600–3200 cm^{-1}. The comparison of the spectra shows that the wide intensive absorption band near to 2800 cm^{-1} is typical for salts of toluene, m-xylene, mesitylene, and pentamethylbenzene, but not for the hexamethylbenzenium ion. Since the former four hydrocarbons take on the proton at the unsubstituted ring positions to form the CH_2 group (see the PMR data), whereas the protonation of hexamethybenzene involves a $CHCH_3$ group the band in question is obviously brought about by the stretching vibrations of the ring CH_2 groups in methylbenzenium ions [50)]. Accordingly the absorption band of the complex 1,3,5-$(CH_3)_3C_6H_3 \cdot HBr \cdot 2\,AlBr_3$ at 2805 cm^{-1} is shifted to 2070 cm^{-1} in passing to the complex 1,3,5-$(CH_3)_3C_6D_3 \cdot DBr \cdot 2\,AlBr_3$ ($v_H/v_D = 1.36$).

The IR spectra of methylbenzenium ions and their deuterated analogues [50)] also revealed the stretching vibrations of the methyl C—H to correspond to the regions of 2890–2920 and 2980–3000 cm^{-1} and the bonds involving single ring hydrogens to the region of 3040–3100 cm^{-1}. The intensity of the absorption resulting from these bonds is by far lower than that of the CH_2 group (see Fig. 11).

The bands of the CH_2 group of methylbenzenium ions are located unusually low for the C—H stretching vibrations, just as for alkylcarbenium ions, since they participate in the σ,p-conjugation with the vicinal carbenium centres. The withdrawing of the electron density from the C—H bonds through the hyperconjugation permits the ring-CH_2-group hydrogens of arenium ions to be involved in the hydrogen bonds with nucleophilic particle of the medium. This is revealed by the large half-width and the high intensity of the bands.

The onium salts of the type $[R_3X^+—CH_2—COR] \cdot Y^-$ where X = N, P and As showed [466, 467)] that the tendency of the CH_2-group hydrogens to form the hydrogen bonds with the Y^- anion (I^-, Br^-, Cl^-, HF_2^-) is especially manifest when the hyperconjugation of the C—H bonds with the onium centre is possible. For instance, the CH_2 group in onium compounds of phosphorus having 3d-orbitals accessible for conjugation gives a wide intensive band greatly shifted to the low-frequency region (for R = C_6H_5 and Y = Br 2638 cm^{-1}). On the contrary, the ammonium group with its inductive mechanism shifts the neighbouring CH_2 group absorption to the region of higher frequencies.

The contribution of σ,π-conjugation of the CH_2 group and of the unsaturated charged part of arenium ions to the stabilization must decrease with the accumulation of substituents efficiently delocalizing the positive charge. Accordingly, in passing from the salt of the benzenium ion[37] to those obtained from toluene, mesitylene, and anthracene the ring CH_2-group absorption shifts to higher frequencies (Fig. 12) with rising hydrocarbon basicity[38]. At the same time its half-width decreases and the integral intensity falls severalfold [21)].

The "acidic" hydrogens of the ring CH_2 in arenium ions interact with the nucleophile particles of the medium. Hence, changing the anion in the salts $AH^+ \cdot M_nY_{3n+1}^-$ causes quite distinct, though small, changes in the CH_2-group absorption. Thus, in the salt of the mesitylenium ion the replacement of the $Al_2Br_7^-$

---

37 The low basicity of benzene prevents to obtain $C_6H_7^+ \cdot Al_2Br_7^-$ under usual conditions (cf. [442)]). The spectrum (1) in Fig. 12 corresponds to the system $C_6H_6 : HBr : Al_2Br_6$ = 2:0.6:1 [21)].

38 The position of absorption band is compared with the ionization potential of the hydrocarbon A [444)].

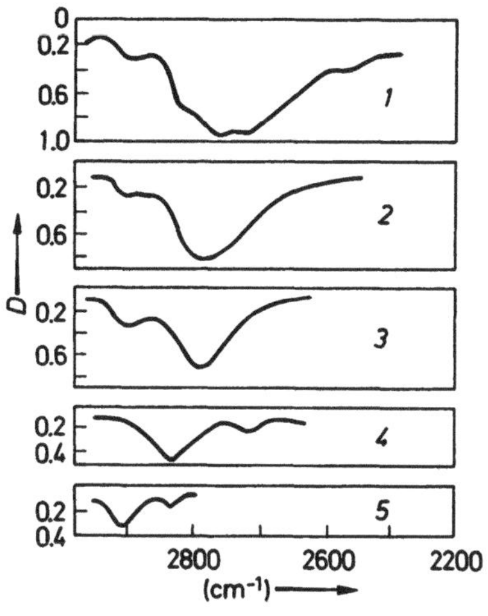

Fig. 12. Infrared spectra of the salts $AH^+ \cdot Al_2Br_7^-$ recalculated to one molar concentration of the absorbent (c = 2.9 M, l = 5.7 · 10^{-2} mm) where A = benzene 1), toluene 2), mesitylene 3), anthracene 4), as well as the infrared spectrum of diphenylmethane 5) [21]

anion by $Al_2Cl_7^-$ leads to a shift of 15 cm^{-1} to high-frequency [21]. To judge by the CH_2-group maximum in the IR spectra of different mesitylenium salts, the aptitude of the anions for interaction with the "acidic" hydrogens of the CH_2 groups diminishes in the series [21,441]: $AlBr_4^- > Al_2Br_7^- > AlCl_4^- > GaCl_4^- > Al_2Cl_7^-$.

The donor ability of the anions follows the same order when they interact with the hydroxy group hydrogen of hydroxybenzenium ions.

The anions of the MY_n^- type are involved in the formation of hydrogen bonds with the "acidic" hydrogens of the ammonium- and hydroxonium-type cations [468–470].

The interaction of anions with the cationic part of the arenium ion salts results in decreasing the anion symmetry to be recognized by the splitting of the $AlCl_4^-$ lines in the Raman spectrum of the mesitylenium tetrachloroaluminate.

Aromatic hydrocarbon molecules may also serve as nucleophilic particles forming hydrogen bonds. Thus, the addition of an equimolar amount of hydrocarbon A to the salts $AH^+ \cdot Al_2Br_7^-$ obtained from toluene or mesitylene increases the integral intensity of the CH_2-group band by 1.5–2 [21] (see also [444]). A less basic benzene changes the intensity of this absorption much less. The IR spectra of ethylbenzene complexes with a less-than-equivalent amount of $AlCl_3$ and HCl are also described [471].

The vibrational spectra of arenium ion salts and respective hydrocarbons show great differences at 600–1600 cm^{-1} as well. This is not surprising since the addition of an aromatic molecule changes its atom symmetry and, hence, the vibrations active in the IR and the Raman spectra. Thus, when passing from benzene to the benzenium ion the symmetry decreases from D_{6h} to C_{2v}. In the IR spectrum of benzenium ion salts an intensive absorption band at 1595 cm^{-1} [16,442] is appeared which is due to the skeleton C—C bond vibrations inactive in the IR spectrum of benzene but manifesting themselves in that of its derivatives. Other examples can be found in [443].

The intensive absorption at 1600 cm^{-1} seems to be characteristic of benzenium ions (cf. the data for other carbenium ions in Table 38). Thus, when 4-methylene-1,1,2,3,5,6-hexamethylcyclohexa-2,5-diene (*30*) is protonated in 95% H_2SO_4 to form the heptamethylbenzenium ion (*16a*) the IR bands at 1670 and 1580 cm^{-1} disappear and an intensive band at 1595 cm^{-1} appears instead [441]. The absorption band resulting from the C⋯C bond vibrations of the mesitylenium ion is located at 1625 cm^{-1}. The respective line in the Raman spectrum, by its position and polarization degree (1625 cm^{-1}, $\varrho = 0.65$ [441]), is similar to those of aromatic C—C bonds (for mesitylene $\nu = 1610\ cm^{-1}$, $\varrho = 0.75$).

This fact agrees with Perkampus's and Baumgarten's suggestion [194,443] that in the vibrational spectra of methylbenzenium ions the two hydrogens of the CH_2 group can, for a first approximation, be regarded as a single substituent H_2 bonded with the aromatic ring and carrying a partial positive charge. The validity of this model follows from the effective participation of the C—H bonds of the ring CH_2 group in the hyperconjugation with the charged dienyl system of benzenium ions. If the model is correct the absorption bands for the out-of-plane deformation of methylbenzenium ions are determined, just as for substituted benzenes [472], by the number and the relative position of the substitutients. Such a correspondence was discovered [16,194,443] (cf. also [444]) by analyzing the low-temperature IR spectra of the salts formed by the interaction of benzene hydrocarbons with hydrogen chloride and gallium chloride. The results are cited in Table 39.

From these data H. Perkampus and E. Baumgarten [443] concluded the salts from m- and p-xylenes, mesitylene, hemimellitene and durene to contain only one of the possible isomeric cations, namely:

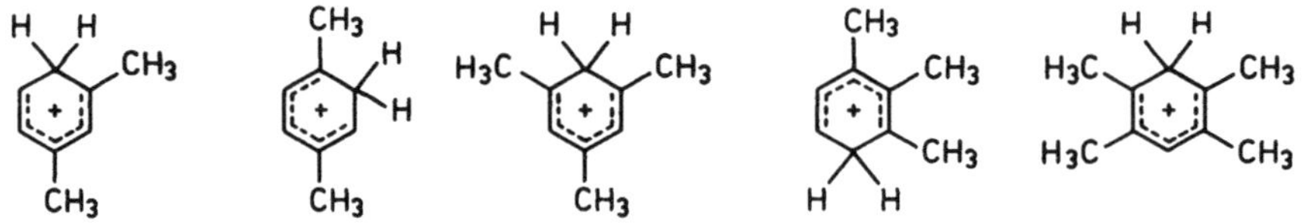

The salts obtained from toluene and pseudocumene consisted, judging by the bands for the out-of-plane deformation modes of C—H bonds, of two isomeric ions:

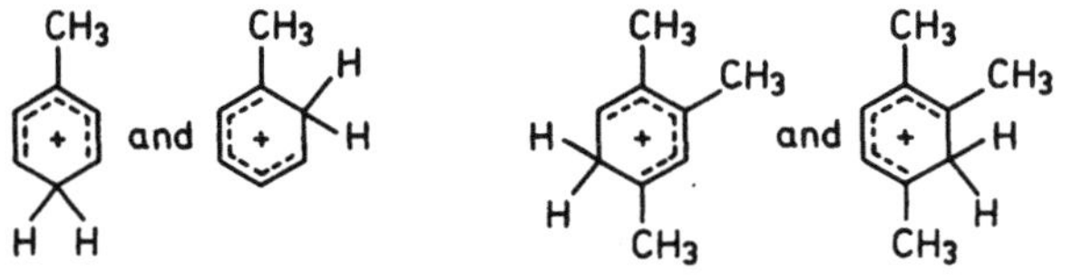

According to PMR in solution at low temperatures the equilibrium between isomeric cations on protonation of toluene and pseudocumene is strongly shifted towards the 4-methylbenzenium and the 2,4,5-trimethylbenzenium ions (see Sect. III.1.A). This discrepancy may be due to the fact that in the samples prepared by "freezing" the components at low temperature from the vapour phase onto the cell windows the salts of isomeric cations could be present in non-equilibrium ratios.

The frequencies of the C_{sp^2}—H stretching vibrations rise as the electron-withdrawing properties of substituents are increased. Therefore for methylbenzenium ions in the approximate model and under the reduced electron density in the ring one could expect the bands of the C_{sp^2}—H stretching vibrations to shift to high

Table 39. Carbon-Hydrogen Out-of-plane Bending Bands of Benzenium Ions in IR Spectra of A + HCl + $GaCl_3$ Systems [16,194,443]

Cation AH^+	The number of adjacent ring hydrogen atoms					Unassigned bands
	5	4	3	2	1	
	The expected position of bands, cm^{-1}					
	690–715 s., 730–770 v.s.	735–770 v.s.	660–725 m., 750–810 v.s.	700–750 m., 800–860 v.s.	860–900 m	
Benzene · H^+	690 v.s.					
Toluene · H^+		742 v.s.		709 v.s., 848 v.s.		898 v.s.
o-Xylene · H^+				710 s. 797 m		
m-Xylene · H^+				732 s., 829 s.	875 s.	697 m
p-Xylene · H^+				720 s., 857 v.s.	873 m	779 s., 830 m
Mezitylene · H^+					867 m	
Hemimellitene · H^+				729 s., 784 v.s.		712 m
Pseudocumene · H^+				734 s., 824 s.	876 m., 888 m	743 m
Durene · H^+					889 m	

frequency as compared to that of hydrocarbons. Such a shift is illustrated by the vibrations in the IR and Raman spectra of the mesitylenium ion and other model compounds [441] (cf. [473]):

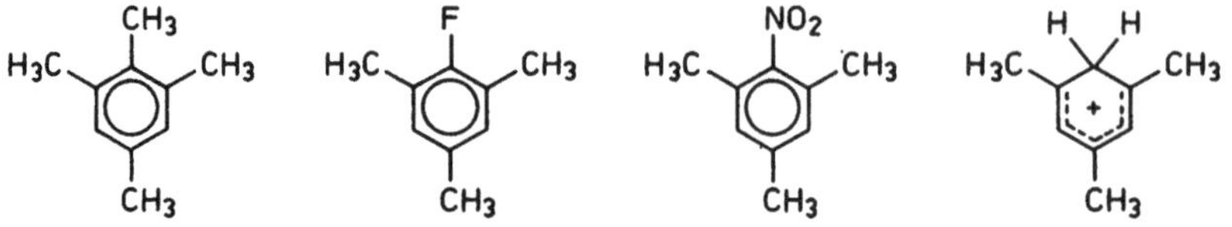

Infrared:	3010	3020	3037	3040 cm^{-1}
Raman:	3011	3022	3035, 3045	3048 cm^{-1}

The approximative modelling of methylbenzenium ions by benzene derivatives containing an extra substituent as compared with the respective methylated benzene was pointed out in the Raman spectra [441, 474] as well. Thus, the Raman spectrum of the mesitylenium ion in the region of 400–1700 cm^{-1} has much in common with that of 1-X-2,4,6-trimethylbenzenes but differs much from those of 1,2,3,4- and 1,2,4,5-tetra-substituted benzenes.

In addition calculating the vibration frequencies of the mesitylenium ion by using the force field of benzene derivatives yielded results that agree well with the observed lines in the IR and Raman spectra [473].

C Benzenium Ions with Heteroatomic Substituents

Vibrational spectra of arenium ions with heteroatomic substituents are very scarce. The protonation of hexamethylcyclohexa-2,5-dienone (*71*) and -2,4-dienone [219], forming the complex (*71*) · HCl · 2 $AlCl_3$ in CH_2Cl_2 shifts the stretching vibration band of the carbon-oxygen bond from 1615 cm^{-1} (C=O) to 1510 cm^{-1} (see also [424]). The band of the OH group at 3450 cm^{-1} and the before mentioned data are in good agreement with those of the PMR, i.e. the complex is a salt of the 4-hydroxy-1,1,2,3,5,6-hexamethylbenzenium ion (*31a*)

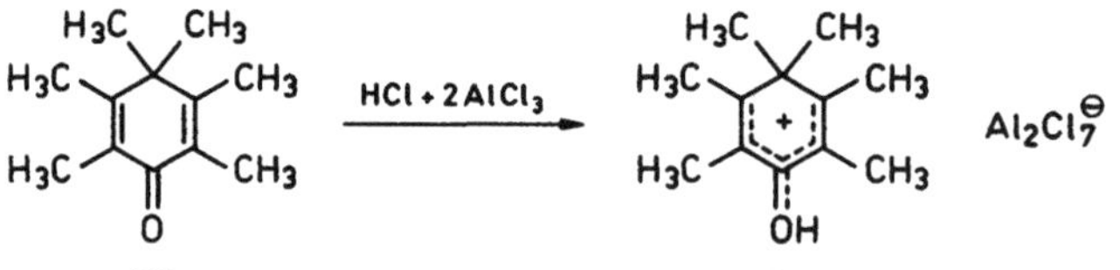

71 *31a*

A significant shift of the hydroxy-group band to low-frequency as compared to the usual position of free OH groups in neutral compounds and its large half-width ($\Delta\nu_{1/2} \approx 200$ cm^{-1}) seem to be due to two factors: the decrease of the force constant of the O—H band by the positive charge of the cation and the participation of this group in the hydrogen bond with the anion. This is confirmed by a 90 cm^{-1} shift to low-frequency observed in passing to the complex (*71*) · HCl · $AlCl_3$ (anion $AlCl_4^-$)[39].

The ring C⋯⋯C bond stretching vibrations of ion (*31a*) give the band at 1630 cm^{-1} whereas the absorption of C=C bond in the parent dienone (*71*) is located at 1660 cm^{-1} [219]. The stretching vibration frequency reflecting the order of C—C bonds decreased when passing from perhalogenated cyclohexadienes to perhalogenated benzenium ions [238].

For the 2,4,6-triaminobenzenium ions (*29*) the C⋯⋯N and C⋯⋯C absorption bands are observed at 1610–1620, 1520–1545 and 1425–1435 cm^{-1} [198].

The IR spectra of 1,3,5-triaminobenzene, its monohydrochloride and trihydrochloride proved the assumption that the monohydrochloride has the structure of the chloride of the 2,4,6-triaminobenzenium ions [124]. The N—H stretching bands of the amino groups (doublet near 3300 cm^{-1}) are similar to those of free amine (3380 and 3200 cm^{-1}) and sharply differing from that of the NH_3^+ group of the trihydrochloride which is shifted to low-frequency (to 2500 cm^{-1}).

F Vibrational Frequencies of Anions

One of the proofs that the complexes of aromatic hydrocarbons with halides of the third-group metals and with hydrogen halides are salts of the hypothetical acids HMY_4 and HM_2Y_7 is the detection of the bands of the anions (MY_4^- and $M_2Y_7^-$) in the low-frequency region.

The tetrahedral anions MY_4^- ($AlCl_4^-$, $AlBr_4^-$, $GaCl_4^-$, BF_4^- etc.) belong to the T_d group of symmetry and have four normal vibrations shown in Fig. 13 [479]. All these vibrations are active in the Raman spectra, the most intensive and the only polarized line being the one assigned to the ν_1 stretching vibration.

The spectral characteristics of the $AlCl_4^-$ anion are most thoroughly studied. Below the positions of the strong absorption band assigned to the ν_3 stretching vibration of inorganic and organic salts containing this anion are listed.

Cation	Band Position, cm^{-1}	Ref.
Na^+	481	460)
	~500	480)
K^+	482	460)
H_4N^+	485	481)
$(CH_3)_4N^+$	495	482)
	496	483)
$(C_2H_5)_2OH^+$	490	484)
	496	483)
PCl_4^+	490	485)
ArN_2^+	480–500	486)
$(C_6H_5)_3C^+$	496	483)
$C_3Cl_3^+$	482, 512	460)
RCO^+	500	480)

39 The hydrogen bond with the hydroxy group hydrogen and different anions has been studied for O-protonated tropone [475–477] and carbonyl derivatives of aromatic systems [478].

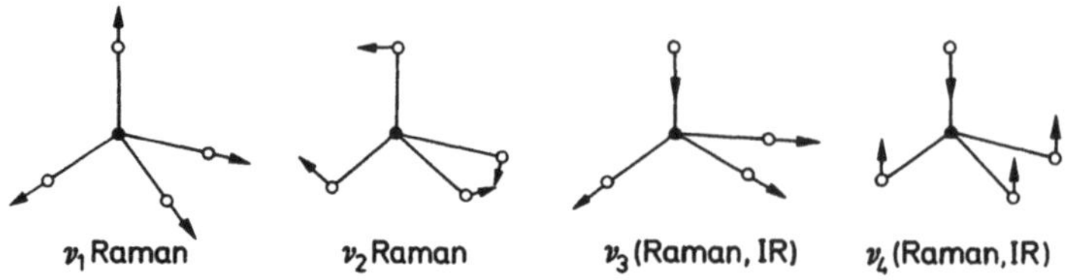

Fig. 13. Normal vibrations of tetrahedral particles [479]

Table 40. Vibrational Frequencies of $AlCl_4^-$ Anion in Raman Spectra of $M^+AlCl_4^-$ Salts

Cation	Conditions	Frequencies (cm^{-1})				Ref.
		ν_1 (p)	ν_2 (dp)	ν_3 (dp)	ν_4 (dp)	
Na^+	melted, 170 °C	349	146	(575)[a]	180	487)
	melted, 200 °C	349	145	(580)[a]	183	488)
	melted, 225 °C	351	121	490	186	489)
K^+	melted, 300 °C	350	122	487	182	490,491)
NH_4^+	solid, 20 °C	356	126	485	184	481)
$(CH_3)_4N^+$	solid	352	125	488	183	482)
NO^+	melted, 190 °C	349	136	...	182	487)
PCl_4^+	solid	352	147	495	...	485)
SCl_3^+	melted, 125 °C	349	130	...	186	492)
	solid	352	138	480	185	492)
	solid, −90 °C	350		481	174	493)
$C_3Cl_3^+$	solution in SO_2	350	...	...		452)

[a] This assignment is questionable

The band of the ν_4 deformation is less accessible for observation in the IR and its evidence is scarce: PCl_4^+ — 180 cm^{-1} [485]; $(C_2H_5)_2OH^+$ — 179 cm^{-1} [484]. Raman spectra of the $AlCl_4^-$ anion are listed in Table 40. An intensive polarized line ν_1 and a depolarized line of mean intensity of ν_4 are independent of the cation at 350 and 180 cm^{-1}, respectively; the ν_3 stretching vibration correlates with a weak line at 490 cm^{-1}. The line of the ν_2 symmetric deformational vibration is less stable; it seems to be sensitive to ambient particles.

Vibrational frequencies of $AlBr_4^-$ in the spectra of various salts are summarized in Table 41.

H. Perkampus and E. Baumgarten [442], when studying the benzenium ion in the C_6H_6—HCl—$AlCl_3$ system by IR, noted in the region of 400–600 cm^{-1} only one new band at 549 cm^{-1} which they tentatively assigned to $AlCl_4^-$. By analogy one could assume the band of the ternary C_6H_6—HBr—$AlBr_3$ system [442] at 450 cm^{-1} to belong to $AlBr_4^-$. The before mentioned data on $AlCl_4^-$ and $AlBr_4^-$, however, preclude such an assignment.

Table 41. Vibrational Frequencies of $AlBr_4^-$ Anion in Spectra of $M^+AlBr_4^-$ Salts

Cation	Conditions	Type of spectrum	Frequencies (cm^{-1})				Ref.
			ν_1	ν_2	ν_3	ν_4	
K^+		IR			401	...	460)
Na^+	melted	Raman	209 v.s.	75 s.	409 w.	114 s.	494)
$(C_2H_5)_4N^+$	solid	Raman	212 v.s.	98 w.	393 m.	114 w.	495)
	solid	IR	(210)		394 m.	114 s.	495)
$(CH_3)_4N^+$	solid	Raman	214	76	400	119	495)
	solid	IR			404		482)
	solid	IR			394		496)
$C_3Br_3^+$		IR			400	...	460)

This discrepancy is explained [53] by the IR spectra of mesitylene and hexamethylbenzene complexes with hydrogen halides and halides of metals $A \cdot HY \cdot nAlY_3$. The IR spectra at n = 1 demonstrate between 400 and 600 cm^{-1} intensive bands at about 490 cm^{-1} (Y = Cl) and at 405 cm^{-1} (Y = Br) corresponding to AlY_4^-. But complexes of the composition $A \cdot HY \cdot 2\,AlY_3$ show the bands of the AlY_4^- anion either very weak or not at all. Instead there are bands at 540–550 cm^{-1} (Y = Cl) and 450 cm^{-1} (Y = Br). These bands seem to have been observed by H. Perkampus and E. Baumgarten [442] who had failed to prepare complexes containing no excess of aluminium halide. These bands are contributed to $Al_2Y_7^-$ [53].

An intensive absorption at about 545 cm^{-1} characterizes the IR spectra of heptachlorodialuminates of the anthracenium ion and its 9,9-dimethyl and 9,9-diethyl derivatives, while bands at 450 cm^{-1} occur in the spectra of the $A \cdot HBr \cdot 2\,AlBr_3$ complexes prepared from toluene, m-xylene and pentamethylbenzene. This band is also observed for the solid KBr—$AlBr_3$ alloy in a 1:2 molar ratio [53] for which a thermal analysis [497] has proved the structure $K^+Al_2Br_7^-$.

The conclusions on the anions in the ternary complexes of aromatic hydrocarbons with hydrogen halides and halides of metals drawn from the IR spectra were later confirmed by the Raman spectroscopy [441,474]. The mesitylene · HCl · $AlCl_3$ complex displays the lines of the $AlCl_4^-$ (350 v.s., ~180 m and ~125 w), the mesitylene · HBr · $AlBr_3$ complex in CH_2Br_2, those of $AlBr_4^-$ anion (396 w, 210 v.s., 114 w, 98 w). The lines of the $AlCl_4^-$ at 183 (ν_4) and 126 cm^{-1} (ν_2) have, respectively, a triplet and a doublet splitting; this indicates a decrease in the symmetry of the anion from T_d to C_{2v} (structure of Z_2XY_2 type [479]) as a result of its interaction with the cation.

As the anion influences the band corresponding to the stretching vibrations of the ring CH_2 group this interaction may be due to the hydrogen bond between the "acidic" hydrogens of this group and the anion.

Recently analyzed Raman spectra of the melted KCl—$AlCl_3$ [490,491] and $NaCl$—$AlCl_3$ [489] systems with a more than equimolecular content of $AlCl_3$ have permitted to identify the line of the $Al_2Cl_7^-$ anion.[40] The calculated vibrational frequencies for the conformation of this anion with a tetrahedral arrangement of chlorine atoms and

40 The ion $Al_3Cl_{10}^-$ may also exist [489].

Table 42. Vibrational Frequencies of $Al_2Cl_7^-$ Anion in Raman Spectra

Vibration modes	Expected frequences (cm^{-1}) and polarization [490,491]	Observed frequencies		
		KCl + $AlCl_3$ [a] [490,491]	NaCl + $AlCl_3$ [a] [489]	Mesitylene + HCl + + 2 $AlCl_3$ [441,474]
A_{1g}	469 (p)	425 w.	435 w. (p)	440 w.
	272 (p)	312 v.s. (p)	313 v.s. (p)	310 s. (p)
	35 (p)	...		...
E_g	470 (dp)	435 w. (dp)	...	440 w.
	166 (dp)	164 m. (dp)	165 s. (dp)	165 m.
				186 m.
	83 (dp)	99 m. (dp)	100 s. (dp)	

[a] melted

with the linear fragment Al—Cl—Al, as well as the observed lines are listed in Table 42. This table also includes the Raman spectrum of the mesitylene · HCl · 2 $AlCl_3$ complex, confirming the presence of $Al_2Cl_7^-$.

For the heptabromodialuminate of the mesitylenium ion see [441,474], data on $Al_2Br_7^-$ are found in [494].

The binary system aromatic hvdrocarbon (A) — gallium chloride [190] is different from the systems A + AlY_3 [190,442,498]. In the latter system, even if interaction occurs, it is restricted to weak π-complexes and is not reflected on the IR spectra of hydrocarbons. By contrast, the IR spectra of the A + $GaCl_3$ binary systems (in some case for the A + $GaBr_3$ as well) at —200 °C show great differences in the ring CH out-of-plane deformational vibrations. According to H. Perkampus and E. Baumgarten [190] these changes are due to bipolar σ-complexes of (*28*). Indeed, the carbon forming a bond with gallium to substitute the aromatic system explains rather well the intensive absorption bands in the out-of-plane deformational vibrations. Attempts to observe bipolar σ-complexes of (*28*) in solutions were without results.

The saturation of A + $GaCl_3$ with hydrogen chloride produces the salt of the arenium ion [443]. Accordingly, the mesitylene · HCl · $GaCl_3$ complex displays the Raman lines of $GaCl_4^-$ (in cm^{-1}): 342 v.s. (p), 142 m and 150 m (dp) [441,474]. The ν_1–ν_4 vibrational frequencies of the $GaCl_4^-$ anion [499,500] correlate with the lines 346, 114, 386 and 149 cm^{-1} in its Raman spectrum. Strangely, mesitylene complex failed to show the ν_3 and ν_4 lines observed in salts with other cations: 369 (m) and 150 cm^{-1} (m) for $(C_2H_5)_2OH^+ \cdot GaCl_4^-$ [484] and 373 cm^{-1} (v.s.) for $(C_2H_5)_4N^+ \cdot GaCl_4^-$ [501].

The vibrational spectra of arenium ion salts with different anions have not been studied yet. Evidence on the vibration frequencies of the other anions, if necessary, can be found in papers for BF_4^- [479], $GaBr_4^-$ [479,499,502], $AlCl_nBr_{4-n}^-$ [482], AlI_4^- [494], $Al_2I_7^-$ [494], $SbCl_6^-$ [503], AsF_6^- [504], as well as in books [479,505].

The structure of the MY_n^- complex anions is also accessible by quadrupole resonance [506] and NMR spectroscopy (see, e.g., AlY_4^- on ^{27}Al [507] and GaY_4^- on ^{71}Ga [508]).

IV Reactions of Arenium Ions

Characteristic of arenium as well as other carbenium ions is their aptitude to rearrange with change of substituent position. In many cases such rearrangements include several stages with consecutive formation of isomeric ions. Therefore, it is reasonable to discuss initially the relative stability of isomeric arenium ions.

1 Effect of Substituents on the Relative Stability of Isomeric Arenium Ions

A Equilibria Between Isomeric Ions Differing in the Site of Proton Attachment

Section III.1.B described that arenium ions differing from one another in the site of proton attachment can first be formed in non-equilibrium ratios. The possibility of such disagreement between kinetic and thermodynamic control was also discovered when protonating some hydrocarbons. This observation was first made for 3,5,8,10-tetramethylaceheptylene (a non-alternant non-aromatic system) [509]; the direction of its initial protonation was effectively explained [510] by considering the map of the electrostatic field. Later some polymethylnaphthalenes were shown to behave in the same manner. Thus, in the case of 1,2,3,4,5,6,7-heptamethylnaphthalene in HSO_3F—SO_2ClF at low temperatures the proton is first attached at the 8-position. But then the ion is rearranged with the proton passing to the 4-position [81]. The rearrange-

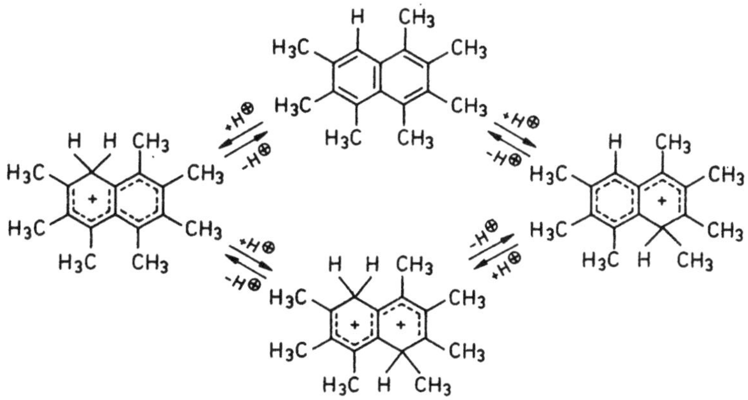

Table 43. Observed and Calculated (in parentheses) Ratio of Protonation at Different Positions of Substituted Benzenes

Substituents						Temp. °C	Acid system	% of protonation at position						Ref.
R_1	R_2	R_3	R_4	R_5	R_6			1	2	3	4	5	6	
CH_3						0	$HF—SbF_5$		12.5		75		12.5	72)
						(0)			(14.5)		(71)		(14.5)	
CH_3		Br				−35	$HF—SbF_5$		3		41		56	363)
						(−35)			(6)		(40)		(54)	
CH_3		Cl				−35	$HF—SbF_5$		2		40		58	363)
						(−35)			(6)		(40)		(54)	
CH_3		F				−35	$HF—SbF_5$		0		11.5		88.5	363)
						(−35)			(2)		(11)		(87)	
Cl		Cl				−10	$HF—SbF_5$		5		47.5		47.5	363)
						(−10)			(7)		(46.5)		(46.5)	
CH_3			OH			−50	HSO_3F	10		45		45		320)
						−50	HSO_3F	11		44.5		44.5		324)
						−50	$HSO_3F—SbF_5$	9		45.5		45.5		324)
						(−50)		(9)		(45.5)		(45.5)		
CH_3		CH_3		Br		0	$HF—SbF_5$		33		33.5		33.5	362)
						(0)			(39)		(30.5)		(30.5)	
CH_3		CH_3		Cl		0	$HF—SbF_5$		36		32		32	362)
						(0)			(39)		(30.5)		(30.5)	
CH_3		CH_3		F		0	$HF—SbF_5$		83.5		8.2		8.2	362)
						(0)			(76)		(12)		(12)	
						−50	$HF—SbF_5$		80		10		10	514)
F		Cl		Cl		−15	$HSO_3F—SbF_5$		12.5		75		12.5	365)
						(−15)			(14)		(72)		(14)	
F		Br		Br		−15	$HSO_3F—SbF_5$		12.5		75		12.5	365)
						(−15)			(14)		(72)		(14)	
Cl		Br		Br		−15	$HSO_3F—SbF_5$		33.5		33		33.5	365)
						(−15)			(33.3)		(33.3)		(33.3)	
Br		Cl		Cl		−15	$HSO_3F—SbF_5$		33.5		33		33.5	365)
						(−15)			(33.3)		(33.3)		(33.3)	
C_2H_5		CH_3		CH_3		−40	$HCl—AlCl_3$		37		26		37	280)

$CH(CH_3)_2$		CH_3		CH_3		−40	$HCl—AlCl_3$	37		26		37	280)
$C(CH_3)_3$		CH_3		CH_3		−40	$HCl—AlCl_3$	41.5		17		41.5	280)
—◁		CH_3		CH_3		−40	$HCl—AlCl_3$			100			280)
$CH_2Si(CH_3)_3$		CH_3		CH_3		−40	$HCl—AlCl_3$			100			66)
OH		OCH_3		OCH_3		−40	$HSO_3F—SO_2FCl$	16.5		67		16.5	354)
						+38	$H_2SO_4—CF_3COOH$	16.5		67		16.5	346)
OH		OC_2H_5		OC_2H_5		+38	$H_2SO_4—CF_3COOH$	16.5		67		16.5	346)
OCH_3		OH		OH		+38	$H_2SO_4—CF_3COOH$	45.5		9		45.5	346)
OC_2H_5		OH		OH		+38	$H_2SO_4—CF_3COOH$	45.5		9		45.5	346)
OH		CH_3	CH_3			−60	$HSO_3F—SbF_5$			10		90	327)
						−50	HSO_3F			10		90	319)
						−50	HSO_3F	3		7		90	324)
						−50	HSO_3F			8		92	320)
						(−50)		(12.5)		(2.5)		(85)	
OH	CH_3		CH_3			−50	HSO_3F			20		80	324)
						−50	HSO_3F			16		84	320)
						(−50)				(17)		(83)	
OCH_3	CH_3	CH_3				−50	HSO_3F			6.5		93.5	324)
OH	CH_3	CH_3	CH_3			−50	HSO_3F			17		83	324)
						−50	HSO_3F			20		80	320)
						(−50)				(3)		(97)	
OH		CH_3	CH_3	CH_3		−50	HSO_3F	28.5		43		28.5	324)
						−50	HSO_3F	25		50		25	319, 320)
						(−50)		(49)		(2)		(49)	
OH	CH_3		CH_3	CH_3		−50	HSO_3F			85		15	324)
						−50	HSO_3F			80		20	320)
						(−50)				(17)		(83)	
OH	CH_3	CH_3	CH_3	CH_3		−50	HSO_3F			94		6	320)
						(−50)				(3)		(97)	
F	CH_3	CH_3	CH_3	CH_3	CH_3	−80	$HF—SbF_5—SO_2ClF$			100			174)
						(−80)		(6.5)		(87)		(6.5)	
Cl	CH_3	CH_3	CH_3	CH_3	CH_3	−80	$HF—SbF_5—SO_2ClF$	12.5	12.5	50	12.5	12.5	174)
						(−80)		(25.5)	(6)	(37)	(6)	(25.5)	

ment can proceed by detachment of the proton and its reattachment at another position or by diprotonation.

The higher thermodynamic stability of the ion with the proton at the 4-position is due to a decrease in the steric-strain on movement of one peri-CH_3-group from the ring plane.

Disagreement between kinetic and thermodynamic control has also been reported for the protonation of 1,2,3,4,6-penta- and 1,2,3,4,6,7-hexamethylnaphthalenes [81].

The possibility of formation of arenium ions in non-equilibrium ratios should be taken into account in studying their relative stability. The proton migration between non-equivalent positions of the aromatic molecule usually occurs rather readily at $> -50 \div -60$ °C; presumably in most publications the authors dealt with equilibrium ratios of ions differing in the sites of proton attachment.

The influence of substituents on equilibrium ratios of arenium ions differing in the site of proton attachment can be described with a modified Hammett equation using σ^+ constants. The experimental data on the basicities of methylated benzenes (see Table 2) has shown [511] (cf. [512]) the reaction constant ϱ (0°) of protonation in HF at 0 °C is −11.3. As the protonation of aromatic hydrocarbons is, for a first approximation, and isentropic reaction [4,83,121,344] it requires [513] the fulfilment of the relationship $\varrho(T) = c/T$. Using the above-mentioned value of ϱ (0°) we have

$$\varrho(T) = -3100/T$$

As a first approximation this equation serves to estimate the reaction constant for the protonation not only of methylbenzenes, but also other substituted benzenes. The ϱ value for strong acid systems is relatively little dependent on the medium.

Assuming the influence of substituents to be additive and applying the above equation for ϱ to the described equilibriums between the isomeric C-protonated substituted benzenes (see Table 43) one can determine, in addition to the known σ^+_{para} and σ^+_{meta} [361], the σ^+_{ortho} and σ^+_{ipso} constants. For example, the σ^+_{ortho} is determined by using the ratios in which the isomeric ions (*72*)–(*74*) are formed on protonation of meta-substituted benzenes with electron-releasing substituents:

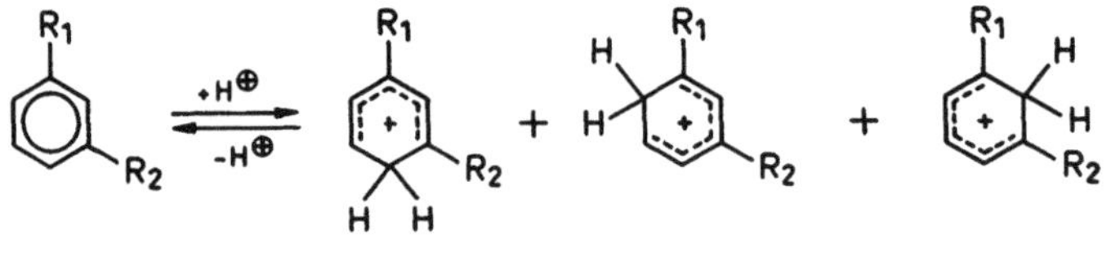

72 73 74

$$\log \frac{[72]}{[74]} = \varrho(\sigma^+_{p-R_1} - \sigma^+_{o-R_1}) \qquad \log \frac{[73]}{[74]} = \varrho(\sigma^+_{p-R_2} - \sigma^+_{o-R_2})$$

as well as ions (*75*) and (*76*) on protonation of 1,3,5-tri-substituted benzenes with identical substituents in two positions:

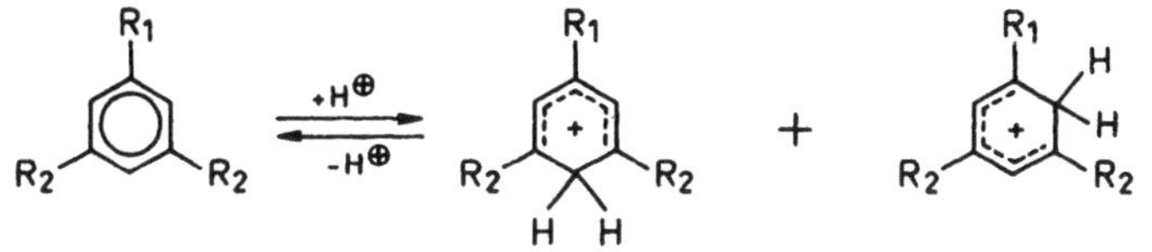

75 76

$$\log \frac{[76]}{[75]} = \varrho(\sigma^+_{o-R_1} + \sigma^+_{p-R_2} - \sigma^+_{o-R_2} - \sigma^+_{p-R_1})$$

In this way the σ^+_{ortho} constants were determined for a number of substituents; the respective data, along with σ^+_{para} and σ^+_{meta} are presented in Table 44. There is a linear correlation between the values of σ^+_{ortho} and σ^+_{para} described [515] by the equation

$$\sigma^+_{ortho} = 0.07 + 0.95\sigma^+_{para} \qquad (r = 0.99, s = 0.05)$$

Table 44. Values of σ^+ Constants

Substituents	σ^+_p	σ^+_m	σ^+_o	σ^+_{ipso}
CH_3	−0.31	−0.07	−0.25[a] [511]	+0.10 [511]
C_6H_5	−0.18	+0.06	(−0.10) [515]	+0.26 [515]
F	−0.07	+0.35	+0.06[b] [531]	
Cl	+0.11	+0.40	+0.18 [515]	+0.60 [515]
Br	+0.15	+0.41	+0.22[c] [515]	≦+0.54 [515]
OH	−0.92	+0.12	−0.80 [296]	

[a] −0.19 [532], −0.27 [533, 534]; [b] +0.31 [532]; [c] +0.184 [533, 534]

The agreement between the ratio of isomeric benzenium ions, calculated from the $\varrho\sigma^+$-approach and that experimentally observed, on protonation of substituted benzenes can be estimated from Table 43. A sharp disagreement of predicted and observed ratios of isomeric ions is only observed for polymethylated phenols.

The determination of σ^+_{ipso} is more complicated. Upon protonation of di-, tri- and tetra-substituted benzenes the PMR spectra show no ions from the addition of the proton to the substituted carbon atom. Therefore, since in routine experiments the lower limit of the detectable equilibrium fraction of ions amounts to 5%, these data are suitable only for estimation of the lower limit of the σ^+_{ipso} values. For instance, 4-chloro-m-xylene is practically protonated only at the 6-position at −40 °C [363]; this indicates that $\sigma^+_{ipso-Cl} \geqq 0.50$.

Greater opportunities are offered by the equilibria of totally substituted benzenium ions. Thus, the 1,2-dimethyl-1,3,4,5,6-pentachlorobenzenium ion turns into the 1,1-dimethyl-2,3,4,5,6-pentachlorobenzenium ion at −50 °C by 85%.

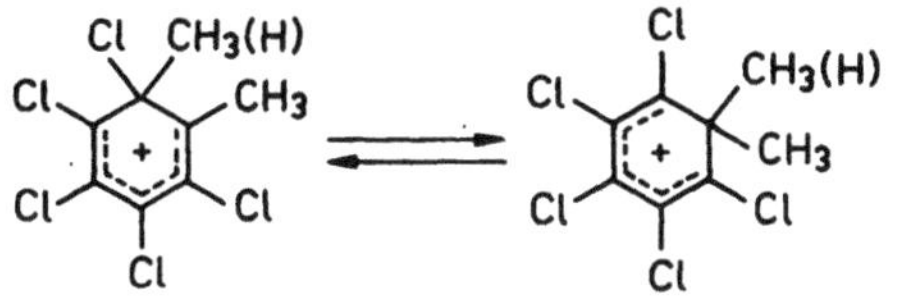

In the additive scheme of substituent effects the substitution of one CH_3 group at the ring sp^3-hybridized carbon of the parent and the formed ions by H should not change the equilibrium. Therefore the equilibria of the above-type benzenium ions are analyzed by the same values of ϱ that were used to describe the equilibria of the C-protonated forms. For the known value of $\sigma^+_{ipso\text{-}CH_3} = +0.10$ [511)] the above equilibrium enables one to find the value of $\sigma^+_{ipso\text{-}Cl} = 0.61$.

Naturally, the relative stability of polysubstituted benzenium ions can be essentially effected, apart from the electronic factors, by some steric ones as well. In the above example of estimating $\sigma^+_{ipso\text{-}Cl}$ the results obtained for tri- and per-substituted benzenium ions agree well; so the value 0.6 was concluded [515)] to correctly reflect the electronic effect of the chlorine atom located in the protonated position.

In this manner one can estimate the σ^+_{ipso} constants of other substituents. They seem to change in parallel with the inductive (polar) constants of the substituents[41].

Positive values of σ^+_{ipso} mean that the substituent at the ring sp^3-hybridized carbon destabilizes the benzenium ion. This is why the proton is seldom attached at the substituted position when unsubstituted positions are available. For example, the equilibrium ratio of the isomeric C-protonated forms of pentamethylbenzene expected for 0 °C is

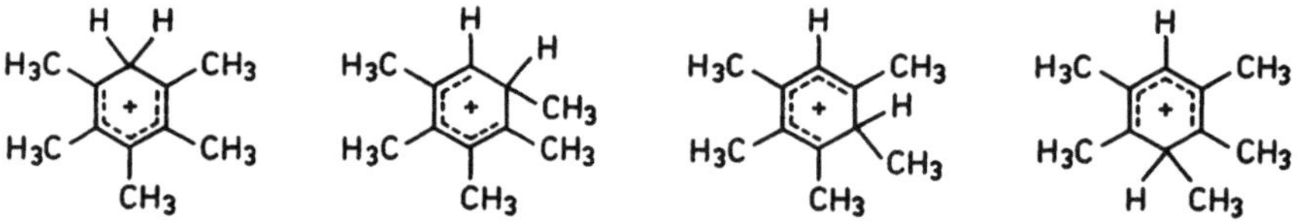

1 $10^{-3.7}$ $10^{-1.6}$ $10^{-4.6}$

Therefore the PMR spectrum shows only the ion with the proton attached at the unsubstituted position.

In some cases the ratio of isomeric ions calculated in terms of the $\varrho\sigma^+$ approach is markedly different from that observed. For example, on protonation of 3,4-dimethyl- and 3,4,5-trimethylphenols in HSO_3F at −50 °C [319)] the observed ratio of protonated forms is as follows:

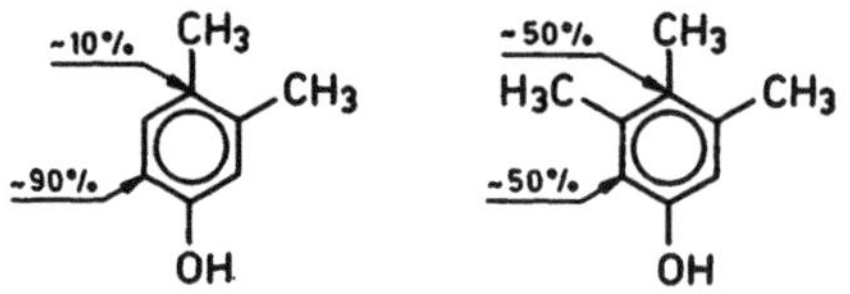

The calculations for the equilibrium give in the first case $C^2:C^4:C^6 = 12.5:2.5:85$ and in the second $C^{2(6)}:C^4 = 98:2$. The abnormally high percentage for the protonation of 3,4,5-trimethylphenol at the 4-position (a similar picture is observed for anisole [319)]) may be due to the decrease in the spatial interaction of the 3-, 4-

41 For the influence of substituents on the basicity of 4-X-4-methylcyclohexa-2.5-dien-1-ones see Table 4 and [231)].

and 5-CH_3 groups upon departure of the 4-CH_3 group from the ring plane. This is usual to explain [4)] a somewhat higher basicity of hexamethylbenzene (calculated per position) as compared with pentamethylbenzene (see Table 2). Before accepting this explanation, however, one should check whether the data on the protonation of polymethylated phenols correspond with the thermodynamic equilibrium (see Sect. III.1.B). A high stability of hydroxy-substituted benzenium ions may lead to a low rate of equilibration between the ions with different sites of proton attachment. If it depended on the steric factor alone one could expect a marked protonation of 1,2,3,5-tetramethylbenzene (isodurene) at the C^2 atom which is not the case.

The efficiency of involving the substituents in the delocalization of the positive charge of arenium ions must in general depend on the acid medium because there may be a change in the extent of the solvation between the ion and the nucleophilic particles of the medium[42]. The change of relative influence of OH and OCH_3 groups with changing medium acidity in the protonation of phloroglucinol and its methyl ethers in 40–70% aqueous $HClO_4$ was established by UV spectroscopy [122,346,356)]. This change seems to be due to the difference in the solvation of the hydroxy group hydrogen with the medium particles whose nucleophilicity also changes with changing medium acidity (cf. [517)]).

Some authors (e.g., [370)]) ascribe this interaction to substituents taking part in delocalizing the positive charge of arenium ions through the σ,π-conjugation, in particular, for the methyl and the other alkyl groups:

Such an interaction is probable. However, according to the available data (see Sect. IV.2.C), a change in the acid medium does not usually cause great changes in the rate constants of carbenium ion rearrangements (cf., however, the rearrangement of hydroxybenzenium ions in aqueous acids, Sect. IV.2.C). One can, therefore, assume the equilibrium between the isomeric ions not to be very sensitive to the nature of acid medium either. The equilibrium constants (K) for isomeric isopropyltrimethylcyclopentenyl cations at 25 °C in different acids [518,519)] are as follows:

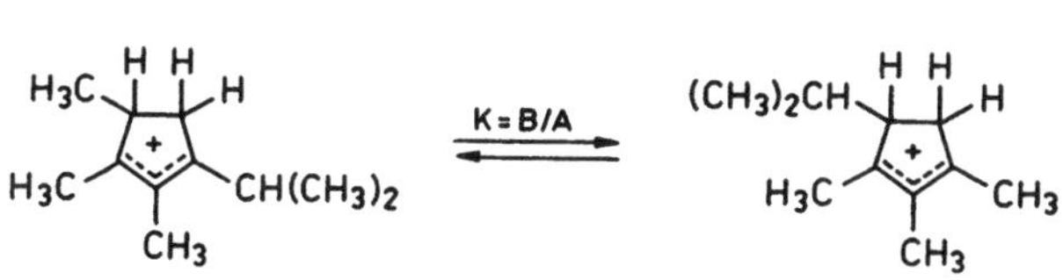

(A) (B)

96% H_2SO_4	2.6	95% CF_3SO_3H	3.0
$H_2SO_4 + CH_3NO_2$	2.5	$CF_3SO_3H + CH_3NO_2$	2.9
HSO_3F	2.5	$CF_3SO_3H + CH_2Cl_2$	2.5
$HSO_3F + CH_3NO_2$	2.7		

42 The solvation enthalpy of the benzenium ion $C_6H_7^+$ in water is estimated 65–82 kcal/mole [516)]; hence the effect of the solvation on stabilizing the ion in weak acids is comparable to that of the delocalizing the charge. One should, however, take into account the data discussed below.

Similarly, the equilibrium ratio of isomeric ions formed on protonation of 9-ethyl-10-methylanthracene is the same for such different systems as liquid HF and $H_2O \cdot BF_3$ (10% m) in CF_3COOH [62].

Different results were obtained by electronic spectroscopy for the protonation of biphenyl and phenanthrene in $HSO_3F—SbF_5$ and in $HF—BF_3$. In $HSO_3F—SbF_5$ at —70 °C biphenyl is mostly protonated at the para-position and phenanthrene at the 9-position [423], whereas in $HF—BF_3$ for biphenyl at 25 °C the spectrum shows the formation of ortho-H-isomer [416] and phenanthrene at —70 °C yields comparable quantities of 4-H- and 9-H-phenanthrenium ions [416, 417]. This difference is due to the fact [423] that the protonating particle in the first case (solvated cation HSO_3^+) is greater than in the second; this is why the proton is preferably attached at sterically better accessible positions. The data discussed are of considerable interest, but it is desirable to verify them by NMR.

On the whole these data testify that in many cases the change of the medium affects comparatively little the physical characteristics, relative reactivity and relative stability of the stable carbocations. Hence, the sensitivity of the PMR and NMR-^{13}C spectra of arenium ions to the medium effects is small[43]; the positions of absorption bands of carbocations including the mesitylenium ion for solutions in acids and for the gas phase are surprisingly close [419]; and the influence of the solvent on the kinetic data on the degenerate rearrangements of arenium ions by 1,2-shifts of CH_3 groups is not appreciable (see Sect. IV.2.B).

Masspectrometric and calorimetric measurements showed no essential change in the difference of enthalpy of isomeric secondary and tertiary carbocations when passing from the gas phase to solutions [520]. For instance, the enthalpy of the second-to-tertiary butyl cation rearrangement in $SbF_5—SO_2ClF$ is 14.5 ± 0.5 kcal/mole while for the gas phase the values are 15 to 17 kcal/mole [520, 521]. Comparisons of the ion-cyclotron resonance with the spectroscopic measurements for solutions have shown [522] that the equilibrium constants for the protonation of α-methylstyrene, 1,1-diphenylethylene, azulene and hexamethylbenzene in passing from the gas phase to water decrease by 25–35 (depending on the proton donor), but the difference between the free energies of formation of the ions $C_6H_5C^+(CH_3)_2$, $(C_6H_5)_2C^+CH_3$, $C_{10}H_9^+$ and $C_6(CH_3)_6H^+$ in gas and water phases is about the same (~35 kcal/mole).

All this confirms the suggestion [523] that electrostatic solvation changes little in the structurally similar ions (e.g., of aliphatic carbocations with a localized charge or of π-delocalized ions). Carbocations are sharply different from the protonated amino and hydroxy compounds characterized by much greater solvation energies. The heat of solvation for a single-charged cation in water by dipoles and induced dipoles in the absence of the specific interaction of the cation with surrounding molecules has been estimated to about 50 kcal/mole [522]. The higher values point to a specific interaction — formation of hydrogen bonds involving the "acidic" hydrogen atoms of the cation and the oxygen atoms of water molecules[44] — or

43 Also the difference between the carbon chemical shifts of the solid sample of heptamethylbenzenium tetrachloroaluminate and of solutions of this ion in different acids is small [269].

44 Or vice versa as with hydroxy- or alkoxy-substituted arenium ions.

interaction as Lewis acids and bases:

$$\left[H_2O\ldots\overset{\vee}{\underset{|}{C}}\ldots OH_2\right]^+$$

The solvation characteristics of aryl-substituted carbocations are comparable to those of tetra-n-butylammonium and tetraphenylphosphonium cations having no specific solvation possibility. Presumably, the specific solvation is negligible for π-delocalized carbocations in a first approximation (see also [524]). For aliphatic carbocations with a localized charge the specific solvation according to the calculation [525] occurs, but for the secondary and tertiary (not the primary) ions the differences in solvation energies are not too great. Actually, between the ionization heats of the R—Cl compounds (R = iso-propyl, tert-butyl, cyclopentyl, 1-methylcyclopentyl-1, norbornyl-2, 2-ethylnorbornyl-2) in SbF_5—SO_2ClF and in the gas phase there exists an approximately linear relation [521]. Linear relations are also observed between ionization heats of chlorides measured for interaction with SbF_5 in four solvents: SO_2ClF, SO_2F_2, SO_2 and CH_2Cl_2. These results again confirm the suggestion that the solvation does not appreciably change the relative stability of carbocations containing no heteroatomic groups.

The effect of temperature on the equilibrium ratios of isomeric arenium ions was not studied, but it cannot be strong. The entropy of protonation of aromatic hydrocarbons and their methylated derivatives in liquid hydrogen fluoride was shown [83] to be practically independent of the structure and basicity of hydrocarbon ($\Delta S_p =$ $= -21$ e.u. [4,512]). This seems to be also valid for other acid systems [62,516], so the equilibrium between isomeric alkylarenium ions differing in the site of proton attachment can be assumed to be mainly determined by the difference in the heats of formation for these ions ($\Delta\Delta F_p \approx \Delta\Delta H_p$)[45]. As the NMR spectra usually show only those ions whose equilibrium portion is no less than 5%, i.e., $K \leqq 20$ and $\Delta\Delta H_p \leqq 1.5$–5 kcal/mole, so a 50 °C change cannot change the equilibrium more than two- or threefold. The same seems to apply for arenium ions with other not very strongly solvatable substituents. Therefore, at small differences in equilibrium concentrations of isomeric ions as, e.g., on protonation of 1,3,5-tri-substituted benzenes containing different alkyl groups or different halogen atoms, the position of equilibrium changes comparatively little after a 40–60 °C change [280,365].

With substituted benzenium ions the effect of temperature on the equilibrium between isomeric ions can be estimated with the mentioned relation between ϱ and T.

Upon protonation of 9-ethyl-10-methylanthracene in liquid HF the equilibrium of 10-H- (*77*) and 9-H-isomers (*78*) changes from 12:88 to 20:80, as temperature rises from −20 to +70 °C [62].

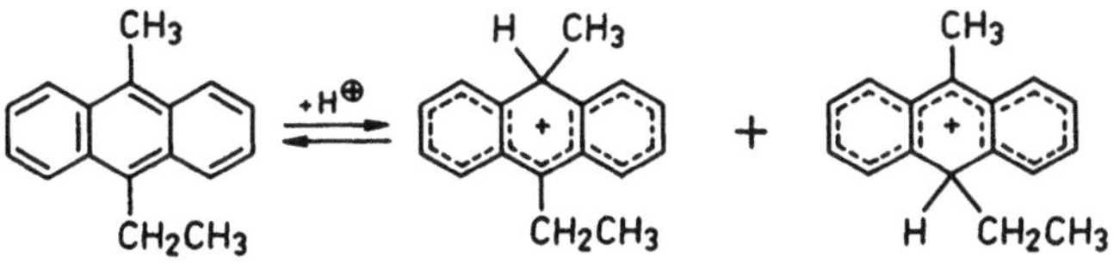

77 *78*

45 This applies also for the equilibria between isomeric alkylcyclopentenyl cations [518].

The equilibrium between the ions (*77*) and (*78*) was studied to elucidate the relative aptitude or methyl and ethyl groups to stabilize carbenium ions.
The strong stabilizing effect of alkyl groups on carbenium ions is usually explained by their participation in delocalizing the positive charge through σ, p-conjugation (hyperconjugation). The +I-effect of alkyl groups is generally believed to rise in the order: $CH_3 < CH_3CH_2 < (CH_3)_2CH < (CH_3)_3C$. This order of electronic effects correlates, e.g., with that of basicities of monoalkylbenzens in the gas phase [85].

Some reactions in solutions, however, follow the reverse sequence (the Baker-Nathan order): $CH_3 > CH_3CH_2 > (CH_3)_2CH > (CH_3)_3C$ which is explained by the σ, p-conjugation of α-C—H bonds being more effective than that of α-C—C bonds (for a critical analysis of these concepts see [528, 529]).

The difficulty of estimating the contribution of the σ, p-conjugation to the electronic effects of alkyl groups is based on the very small difference between the effect of CH_3 group and that of $(CH_3)_3C$ group in reactions for which the Baker-Nathan order is characteristic. As a rule, the difference between the changes of free energy of activation does not exceed 0.5 kcal/mole. In the carbenium ions where the "response" of the substituents to a positive charge must be particularly strong one could expect this difference to be much greater. The results obtained, however, proved to be contradictory.

To judge by the equilibrium between the isomeric tertiary hexyl cations

$CH_3-\overset{+}{C}(CH_3)-CH_2CH_2CH_3$ (32 %) $CH_3CH_2-\overset{+}{C}(CH_3)-CH_2CH_3$ (38 %)

$CH_3-\overset{+}{C}(CH_3)-CH(CH_3)_2$ (30 %)

and the tertiary heptyl cations

$CH_3-\overset{+}{C}(CH_3)-CH_2CH(CH_3)_2$ (83 %) $CH_3-\overset{+}{C}(CH_3)-C(CH_3)_3$ (17 %)

the difference between the alkyl groups on the stability of alkycarbenium ions is small ($\Delta\Delta F \leqq 1$ kcal/mole) [62]. In alkyl-substituted cyclopentyl cations [518] this difference proved to be more marked (in 96 % H_2SO_4 at 35 °C):

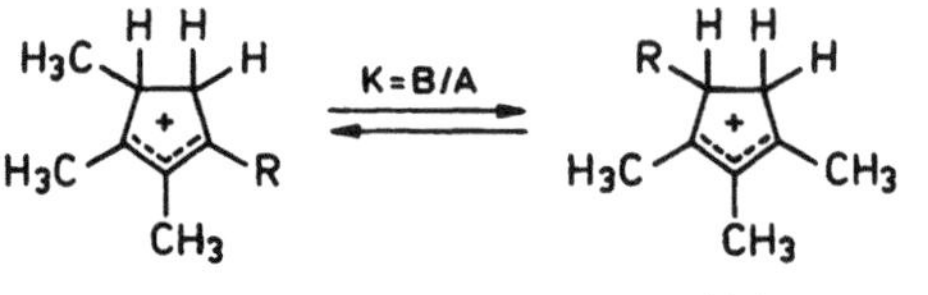

(A)   (B)

R = CH_2CH_3, K = 1.6
R = $CH(CH_3)_2$, K = 3.0
R = $C(CH_3)_3$, K = 18

When passing from an ion with the CH_3 group at the ring sp^2-hybridized carbon atom of the allyl system to an isomeric ion with the $(CH_3)_3C$ group at the same position the free energy is changed by 1.8 kcal/mole.

A still greater effect was reported by A. Arnett and J. Larsen [526, 527]. From the heats of formation for alkylbenzenium ions in HSO_3F—SbF_5 at —65 °C they concluded that the difference in the free energies of ion formation from toluene and ethylbenzene, as well as from toluene and tert-butylbenzene amounts to 2.5 and 3.8 kcal/mole.

D. Brouwer and J. van Doorn [62], however, doubted the reliability of estimating the stabilizing effects of alkyl groups by the Arnett and Larsen method[46]. They find it difficult to correlate the results of these authors with the small difference in the basicity constants of hexamethylbenzene and hexaethylbenzene ($10^{1.4}$ and $10^{2.0}$), as well as 9-methyl- and 9-ethylanthracenes ($10^{5.7}$ and $10^{5.4}$) in liquid HF [83]. Moreover, the methyl and the ethyl groups exert practically equal effects on the basicity of cyclohexa-2,5-dien-1-ones and cyclohexa-2-en-1-ones (from the 3-positions) [227].

The equilibrium between the 9-ethyl-9,10-dimethyl- and 10-ethyl-9,9-dimethylphenanthrenium ions (*79*) and (*80*) can be interpreted as a far greater stabilizing effect by the CH_3 group than by the C_2H_5 group [212] since at —70 °C the equilibrium fraction of the ion (*80*) does not exceed 1 %.

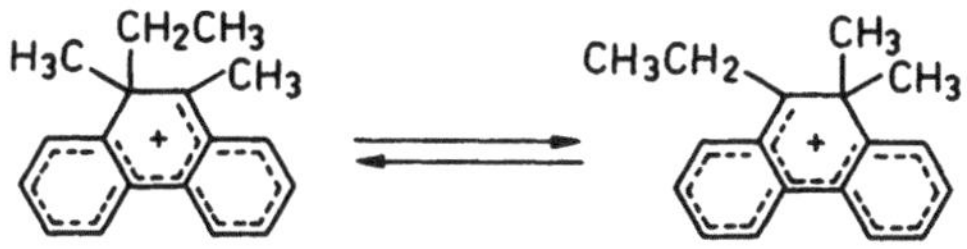

79 *80*

The 9-methyl-9,10-diethylphenthrenium ion converts almost completely into the 9,9-diethyl-10-methylphenanthrenium ion (at 25 °C the equilibrium ratio ist ~1:9) [215) 47]

Similarly, among the ions of acenaphthylenium given below the equilibrium is practically completely shifted to the ion (*81*) with a CH_3 group at the carbenium centre [315]:

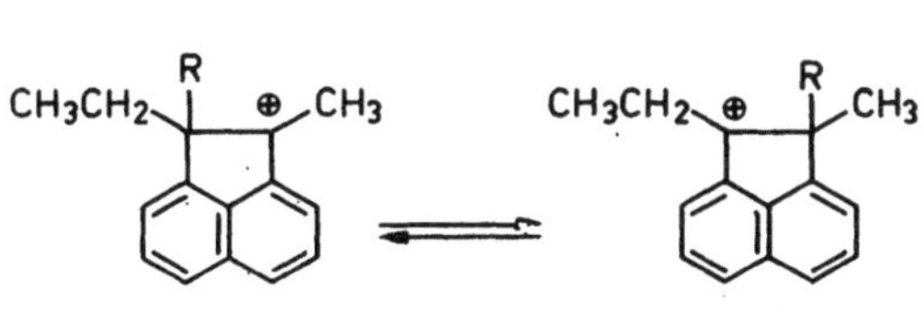

81 *82*

R = H, CH_3

46 The differences in the heats of formation for benzenium ions from C_6H_5R (R = CH_3, C_2H_5, n-C_3H_7, t-C_4H_3) were later shown to be largely due to solvational factors of steric nature (cf. [85,119]).

47 The equilibria between 9,9,10-alkyl-diaryl- and the 9,9,10-dialkyl-arylphenanthrenium ions are described in [309,217], for the 9,9,10-chlorodimethyl- and 9,9,10-chlorodiethylphenanthrenium ions, in [177].

These data permit the conclusion that according to their stabilizing effect on carbenium, and, in particular, on arenium, ions the alkyl groups are arranged in the Baker-Nathan order, the difference in the effect depending on the ion structure. However, the difficulty of estimating the relative role of electronic and steric effects requires much care, so one can hardly answer the question as to which of these effects actually determines this sequence (see [519, 530]).

By describing the electronic effect of the methyl groups in benzenium ions as a combination of the inductive and the hyperconjugative effects the relative basicity of methylbenzenes is interpreted rather successfully by quantum chemical calculations [16, 93, 94] (see also ref. in Sect. II).

By protonating naphthalene derivatives, 1,4,5-trimethylnaphthalene in HSO_3F—SO_2ClF at —80 °C adds the proton at 8- and 4-positions in the ratio of 1:4 [302]. 1,6-Dimethylnaphthalene in HSO_3F—SbF_5—SO_2ClF is mostly protonated at the 4-position; however, two more ions are formed one of which seems to have the proton attached at the 5-position [301]. 1,7-Dimethylnaphthalene yields ions with the proton attached to the 8- and 4-positions (85:15 at —40 °C) [301], while 1,4-dimethylnaphthalene is protonated at the 1-, 2- and 5-positions, the equilibrium concentrations of the three formed ions being comparable [301] (cf. [300]).

1- and 2-halonaphthalenes add the proton both at α- and at β-position, the equilibrium ratios of isomeric ions at —80 ÷ —20 °C being as follows:

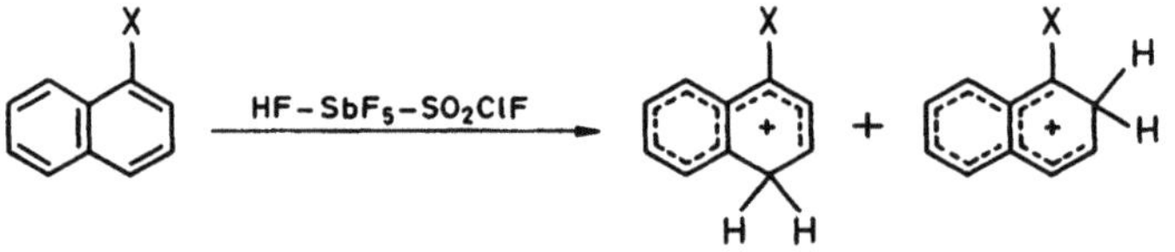

X		
F	100%	0%
Cl	90%	10%
Br	60%	40%
I	0%	100%

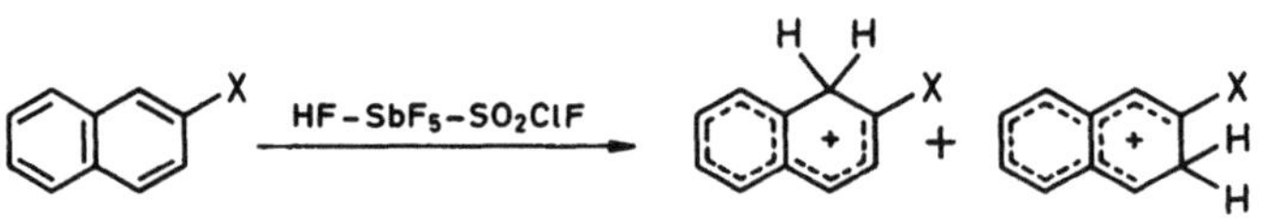

X		
F	100%	0%
Cl	95%	5%
Br	90%	10%
I	60%	40%

However, the data on the chloro-, bromo- and iodonaphthalenium ions should be considered with caution as their PMR spectra are not well enough resolved to draw unequivocal conclusions.

The electronic effects of substituents in arenium ions deserves quantitative estimation on a wider scale because a comparison of the data with the accumulated data on the effect of substituents in electrophilic substitutions will help to judge how close the transition states of these reactions are to the structure of arenium ions.

B Isomeric Conversions of Arenium Ions Due to Substituent Transfer

A characteristic property of carbenium ions is their aptitude to rearrangements due to the shift of the substituent R from α-sp^3-hybridized carbon atom to the positively charged sp^2-hybridized atom (1,2-shift of the R substituent):

Similar rearrangements are observed with arenium ions. In particular, the equilibria (see Section IV.1.B) between isomeric ions differing in the site of proton attachment imply the possibility of their interconversions. These conversions will be shown below to be feasible not only through intermolecular proton transfer but also as a result of intraionic 1,2-shifts (R = H). The transfer of the substituents, other than hydrogens, in arenium ions, results in isomeric ions differing in the position of substituents in the ring. Rearrangements of this type underlie the intramolecular acid-catalized isomerization of aromatic compounds [13].

Isomeric conversions of xylenes: According to the physical data (see Sect. III), o-, m- and p-xylenes are protonated with strong acids to yield ions (*83*), (*84*) and (*85*). Very small amounts of ions differing in the site of proton attachment must be in equilibrium with them, i.e. ions (*86*), (*87*) and (*88*) in which one of the methyl groups is located at the sp^3-hybridized carbon atom. The rearrangements of these ions due to the 1,2-shifts of CH_3 groups bring about the isomeric conversions of xylenes.

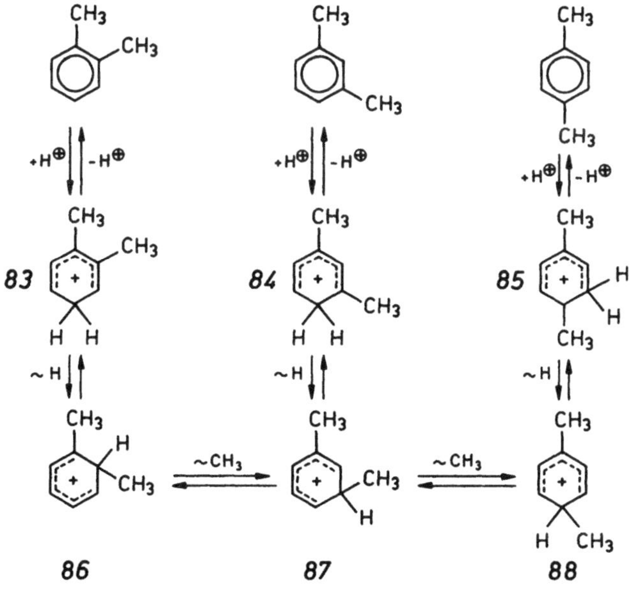

The intramolecular isomerization of aromatic compounds involving a change of position in the aromatic ring for the primary alkyl groups, aryl fragments, chlorine atoms and CH_3SO_2 group is similar [13].

There is another mechanism for the acid-catalysed isomerization of aromatic compounds connected with the reversibility of some types of electrophilic substitutions:

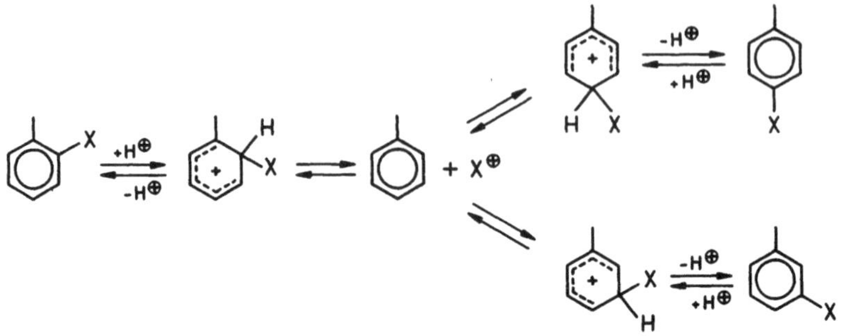

Aromatic sulphonic acids ($X = SO_3H$) [13] for example, are isomerized in this way. The change of position in the aromatic ring of the tertiary alkyl groups and bromine likewise cannot be confined as intramolecular mechanism with successive 1,2-shifts of the migrating group. Thus, the interconversions of ortho- and para-disubstituted benzenes with substituents of the above type proceed faster than their conversion into a thermodynamically more stable meta isomer [13, 65].

When the protonation of the parent compound and the product is low (insufficient medium acidity, limited amounts of strong acid, high temperature etc.) the equilibrium between the isomers is determined by their own thermodynamics. For example, for xylenes the thermodynamic equilibrium in the vapour phase corresponds to the following ratio ortho:meta:para at 300 K — 16:60:24; at 700 K — 24:52:24 [13]. Table 45 shows that the equilibrium xylene mixtures obtained by isomerization under conditions failing to provide their complete protonation have a composition close to the one calculated from their thermodynamics.

By contrast, when aromatic compounds isomerize under conditions providing their complete conversion into arenium ions the final composition is determined by the equilibrium between isomeric ions with different relative arrangement of substituents. As substituents have a strong effect on the stability of arenium ions, the differences in the thermodynamic stability of isomeric arenium ions differing in the position of substituents are expected to be far greater than for unprotonated compounds and, consequently, the equilibrium between them must be shifted more in favour of one of the isomers. Obviously among the isomeric dimethylbenzenium ions the most

Table 45. Observed Equilibrium Concentrations of Isomeric Xylenes

Temp.		Catalyst	Molar ratio MX_3: xylene	Equilibrium composition (%)			Ref.
°C	K			ortho	meta	para	
50	323	$HBr + AlBr_3$	0.1	12 ± 3	71 ± 5	17 ± 2	535)
50	323	$HCl + AlCl_3$	0.5	17	62	21	536)
100	373	$HF + BF_3$	0.1	19	61	20	537)
450	723	alumosilicate	—	28	50	20	538)

stable one must be the ion (*84*). Accordingly, the isomerization of xylenes providing their complete protonation (e.g., in liquid HF with 3 moles of BF_3 [537]) yields (after decomposition of the σ-complex by adding a base), practically pure m-xylene.

A marked shift in favour of 1,3-isomer upon isomerization under complete protonation must also be observed with other disubstituted benzenes with substituents capable of taking part in delocalizing the positive charge of benzenium ions. Similarly, in the case of tri- and tetra-substituted benzenes the 1,3,5- and 1,2,3,5-isomers must be sharply predominant in the equilibrium. Some examples of such isomerization of aromatic compounds are of synthetic interest (see [539]) and Table 46).

The conversions of isomeric benzenium ions due to the 1,2-shifts of various substituents have been observed by NMR. In particular, the rearrangements of completely substituted benzenium ions were studied. By generating individual isomers of these ions at low temperatures when the rearrangements rates are low, and then gradually raising the temperature on can observe on the NMR the conversions leading to more stable ions.

The 1,2-shifts of X-substituents in the 1-X-1,2,3,4,5,6-hexamethylbenzenium ions (*22*) yielding ions of the same structure correspond with degenerated rearrangements to be discussed in Sect. IV.2.B.

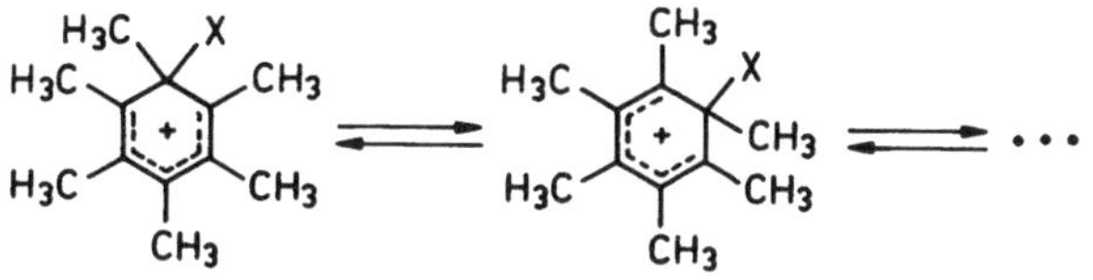

22 *22′*

The shifts of CH_3 groups in these ions result in isomeric cations.

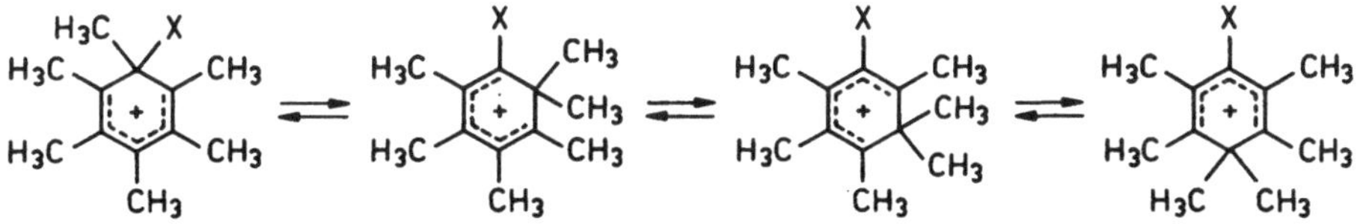

The equilibrium between X-hexamethylbenzenium ions must evidently be determined by the X substituent.[48] When the X group, if located at the sp²-hybridized carbon atom, acts more effectively in delocalizing the positive charge than the CH_3 group, the 4-X-isomer must, according to the results of the preceding Section, predominate. If the X substituent is a weaker electron donor than the CH_3 group, then the most favourable, from the stabilizing effect, are likely to be the 1-X- and 3-X-isomers.

For the 1-X-1,2,3,4,5,6-hexamethylbenzenium ions with such electronegative X substituents as NO_2 [166,169]), SO_3H [171]) and $CHCl_2$ [155]), no isomeric ions are observed

48 In further discussion no account is taken of the steric factors which may also affect the equilibrium between ions.

Table 46. Isomerization of Substituted Benzenes when Fully Protonated

Starting compounds (A)	Acid system	Molar ratio MX_n:A	Temp. °C	Product composition		Ref.
1,2-$C_6H_4(CH_3)_2$	HF + SbF_5	10	25	1,3-$C_6H_4(CH_3)_2$	(100%)	[512]
1,4-$C_6H_4(CH_3)_2$	HF + BF_3	3	30	1,3-$C_6H_4(CH_3)_3$	(100%)	[537]
	HF + SbF_5	10	25	1,3-$C_6H_4(CH_3)_2$	(100%)	[512]
1,2-$CH_3C_6H_4Cl$	HF + SbF_5	>1	0	1,3-$CH_3C_6H_4Cl$		[363]
1,2-$CH_3C_6H_4Br$	HF + SbF_5	>1	0	1,3-$CH_3C_6H_4Br$		[363]
1,2-+1,4-$C_6H_4Cl_2$	HF + SbF_5	5	20	1,3-$C_6H_4Cl_2$	(100%	[540]
1,2-+1,4-C_6H_4ClBr	HF + SbF_5	5	−2	1,3-C_6H_4ClBr	(91%)	[540]
1,2-+1,4-$C_6H_4Br_2$	HF + SbF_5	~8	−2	1,3-$C_6H_4Br_2$	(92–96%)	[540]
1,4-$CH_3C_6H_4OH$	HCl + $AlCl_3$	2	120	1,3-$CH_3C_6H_4OH$	(85%)	[539]
1,4-BrC_6H_4OH	HCl + $AlCl_3$	2	60	1,3-BrC_6H_4OH	(86%)	[539]
1,2,3-$C_6H_3(CH_3)_3$	HF + SbF_5	10	25	1,3,5-$C_6H_3(CH_3)_3$	(100%)	[512]
1,2,4-$C_6H_3(CH_3)_3$	HF + BF_3	>1	82	1,3,5-$C_6H_3(CH_3)_3$	(100%)	[537]
	HF + SbF_5	10	25	1,3,5-$C_6H_3(CH_3)_3$	(100%)	[512]
1,2,4-$C_6H_3(C_3H_7$-n$)_3$	HCl + $AlCl_3$	1	40	1,3,5-$C_6H_3(C_3H_7$-n)	(94%)	[541]
2-+4-Br-1,3-$C_6H_3(CH_3)_2$	HF + SbF_5	2	0	1,3,5-$BrC_6H_3(CH_3)_2$	(>99%)	[362]
1,2,3-+1,2,4-$C_6H_3Cl_3$	HF + SbF_5	5–10	70	1,3,5-$C_6H_3Cl_3$	(100%)	[540]
1-F-2,4-$C_6H_4Cl_2$	HF + SbF_5	5	70	1,3,5-$C_6H_3FCl_2$	(98,5%)	[540]
1-F-2,5-$C_6H_4Cl_2$	HF + SbF_5	5	70	1,3,5-$C_6H_3FCl_2$	(99.5%)	[540]
1,2,3-+1,2,4-$C_6H_3Br_3$	HF + SbF_5	5	0	1,3,5-$C_6H_3Br_3$	(100%)	[540]
1,2,3,4-+1,2,4,5-$C_6H_2(CH_3)_4$	HF + SbF_5	1	70–80	1,2,3,5-$C_6H_2(CH_3)_4$	(100%)	[537]
	HF + SbF_5	10	25	1,2,3,5-$C_6H_2(CH_3)_4$	(100%)	[512]
1,2,4,5-$C_6H_2(CH_3)_4$	HCl + $AlCl_3$	1	5	1,2,3,5-$C_6H_2(CH_3)_4$	(93.5%)	[274]
2,4,5-+2,4,6-+3,4,5-$(CH_3)_3C_6H_2OH$	HCl + $AlCl_3$	[illegible]	105	2,3,5-$(CH_3)_3C_6H_2OH$	(100%)	[539]

up to decomposition temperatures; this indicates a greater stability of 1-X-isomers than of 3-X-isomers. It is not excluded, however, that in these examples the equilibrium between isomeric ions was not reached since the 1,2-shift of a CH_3 group is too difficult for the high barrier separating the 1-X-isomers from far less stable 2-X-isomers. Therefore, it would be helpful to generate the other isomers and to follow their transformations.

1-H-1,2,3,4,5,6-hexamethylbenzenium ion does not form isomeric ions either. In this case the equilibrium should be quickly established; hence, the 1-H-isomer is thermodynamically more stable than the 3-H-isomer.

For the ions with the substituents X = CH_2Cl, CH_2CH_3, C_6H_5 (with σ^{+}_{p-X} — 0.01, —0.295 and —0.179, respectively [361]) the equilibrium mixtures contain 1-X- and 3-X-isomers. Thus, dissolving 4-methylene-1-chloromethyl-1,2,3,5,6-pentamethylcyclohexa-2,5-diene in HSO_3F at —80 ÷ —50 °C yields the 1-chloromethyl-1,2,3,4,5,6-hexamethylbenzenium ion which at —20 °C turns in about 20% into the 3-chloromethyl isomer [153,155]. Similarly, about 15% of 1-ethyl-1,2,3,4,5,6-hexamethylbenzenium ion in HSO_3F converts into the 3-ethyl isomer [153,155]. For the conversions of the 4-ethyl-1,1,2,3,5,6-hexamethylbenzenium ion into the 1-ethyl isomer see [209]. An equilibrium mixture of phenylhexamethylbenzenium ions was obtained from three individual isomers containing the phenyl group at C_1, C_3 and C_4 [208].[49] At 0 °C the equilibrium only contains 1-phenyl- and 3-phenyl isomers in about a 20:80 ratio [208,242]. In the case of X = $CH_2CH_2N^+(CH_3)_3$ the equilibrium ratio of 1-X- to 3-X-isomers is 60:40 [209].

4-Benzyl-1,1,2,3,5,6-hexamethylbenzenium ion generated at —70 °C in HSO_3F converts, at —55 °C, completely into the 3-benzyl isomer [155]. The further rise of the temperature to —40 °C decomposes the 3-benzyl isomer; after its rearrangement into a 1-benzyl-substituted ion the benzyl group splits off and leaves hexamethylbenzene.

Protonation of 4-formylmethylene-1,1,2,3,5,6-hexamethylcyclohexa-2,5-diene in HSO_3F—SO_2ClF at —115 °C results in the cation (*89*) which at —90 °C is converted into a 4-formylmethyl-1,1,2,3,5,6-hexamethylbenzenium ion (*90*); at —80 °C this is completely rearranged into the 3-formylmethyl isomer (*91*):

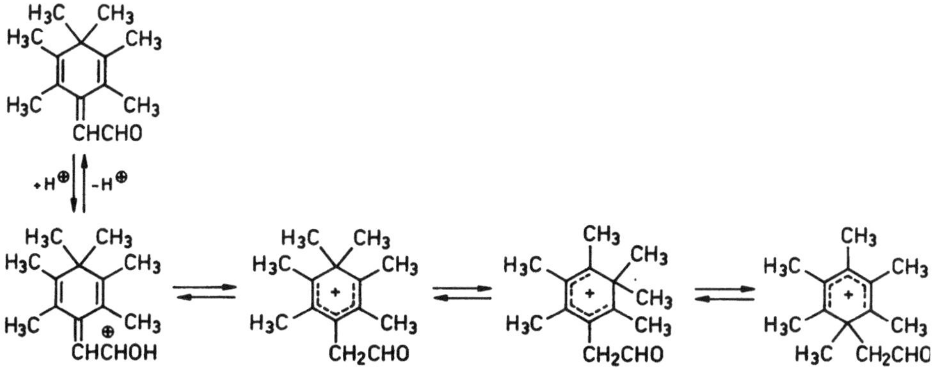

89 *90* *91* *92*

49 For comparing the effects of CH_3 and C_6H_5 groups on the stability of carbocations see [542].

At above —40 °C an equilibrium is established between the 3- and 1-formylmethyl isomers, the ratio between (*91*) and (*92*) being 65:35 [155]. In more acidic media the formylmethylbenzenium ion is additionally protonated at the carbonyl group.

Ions with X substituents characterized by a stronger stabilizing effect than that of the CH_3 group are those with X = OCH_3 (σ_p^+ = —0.778) and OH (σ_p^+ = —0.92 [361]).

The 1-methoxy-1,2,3,4,5,6-hexamethylbenzenium ion (*93*) generated by dissolving 1,4-dimethoxy-1,2,3,4,5,6-hexamethylcyclohexa-2,5-diene in HSO_3 at —75 °C, quickly rearranges at —30 °C into the 2-methoxy-1,1,3,4,5,6-hexamethylbenzenium ion (*94*) which at —10 °C is completely converted into the 4-methoxy isomer (*95*) [243].

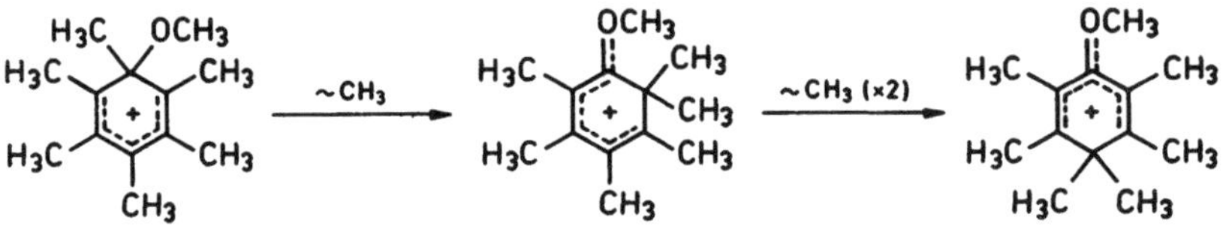

93 94 95

Quite unexpected, at first sight, is the rearrangement of the 2-methoxy-1,3,4,5,6-pentamethylbenzenium ion into the 4-methoxy-1,1,2,3,6-pentamethylbenzenium ion (in CF_3SO_3H at 95 °C [265]):

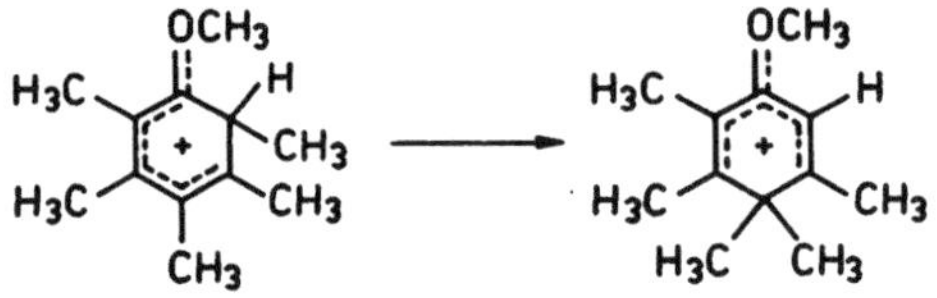

Here the methyl group is probably shifted since the OCH_3 group located between two CH_3 groups is removed from the ring plane and looses its efficiency in delocalizing the positive charge.

An attempt to generate the 1-hydroxy-1,2,3,4,5,6-hexamethylbenzenium ion (*96*) by dissolving 1-methoxy-4-hydroxy-1,2,3,4,5,6-hexamethyl-cyclohexa-2,5-diene in HSO_3F at —80 °C resulted [243] in the formation of the ion (*93*) and the 2-hydroxy-1,1,3,4,5,6-hexamethylbenzenium ion (97). The latter is also generated by protonating 2,3,4,5,6,6-hexamethylcyclohexa-2,5-dienone [218,219,329,332].

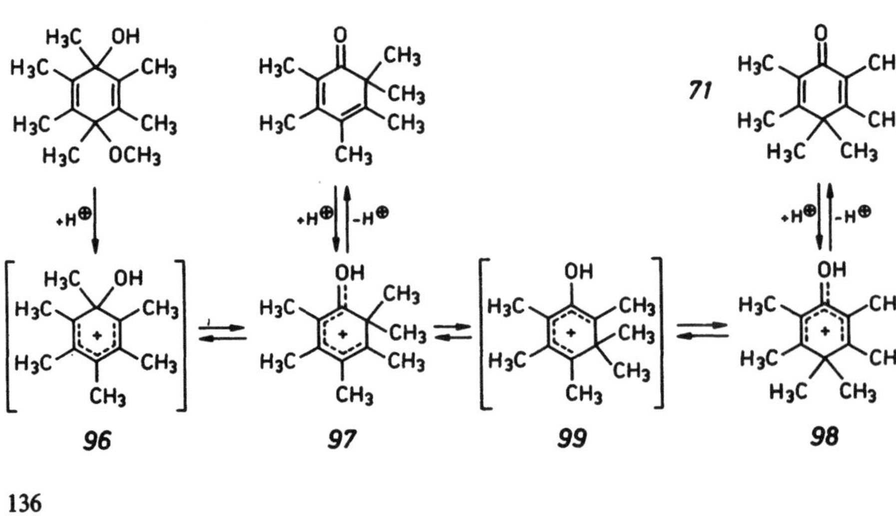

In comparison with its isomers the ion (*96*) seems to be far less stable than the ion (*93*), and under the condition of generation is quickly rearranged, by the 1,2-shift of a CH_3 group, into the ion (*97*). With rising temperature this is practically completely converted into the 4-hydroxy-1,1,2,3,5,6-hexamethylbenzenium ion (*98*) [543–545] which is a conjugate acid of the 2,3,4,4,5,6-hexamethylcyclohexa-2,5-dienone (*71*). By deuterium labelling the rearrangement of ion (*97*) into ion (*98*) was observed to be reversible by 1,2-shifts of CH_3 groups through the intermediate ion (*99*). The diene (*100*) and the triene (*101*), possible precursors of the ion (*99*), are rearranged with acids into the dienone (*71*) [544].

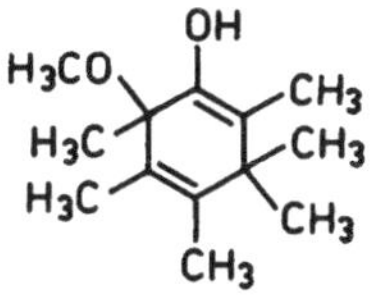

100

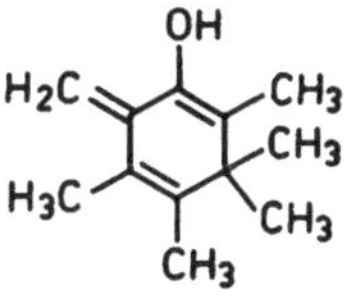

101

The easy conversion of the ion (*96*) into the ion (*97*) accounts for the formation of 2,3,4,5,6,6-hexamethylcyclohexa-2,4-dienone upon electrophilic hydroxylation of hexamethylbenzene [546, 547].

The rearrangement of the ion (*96*) into the ion (*97*) must nevertheless, just like other isomeric conversions of arenium ions, be reversible. This is indirectly indicated by the label migration over the ring in 2,3,4,4,5,6-hexamethylcyclohexa-2,5-dienone-1-^{13}C when heated with the $(CH_3)_2O \cdot BF_3$ complex [548]. These data and those on the isomeric conversions of phenylpentamethylbenzenium ions [549] indicate that the migrating group may also appear at the carbon bonded with oxygen.

The 1-chloro-1,2,3,4,5,6-hexamethylbenzenium ion (*22c*) is reported [172, 259] to be completely converted at —15 °C into a 2-chloro isomer (*102*):

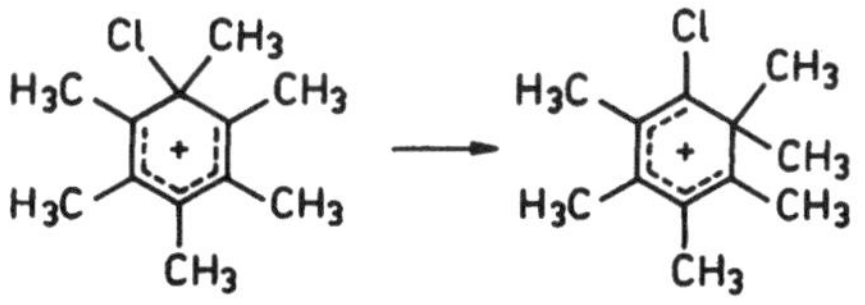

22c *102*

In this case the chlorine, being at the 1-position in the ion (*22c*) seems to exert only an unfavourable inductive effect, whereas in ion (*102*) the +M-effect of the chlorine may be switched on and +I-type substituent (CH_3 group) appears simultaneously at the 1-position instead of the —I-substituent (Cl).

Similarly, 1-chloro-5-X-1,2,3,4,6-pentamethylbenzenium ions (X = NO_2, $CONR_2$) are rearranged into 2-chloro-5-X isomers [176].

An interesting observation [332] was made in the rearrangement of the 2-hydroxy-1,1,3,4,5,6-hexamethylbenzenium ion (*97*) into a 4-hydroxy isomer (*98*) in different acids. The rate of this rearrangement proved to be the same ($k \approx 10^{-3}\ s^{-1}$ at 60 °C) in 85 and 95% H_2SO_4 and in HSO_3F containing some HF. When HSO_3F

is used without HF the conversion rate rises 10^7 times. R. Childs [332] explained this effect by SO_3 usually present in HSO_3F and bonded by HF. The acceleration is effected by the sulphonation of the hydroxy group in the ion (*97*) which decreases its electron-releasing influence and facilitates the conversion into the ion (*49*) with a meta-positioned acid function.

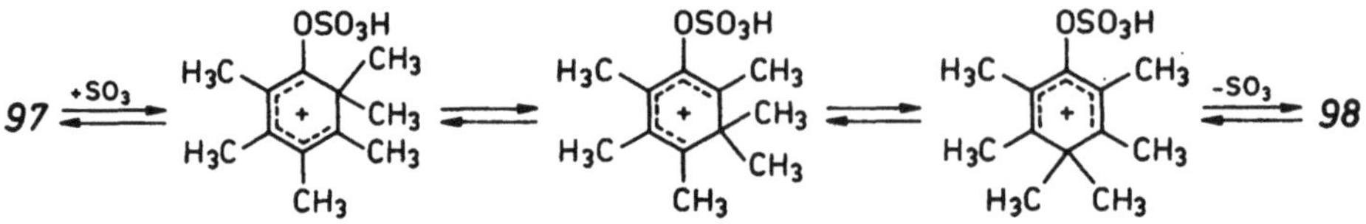

49

At low concentrations of SO_3 the ion (*49*) is an intermediate particle between the ions (*97*) and (*98*). As the concentration of SO_3 in HSO_3F, however, rises to 20% the equilibrium at −50 °C is completely shifted toward this ion. Assumably, OSO_3H, as distinct from OH and OCH_3, takes part in delocalizing the positive charge of the ion less effectively than the CH_3 group.

Some substituted 2-hydroxybenzenium ions were observed to rearrange into 4-hydroxy isomers by the migration of substituents different from CH_3:

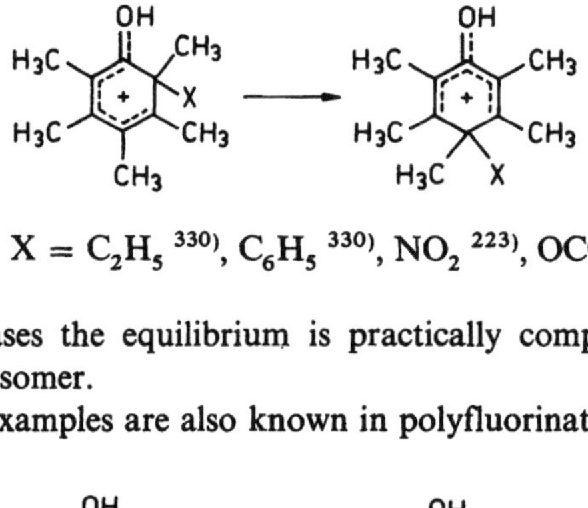

X = C_2H_5 [330], C_6H_5 [330], NO_2 [223], $OCOCH_3$ [222]

In all cases the equilibrium is practically completely shifted in favour of the 4-hydroxy isomer.

Similar examples are also known in polyfluorinated arenium ions [235]:

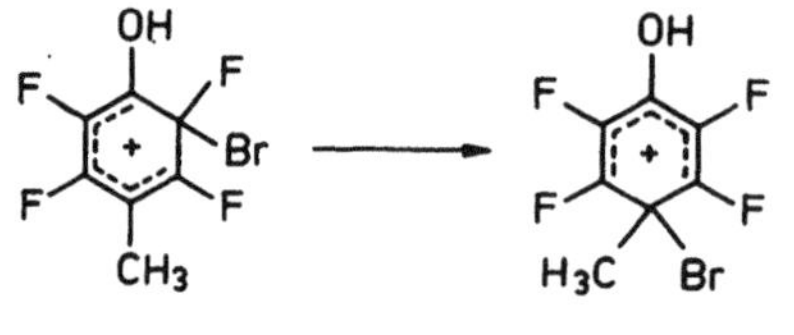

If there is an unsubstituted position in the ion, a transfer to this position of an electronegative substituent leads to aromatization [202]:

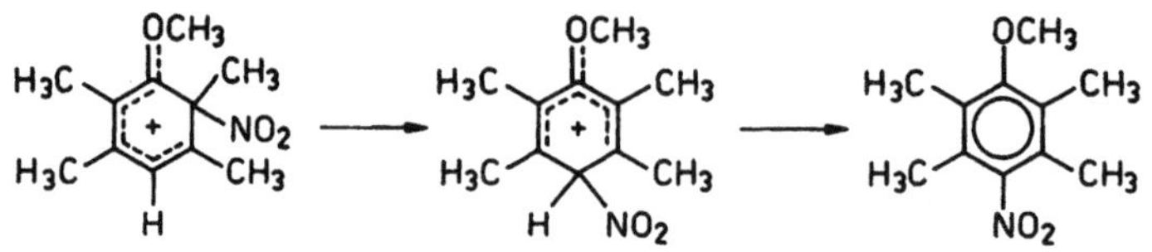

Shifts of substituents in hydroxybenzenium ions form the basis of dienone-phenol rearrangements observed in many cases (see surveys [550-552]), e.g. 2,4,4-trimethylcyclohexa-2,5-dienone into 2,4,5-trimethylphenol under the influence of acids:

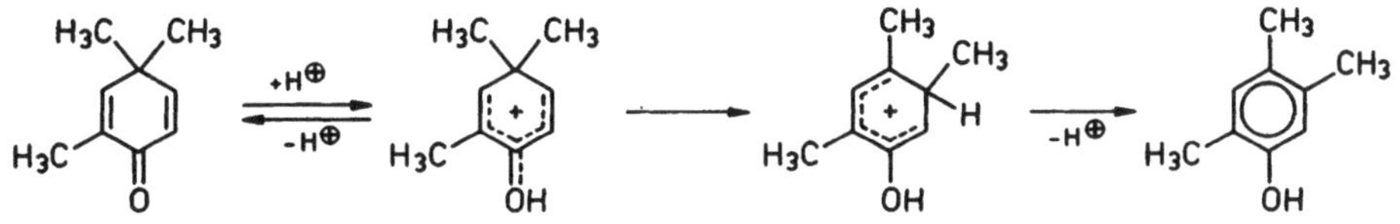

Among the possible directions of the 4-CH_3-group migrations (into the 5- and 3-positions) the one realized leads to a more stable hydroxybenzenium ion. Additional evidence on the rearrangement of this type is given in Sect. IV.2.C.

For the ions with electronegative substituents at the ring sp^2 hybridized carbon, the isomerization can proceed via the "addition-elimination" scheme. In this manner the polyfluorinated arenium ions are likely to be rearranged in acid media containing the donors of the fluoride-ion (HF, SbF_6^- etc.) [178, 235, 238, 239, 413, 414]:

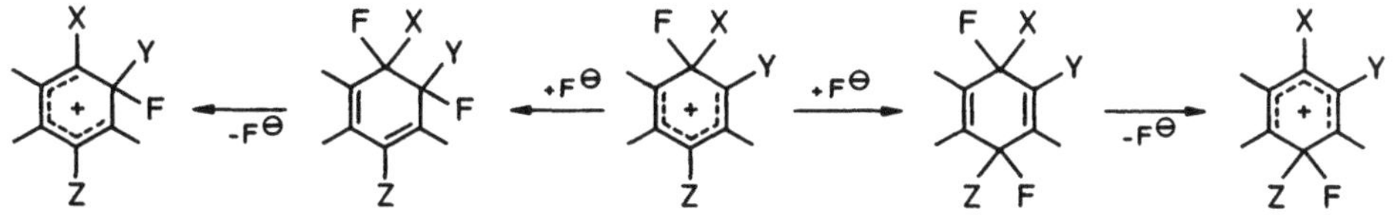

The simultaneous addition-elimination in the cyclic transition state with, say, HF taking part in the process is neither excluded [235, 239].

The isomeric conversions of polyfluorinated arenium ions can inform on the thermodynamic equilibrium between the isomeric ions and establish the ratio of the stabilizing effects for different substituents in these ions. The interconversions of isomers is essential since under the conditions of generation they can be formed in non-equilibrium ratios. For example, 1-haloheptafluorocyclohexa-1,4-dienes interact with SbF_5 in SO_2ClF at —80 °C to form a mixture of 2- and 3-haloheptafluorobenzenium ions, the latter being predominant (66 and 75% for X = Cl and Br). But as the temperature rises to —60 °C the amount of this ion falls to 30—40% and at —20 °C an equilibrium is established among 2-, 3- and 4-haloheptafluorobenzenium ions (7.5:4:1 for X = Cl; 1.5:1:0 for X = Br) [414].

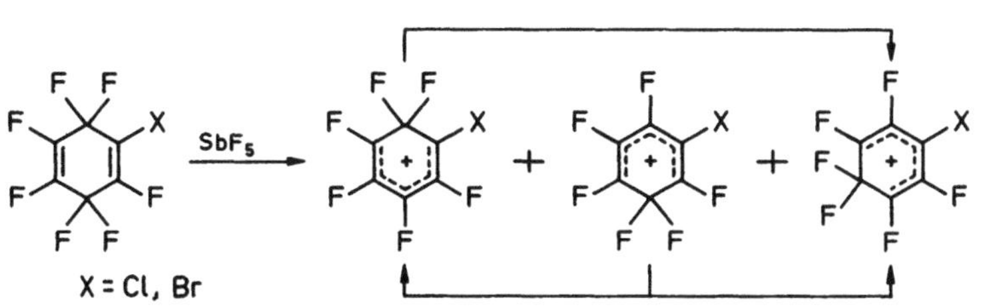

The electronic effects of the F, Cl and Br atoms in polyhaloarenium ions are discussed in [19].

The rearrangement of polyhalogenated arenium ions due to the 1,2-shift of a substituent is rather interesting. At present only the conversion of 1,4-dichloropentafluorobenzenium ion into 1,3- and 1,2-dichloro isomers:

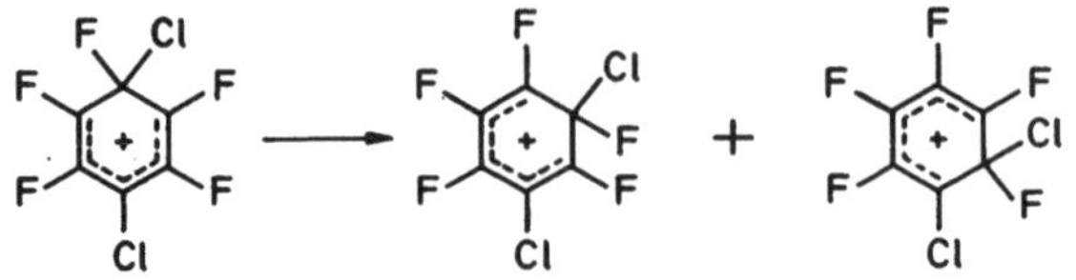

and the conversions of trichlorotetrafluorobenzenium ions can be considered as examples.

2 Kinetics and Mechanism of Isomeric Conversions of Arenium Ions

A Processes of Intermolecular Proton Transfer

One of the possible interconversions of arenium ions differing in the proton site is the reversible proton transfer to the nucleophilic particles in the solution. Such particles can be acid anions, free hydrocarbon molecules and non-dissociated acid molecules:

$$AH^+ + Y^- \rightleftarrows A + HY \qquad (1)$$

$$AH^+ + A \rightleftarrows A + AH^+ \qquad (2)$$

$$AH^+ + HY \rightleftarrows A + H_2Y^+ \qquad (3)$$

The intermolecular proton transfers reduce the lifetime of arenium ions and at their high rates lead to the broadening and coalescence of signals in the NMR spectra. The changes in the spectra with changing temperature and medium acidity indicates the predominant direction and rate of proton transfer.

By dissolving e.g. mesitylene in liquid HF with a limited amount of BF_3 C. MacLean and E. Mackor [34] prepared solutions containing comparable quantities of mesitylene and mesitylenium ion. In the PMR spectrum at —100 °C the mesitylene and its protonated form gave separate sets of signals. A rise in temperature to —30 °C (Fig. 14) resulted in the broadening and merging of the signals for the ion CH_2 group, the ring protons of the ion and the mesitylene, as well as the coalescence of the signals for the CH_3 groups of the ion with that of the free hydrocarbon. Since the HF signal was not appreciably broadened the coalescence of the above signals reflects the proton transfer between the mesitylenium ion and mesitylene. This transfer average the state of the protons bonded with the ring and simultaneously the states of the protons of the ion methyl groups and of the free hydrocarbon, but it does not affect those of the acid.

A further rise of temperature to +50 °C broadened and merged the signal of the ring protons from the mesitylenium ion and mesitylene with that of HF by the exchange process (1) with the HF_2^- anion. The temperature dependence of the PMR

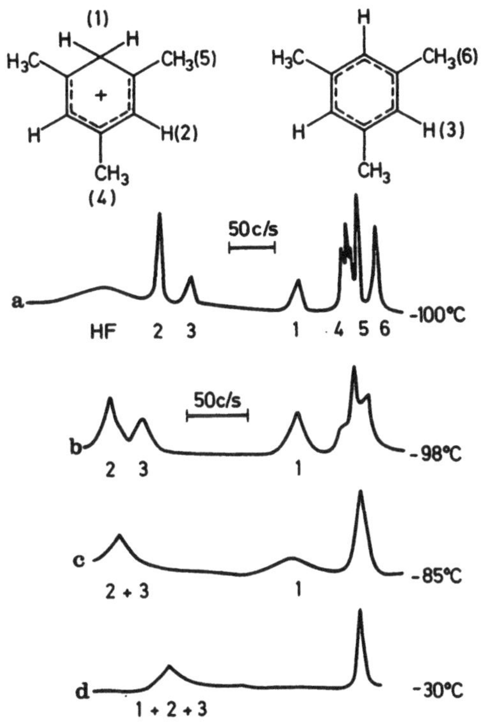

Fig. 14. PMR spectra (60 MHz) of mesitylene and tetrafluoroborate of the mesitylenium ion in HF at various temperatures (the curves b, c and d are presented in a different scale excluding the HF signal)

spectra in terms of the Bloch equations modified for the exchange by H. McConnell (see [553]) made it possible to determine the kinetic characteristics of the proton transfer from the mesitylenium ion to mesitylene and the HF_2^- anion. For similar experiments with m-xylene and anisole, the data obtained were less complete. The results of the kinetic measurements are presented in Table 47 (as reported later [4,415] the activation energies seem to be underestimated).

The rate coefficients of proton transfer from the mesitylenium ion to the HF_2^- anion are of the same order of magnitude (10^4 l · mole^{-1} · s^{-1} at 0 °C). The lower rate of the second described process (at below 0 °C it is practically suppressed), is attributed to the low concentration of the HF_2^- anion being transformed by the reaction

$$HF_2^- + BF_3 \rightleftarrows HF + BF_4^- .$$

A similar situation occurs with protonated m-xylene. But in passing to the 4-methoxybenzenium ion the rate of proton transfer to the conjugate base drops by two orders of magnitude.

The proton transfer from the arenium ion to aromatic hydrocarbon is likely to occur in "cation-molecule" associated pairs of charge transfer complexes [34].

Later T. Birchall and R. Gillespie [271] studied proton exchange reactions for methylbenzenium ions in HSO_3F. As the protonation of m-xylene, mesitylene, durene, pentamethylbenzene and hexamethylbenzene in an excess of HSO_3F is strong enough, the proton transfer by reaction (2) is suppressed, and the exchange of a

Table 47. Arrhenius Parameters for Intermolecular Proton Exchange [34)]

Ion AH^+	Transfer to hydrocarbon A	Transfer to anion HF_2^-
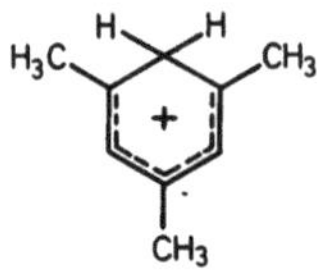	$10^{11.0}$ exp (—8000/RT)	$10^{10.3}$ exp (—7400/RT)
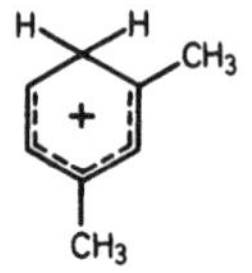	10^{10} exp (—7500/RT)	—
	$\leqq 10^{8.5}$ exp (—7700/RT)	$\geqq 10^{10}$ exp (—7700/RT)

proton with the solvent by reaction (1) turned out to be predominant. Accordingly, an addition of KSO_3F to the system increases the exchange rate, and that of SbF_5 decreases it.

The authors determined the lifetimes of methylbenzenium ions corresponding to the mentioned hydrocarbons in HSO_3F and HSO_3F—SbF_5 at different temperatures; from these they estimated (the absence of information on the SO_3F^- anion did not allow them to calculate the rate coefficients) the activation energies of proton transfer to the solvent anion. The values obtained (13–15 kcal/mole) proved to be much higher than those found by C. MacLean and E. Mackor. This seems to reflect the lower basicity (higher acidity) of the medium in the mentioned experiments [271)]. Their conclusion that for conjugate acids of durene and hexamethylbenzene the proton exchange with the solvent in HSO_3F predominates over the intramolecular 1,2-shifts of hydrogen contradicted the data obtained for HF [34,275,415)] and proved to be erroneous [4)].

On the whole the above data indicate rather a high activation barrier for the proton transfer from arenium ions to nucleophilic particles of the medium. The proton transfer between O- and N-protonated particles and respective conjugate bases or anions usually proceeds at far higher rates [554)] often determined by the probabilities for the two particles to meet (the rate coefficients are of the order of 10^{10}–10^{11} $l \cdot mole^{-1} \cdot s^{-1}$)[50]. In such cases the Arrenius energies of activation amount to 2–3 kcal/mole just as expected for diffusion controlled processes. These reactions proceed easily since they are effected by the transfer of proton along the strong hydrogen bond whose preliminary formation facilitates the subsequent cleavage of the initial bond.

50 The rate coefficient of the proton transfer from H_3O^+ to HF_2^- amounts to about 10^{12} $l\ mole^{-1}\ s^{-1}$ [34)].

In the case of arenium ions the efficiency of interaction between the hydrogen atoms of the ring CH_2 or CHR group and the ambient nucleophilic particles on the pattern of hydrogen bonding is far lower than with O- or N-acids (for such interaction see Sect. III.5).

The marked activation barrier for the intermolecular proton transfer from arenium ions to conjugate bases and anions facilitates the suppression of this process hindering the observation of the fixed structure of ions by the NMR method. This suppression is usually achieved by raising the acidity, lowering the temperature and the concentration of the protonated compound in the solution (in this case the anion concentration falls too). Since in strong acid media the signals of the ring CH_2 and CHR groups of arenium ions are sufficiently sharp the rate coefficients of reactions (3) under direct observation of arenium ions are very small.

However, in strong acids one should not leave out the probability of intermolecular proton exchange with diprotonated forms of aromatic compounds:

$$AH_i^+ + HY \rightleftarrows AH_iH_j^{2+} + Y^- \rightleftarrows AH_j^+ + HY$$

Diprotonation was observed, for instance, for polymethylated derivatives of biphenyl and naphthalene in the $HSO_3F{-}SbF_5$ and $HF{-}SbF_5$ systems [79-82].

Suppression of intermolecular proton transfer enables one in many cases to observe intramolecular (intraionic) migrations which are discussed in the next section.

B Regularities of Intermolecular 1,2-Shifts of Hydrogen and Other Migrants in Arenium Ions[51]

A characteristic property of carbenium ions in general and arenium ones in particular is a shift of one of the groups located at the sp^3-hybridized carbon atom to the neighbouring positively charged carbon atom (carbocationic centre):

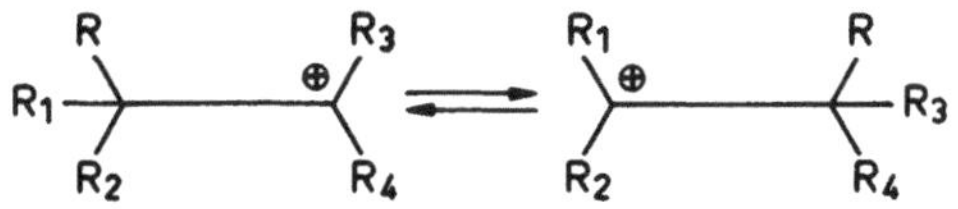

If $R_1 = R_3$ and $R_2 = R_4$ then the structure of the ion before and after the rearrangement is the same, so such processes are termed degenerate. The principal tool for fast degenerate processes is at present the dynamic NMR. Its application to ionic systems started from the study of fast 1,2-hydrogen shifts. Consider, e.g., the changes in the spectrum of the 1-H-1,2,3,4,5,6-hexamethylbenzenium ion generated in $HF{-}BF_3$ [4,34,415]. At −85 °C the PMR spectrum of this cation (see Fig. 1a) displays separately the captured proton and the four types of CH_3 groups. With rising temperature the signals of the CH_3 groups are broadened and merged and at 0 °C (Fig. 15) are displayed as a well resolved doublet; simultaneously, the quartet signal of a single hydrogen atom converts into a complicated multiplet. With decrease in temperature the original PMR spectrum is restored.

51 See also [555].

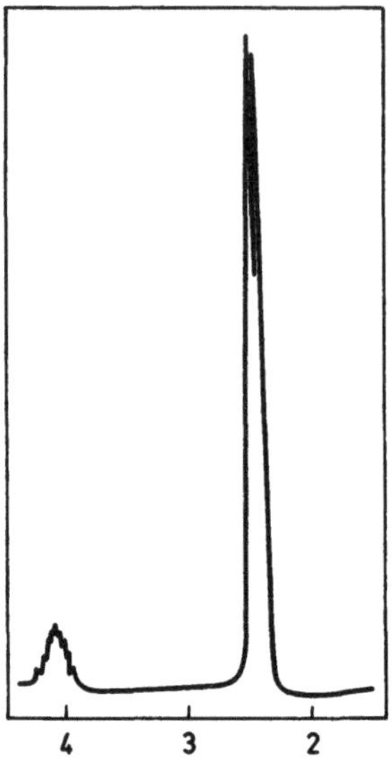

Fig. 15. PMR spectrum of 1-H-1,2,3,4,5,6-hexamethylbenzenium ion tetrafluoroborate in HF at 0 °C [4]

The coalescence of the hexamethylbenzenium ion CH_3 signals with rising temperature reflects the averaging of these groups due to the change of the site of proton attachment. If the change in the proton position were a consequence of its intermolecular transfer to the acid anion and back, then simultaneously with the coalescence of the CH_3 signals a coalescence of the signals of the ion's single hydrogen atom and of acid protons would occur; this is not observed. The mechanism and the possibility of explaining the observed changes in the spectrum by the proton transfer to the free hexamethylbenzene (which must be present in equilibrium if only in insignificant amounts) are also rejected since the increase in the medium acidity by introducing additional BF_3 and the decrease in the concentration of the substrate (and hence of the HF_2^- anion) does not influence the temperature changes of the PMR spectrum of the hexamethylbenzenium ion. Also, the doublet splitting of the merged signal indicate the spin coupling of the migrating proton with the protons of the same signal indicate the same spin coupling of the migrating proton with the protons of the six CH_3 groups. The observed signal splitting of the CH_3 groups (2.1 cps) correlates with the averaged values of the spin-spin coupling constant of the proton with the protons of the CH_3 groups for a fixed ion structure:

$$(J_{1\text{-H},1\text{-CH}_3} + 2\,J_{1\text{-H},2\text{-CH}_3} + 2\,J_{1\text{-H},3\text{-CH}_3} + J_{1\text{-H},4\text{-CH}_3}):6$$
$$= (6.8 + 2.1 + 2.0 + 2.0):6 \approx 1.8 \text{ cps.}$$

In this case the averaging of the CH_3-groups is due to the intraionic shifts of hydrogen:

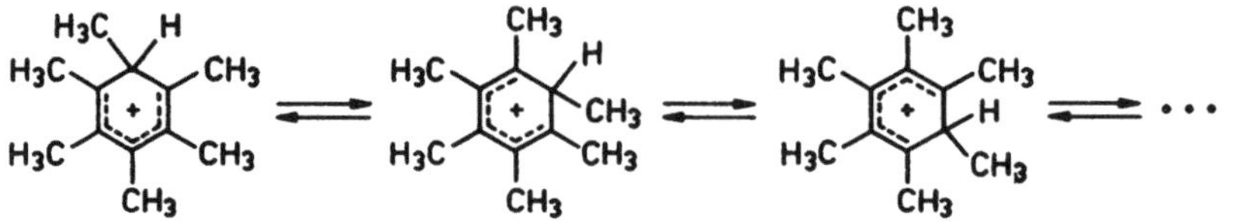

Using the chemical shifts and the half-widths of signals for various protons in the low-temperature spectrum corresponding to the 1-H-1,2,3,4,5,6-hexamethylbenzenium ion one can, by the Bloch equations modified for exchange processes by H. McConnell [553)], calculate the spectrum shape for different times of its life and, by comparing the calculated spectra with the experimental ones, determine the rate coefficients of the degenerate rearrangements at different temperatures.

With this technique C. MacLean, E. Mackor and D. Brouwer [34,275,415)] determined the kinetics of the degenerate rearrangements of benzenium ions realized by 1,2-hydrogen shifts. The results obtained from sufficiently complete kinetic analyses are presented in Table 48. The data on the 1,2-hydrogen shifts in aliphatic ions are given in [555)].

In cases listed in Table 48 the hydrogen at each 1,2-shift migrate to a carbon atom of the same basicity as the one it comes from. Hydrogen migration to a less basic carbon atom must overcome a higher activation barrier; it is not always possible to observe such rearrangements by the temperature changes of the NMR spectra. Thus, the intramolecular degenerate rearrangement of the 2,4,6-trimethylbenzenium (mesitylenium) ion is associated with the intermediate formation of the extremly unstable 1,3,5-trimethylbenzenium ion and therefore proceeds so slowly that it is not detected on the background of the intermolecular proton transfer [34)].

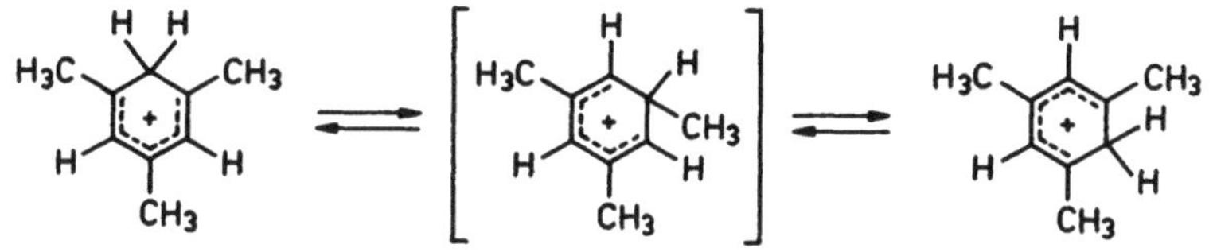

An intramolecular rearrangement with 1,2-shifts of the hydrogen between non-equivalent positions is the degenerate conversion of the 2,3,5,6-tetramethylbenzenium ion (*103*).

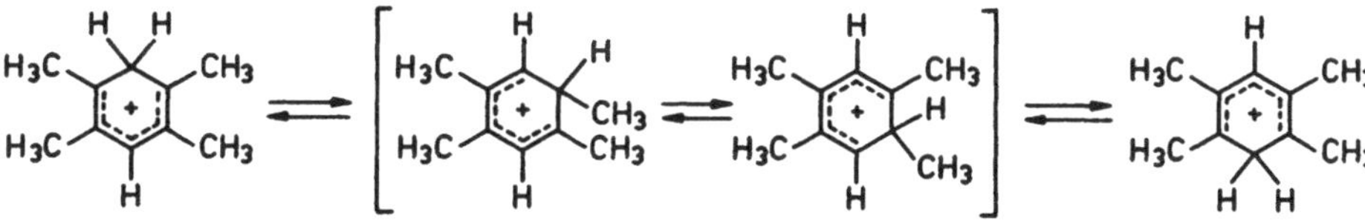

103 *103a* *103a'* *103'*

The rate of this rearrangement is of the same order of magnitude as for the ions presented in Table 48 [34)] (see also [4,415)]). The difference in the stability of isomeric ions (*103*) and (*103a*) though appreciable (the latter ion is not detected in equilibrium with the former by the PMR spectra) is not so great as for the trimethylbenzenium ions.

The rate coefficient of the degenerate two-stage rearrangement of the 2,3,4-trimethylbenzenium ion is about $10^2\ s^{-1}$ at 0 °C [4)].

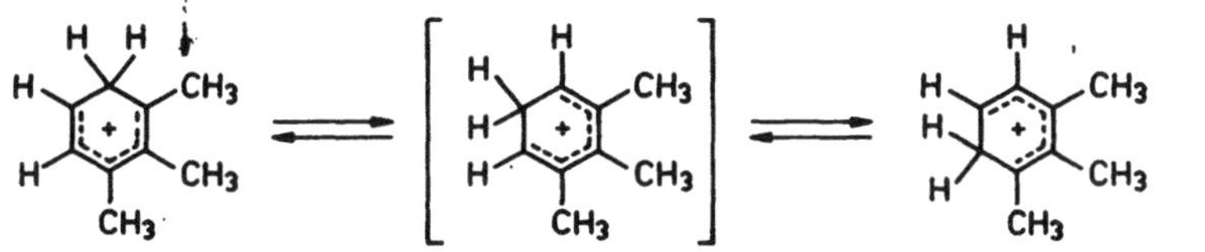

Table 48. Kinetic Parameters of Hydrogen 1,2-Shifts[a] in Arenium Ions

Ions	Acid system	log A	E_a kcal/mole	log k (25 °C) (k, s^{-1})	ΔG* (25 °C) kcal/mole	Ref.
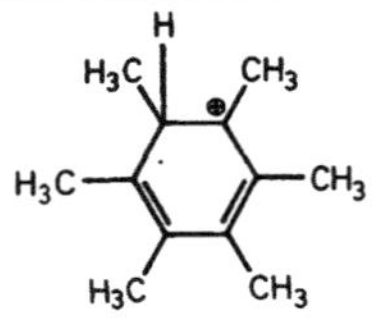	$HF—BF_3(HF—SbF_5)$ $HSO_3F—SbF_5—SO_2ClF$	13.3 13.5	11.3 11.8	5.01 4.85	10.6 10.8	4,275) 308)
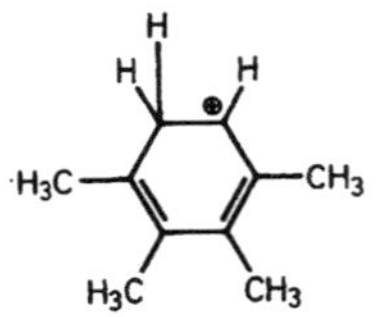	$HF—BF_3$	13.0	11.6	4.50	11.3	4,275)
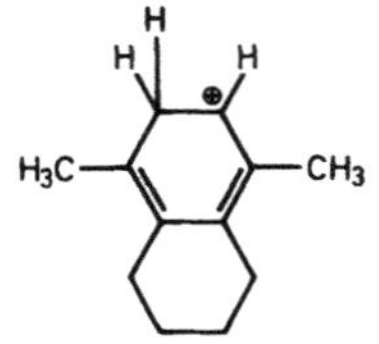	$HF—BF_3$	13.0	11.2	4.79	10.9	275)
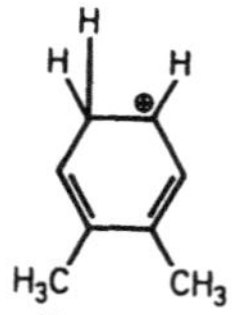	$HF—SbF_5—SO_2ClF$	8.5	5.3	4.62	11.2	283,367)
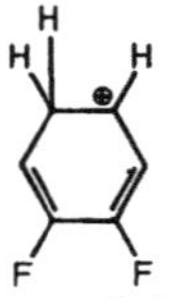	$HF—SbF_5—SO_2ClF$	8.6	5.7	4.42	11.4	283,367)

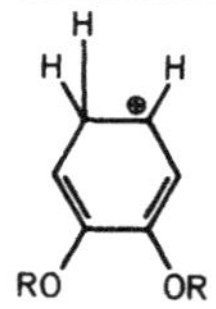	R = H	$HSO_3F—SbF_5—SO_2ClF$			<1 (—20 °C)	>13.6	[333,555]
	R = CH_3	$HSO_3F—SbF_5—SO_2ClF$			<1 (—40 °C)	>12.4	[275,333,555]
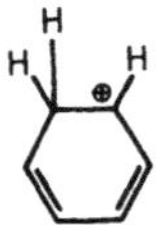		$HF—SbF_5—SO_2ClF—SO_2F_2$	15.3	10	7.97	6.6	[65,139,282]
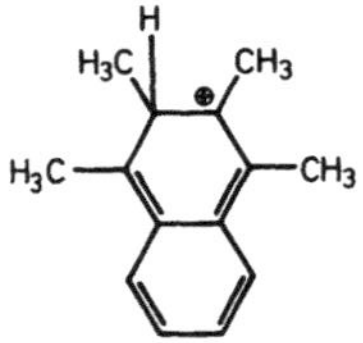		$CF_3SO_3H—SbF_5$				12.8 (62 °C)	[556]
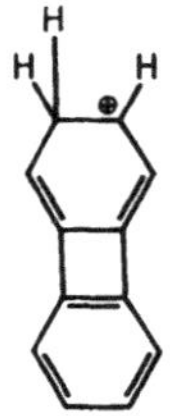		$HSO_3F—SbF_5—SO_2ClF$			~1.4 (—32 °C)	~12.4 (—32 °C)	[314]

Table 48. (continued)

Ions	Acid system	log A	E_a kcal/mole	log k (25 °C) (k, s^{-1})	ΔG* (25 °C) kcal/mole	Ref.
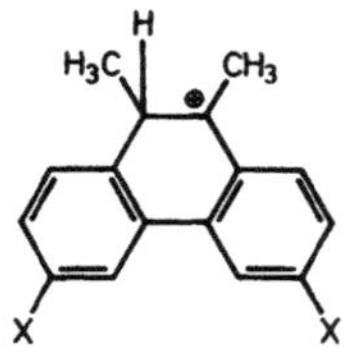 X = H	$HSO_3F—SbF_5—SO_2ClF—SO_2F_2$	12.8	8.7	6.42	8.7	308)
X = Br	$HSO_3F—SbF_5—SO_2ClF—SO_2F_2$	12.8	8.8	6.35	8.8	308)
X = CH_3	$HSO_3F—SO_2ClF—SO_2F_2$	12.8	9.8	5.62	9.8	308)
X = CF_3	$HSO_3F—SbF_5—SO_2ClF—SO_2F_2$	13.5	7.5	8.00	6.5	308)
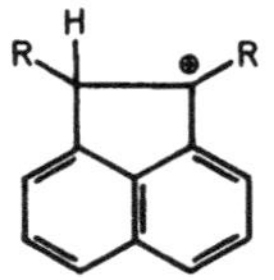 R = CH_3	$HSO_3F—SbF_5—SO_2ClF$	13.3	13.8	3.2	13.1	316) cf. 315)
R = H	$HSO_3F—SbF_5—SO_2ClF$			<1 (40 °C)	>16.9 (40 °C)	315)
$(CH_3)_2CH—C^{\oplus}(CH_3)_2$	$SO_2ClF—SO_2F_2$			7.5 (—138 °C)	3.1 (—138 °C)	557)
$CH_3(H)CH—C^{\oplus}(H)CH_3$	$SO_2ClF—SO_2F_2$			≧9.3 (—140 °C)	≦ 1.9 (—140 °C)	557)

[a] All data are related to the shift of one hydrogen atom in one direction. The statistical factors 2 and 4 are taken for hexamethylbenzenium and unsubstituted benzenium ions.

Rearrangements with 1,2-hydrogen shift were also observed for the ions from the protonation of toluene [4,65], xylenes [4,367], halogenated toluenes [363], fluorobenzene [283], difluorobenzenes [283,367], 1,2,3,4-tetrafluorobenzene [283], methylnaphthalenes [301].

Factors determining the rate of 1,2-hydrogen shifts: The Dutch researchers suggested that the 1,2-hydrogen shift in carbenium ions is realized via a transition state of the π-complex (*104*) in which the migrating proton interacts with the π-bond between two carbon atoms [34,415]:

104

Since the activation energy of the 1,2-hydrogen shift for the hexamethylbenzenium ion is only 3–4 kcal/mole lower than the enthalpy of the proton transfer from this ion to the medium [415], the bond of the migrating proton with the ion skeleton in the transition state must be weak. Accordingly the difference in the activation energies of the 1,2-shifts of protium and deuterium in the 1-H- and the 1-D-1,2,3,4,5,6-hexamethylbenzenium ions ($k_H/k_D = 0.35\ e^{+1200/RT} = 5$–$10$) proved to be close to that of the vibrational zero-point energies of the C—H and C—D bonds (1150 cal/mole).

If the migrating proton is linked in the transition state by a weak π-bond, then its 1,2-shift should be realized the more readily the less basic the conjugate base. On some examples (see Table 48) such a correlation is actually traced[52]:

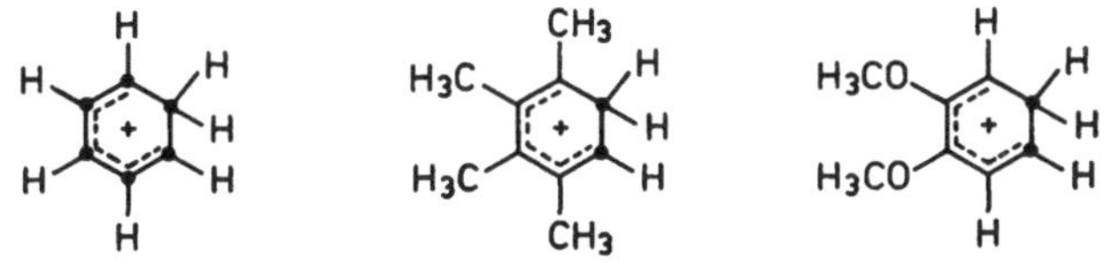

$k_{25°} \approx 10^5\ s^{-1}$ [65] $k_{25°} = 10^{4.5}\ s^{-1}$ [275] $k_{-20°} < 1\ s^{-1}$ [275]

For the 1,2-hydrogen shift in the hexamethylbenzenium ion the rate coefficient turned out to be somewhat higher than that for the protonated prehnitene (see Table 48), although the basicity of hexamethylbenzene exceeds that of prehnitene by more than 3 orders of magnitude[53]. This discrepancy and some others made the Dutch scientists doubt the adequacy of the transition state model of the π-protonated double bond and assume that a large part of the positive charge in the transition state remains on the carbons; therefore, the factors stabilizing the parent ions act

52 The basicity of o-dialkoxybenzenes is higher than that of polymethylated benzenes [275].

53 In protonated o-xylene and o-difluorobenzene, the degenerate 1,2-hydrogen shifts have about the same rates [283].

likewise in the transition state [4,275]. They present the transition state of the 1,2-hydrogen shift as a resonance hybrid of the canonical structures (A) and (B)

(A) (B)

This description implies that the three-centre bond of the migrating hydrogen with the carbon skeleton partially preserves the σ-character.

G. Olah suggested that for denoting the transition state with the three-centre two-electron bond the following symbolics should be used [1,65]:

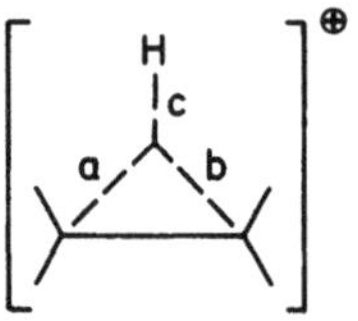

He believes the three-centre bond in the transition state to be broken in to three directions: two of them (a and b) result in the formation of carbenium ions, and one (c), in the loss of the proton and formation of the C=C bond. The preference of one or the other direction depends on both internal and external (solvation) factors.

The symmetric arrangement of the migrating group (in this case hydrogen) may be an unstable intermediate rather than the transition state (cf. [558]).

By doubting that the rate coefficients of carbenium ions rearrangements due to a 1,2-hydrogen shift depend on the basicities of the conjugate bases, the Dutch scientists suggested that the factor determining the rate may be the positive charge on the carbon atom to which the proton is shifted [4,275]. The larger the positive charge on the neighbouring carbon atom, the stronger the C—H bond is polarized and the more readily the three-centre-bond configuration is attained.

Relationship between the activation parameters of the 1,2-hydrogen shift and the deficiency of electron density [308]: Characteristing the electron deficiency with the chemical shift of the carbon atom (δ_c) to which the hydrogen migrates, the authors found for the free activation energies of the 1,2-hydrogen shifts in the 1-H-1,2,3,4,5,6-hexamethylbenzenium and 3,6-di-substituted 9-H-9,10-dimethylphenanthrenium ions the simple linear relationships.[54]

$$\Delta G^{\ddagger}_{H}\ (-110\ ^{\circ}C) = 26.8 - 0.080\ \delta_c; \quad r = -0.961; \quad s = 0.51;$$

$$\Delta G^{\ddagger}_{H}\ (25\ ^{\circ}C) = 25.9 - 0.076\ \delta_c; \quad r = -0.901; \quad s = 0.80;$$

$$E_a = 28.2 - 0.084\ \delta_c; \quad r = -0.992; \quad s = 0.24\,.$$

54 This relationship was first established for the CH_3 group migration (see below).

Later it was shown [556] that taking in addition the data on the degenerate 1,2-hydrogen shift in the 2-H-1,2,3,4-tetramethylnaphthalenium ion and in the 1-H-1,2-dimethyl-3,4,5,6-tetrachlorobenzenium ion yields the following expression:

$$\Delta G^{\ddagger}_{H}\ (25\ ^\circ C) = 24.6 - 0.07\ \delta_c;\quad r = -0.954;\quad s = 0.66\,.$$

Just as in degenerate rearrangements due to the 1,2-hydrogen shifts, the temperature changes in the NMR spectra reveal the fast reversible rearrangements of ions connected with the shifts of other substituents. The first to be studied was the degenerate rearrangement of the heptamethylbenzenium ion [154,279] (see also [559]).

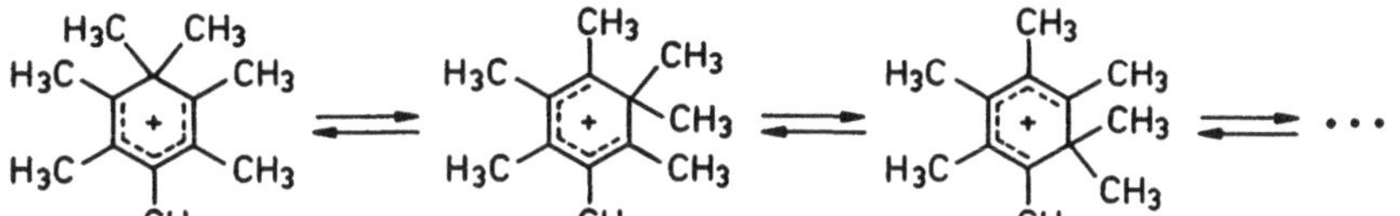

Figure 16 depicts the PMR spectra of heptamethylbenzenium ion chlorosulphonate at different temperatures and the curves (indicated by circles), closest to them, from the series of calculated for different rates of degenerate rearrangement realized by the 1,2-shift of CH_3 groups. That the rearrangement of the heptamethylbenzenium ion is due to the 1,2-shifts of CH_3 groups, i.e., that at each migration the CH_3 group shifts no farther than to the neighbouring ring carbon atom has been proved by experiments using the "labelling via saturation" [560,561] 55 and corroborated by the "two-dimensional" NMR spectroscopy [765]. The kinetics of this rearrangement obtained from the temperature dependence of the PMR spectra are listed in Table 49.

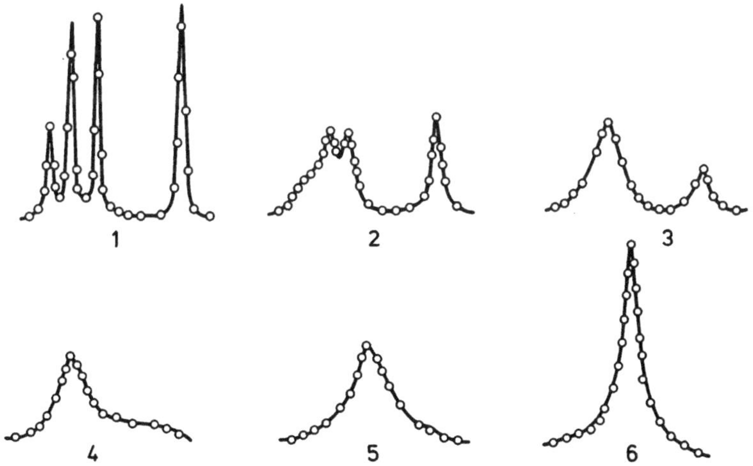

Fig. 16. PMR spectra (40 MHz) of heptamethylbenzenium ion chlorosulphonate at different temperatures

Curves:	1	2	3	4	5	6
t, °C:	−10	49	68	77	96	113
$\tau = {}^1/k$, sec:	10^2	$8.0 \cdot 10^{-2}$	$2.2 \cdot 10^{-2}$	$5.4 \cdot 10^{-3}$	$1.7 \cdot 10^{-3}$	$7.6 \cdot 10^{-4}$

55 For other examples see [244,252,314,371,372,556].

Table 49. Kinetic Parameters of 1,2-Methyl Shifts[a] in Arenium Ions

Ions		Acid system	log A	E_a kcal/mole	log k (25 °C) (k, s^{-1})	$\Delta G^{\neq}$ (25 °C) kcal/mole	Ref.
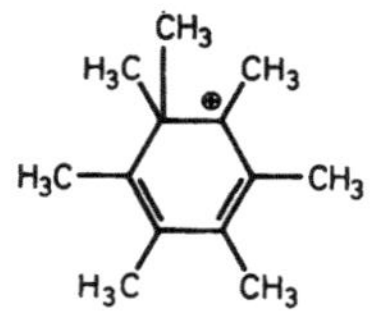		HSO_3Cl	12.9	18.2	—0.44	18.1	279)
		H_2SO_4	11.7 (13.0)	15.2 (18.3)			555, 559)
		CF_3COOH	13.1	18.4	—0.38	17.9	560, 561)
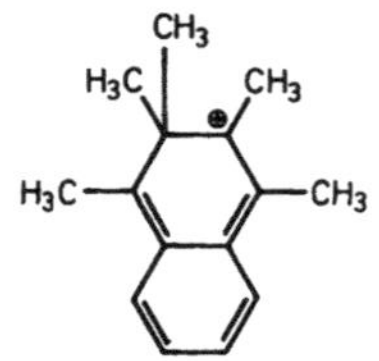						22.7 (—125 °C)	210)
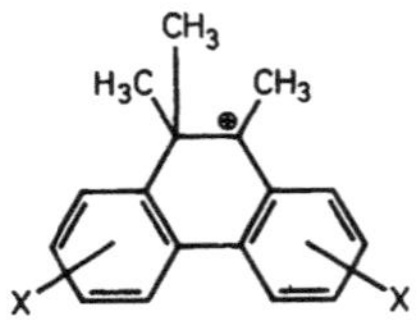	X = H	$HCl-2\,AlCl_3-CH_2Cl_2$	13.2	12.1	4.34	11.5	211)
		HSO_3F-SO_2ClF	13.3	12.2	4.36	11.5	410)
	X = Br (3,6)	HSO_3F-SO_2ClF	12.5	11.2	4.29	11.6	410)
	X = CH_3 (3,6)	HSO_3F-SO_2ClF	12.8	13.4	2.98	13.4	410)
	X = CF_3 (3,6)	HSO_3F-SO_2ClF	11.5	7.9	5.71	9.7	410)
	X = Br (2,7)	HSO_3F-SO_2ClF	12.3	10.2	4.82	10.9	410)

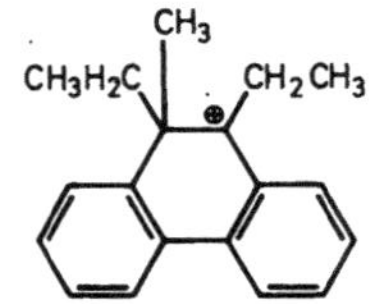	$HCl-2\,AlCl_3-CH_2Cl_2$			~4	~12	[215]
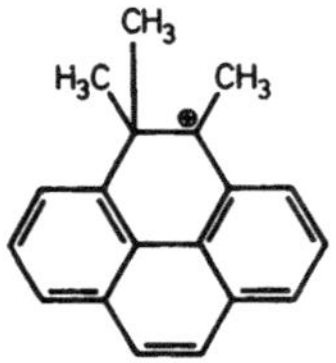	HSO_3F-SO_2ClF	12.8	11.7	4.22	11.7	[410, 562]
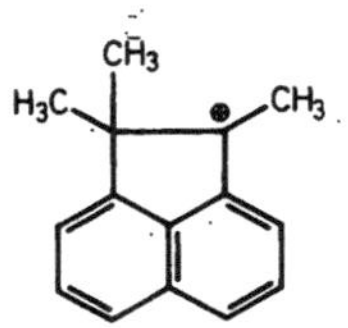	CF_3SO_3H	12.9	18.5	—0.66	18.4	[316]
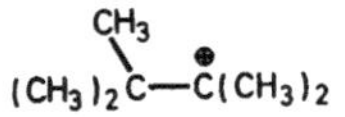	$SO_2ClF-SO_2F_2$			6.45 (—136 °C)	3.8 (—136 °C)	[557]

[a] All data are related to the shift of one CH_3 group in one direction; see table 48

The data characterizing the 1,2-shift of the CH_3 group in other arenium ions are also presented. Some of these data (for slow processes) are obtained by using labeled atoms (see, e.g., [210]).

The rate of the heptamethylbenzenium ion rearrangement is practically the same in the following media: HSO_3F—SbF_5 (9:1), HSO_3F, HSO_3Cl, H_2SO_4, HCOOH, $C_6H_5NO_2$ and CH_2Cl_2 [563]. No solvent effect has been observed on the 1,2-shift of the C_6H_5 in the 1-phenyl-1,2,3,4,5,6-hexamethylbenzenium ion [242] or on the chlorine migration for the degenerate rearrangement of the 1-chloro-1,2,3,4,5,6-hexamethylbenzenium ion [564].[56]

A similar extremely low sensitivity to the acid medium is demonstrated for the rearrangements of the 6-hydroxy-1,1,2,3,4,5-hexamethylbenzenium ion [332], substituted cyclopentyl cations [518] and alkylcarbenium ions [143] (see also [558]). These observations allow us to compare the carbenium ion rearrangements obtained in different acid media.

The 1,2-shifts of CH_3 groups were studied in greatest detail to estimate carefully how the electron deficiency of the carbocationic centre (the carbon atom the migrant shifts to) affects the activation parameters of arenium ion rearrangements. Thus, for structurally similar ions including the heptamethylbenzenium ion, the 9,10,10-trimethylphenanthrenium ion and its disubstituted derivatives, as well as the 9,9,10-trimethylpyrenium ion the following linear relations are found [409,410]:

$$\Delta G^{\ddagger}_{CH_3}\ (-110\ ^\circ C) = 55.8 - 0.192\ \delta_c; \quad r = -0.993; \quad s = 0.39;$$

$$\Delta G^{\ddagger}_{CH_3}\ (25\ ^\circ C) = 53.7 - 0.181\ \delta_c; \quad r = -0.995; \quad s = 0.29;$$

$$E_a = 59.4 - 0.207\ \delta_c; \quad r = -0.977; \quad s = 0.74\,.$$

Taking in addition the more recent data on the rearrangement of the 1,2,2,3,4-pentamethylnaphthalenium ion [210] we have

$$\Delta G^{\ddagger}_{CH_3}\ (-110\ ^\circ C) = 53.7 - 0.183\ \delta_c; \quad r = 0.997; \quad s = 0.39$$

Graphically this dependence is represented in Fig. 17.

The value of $\Delta G^{\ddagger}_{CH_3}$ equals zero at $\delta_c \approx 290$ ppm. This value corresponds to the chemical shift of the sp^2-hybridized carbon atom bearing a single positive charge (see Sect. III.2); hence, the 1,2-shifts of the CH_3 in the simplest aliphatic carbocations with limited possibilities for delocalizing the charge must be very fast. This is in good agreement with data obtained for the 1,1,1,2,2-pentamethylethyl cation (see Table 49).

According to above mentioned data the activation parameters of the 1,2-shifts in arenium ions are related to the deficiency of electron density on the atom (C_t) the migrant is shifted to [410]. Let us present the rearrangement in two stages: first localize the positive charge on the C_t atom spending on it the energy ΔE^{+}_{loc} and then shift the migrating group to C_t. If the energy expenditures at the second stage for the same

56 This is rather unexpected since the equilibrium between the aliphatic halosubstituted carbocations and the respective cyclic halogenium ions depend on the solvent polarity [565,566].

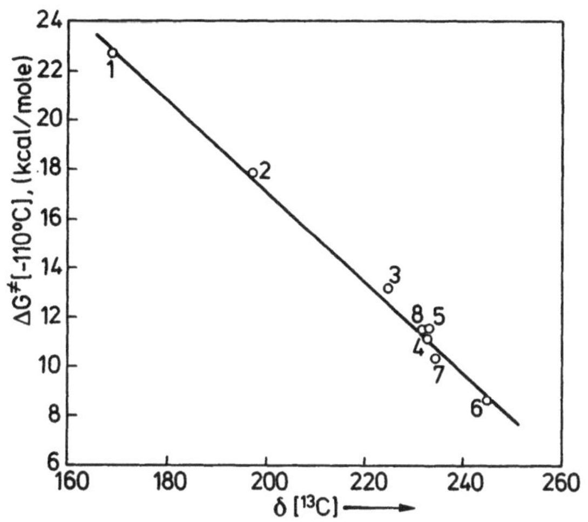

Fig. 17. Dependence of the free activation energy of the CH_3 group 1,2-shift on the extent of electron deficiency characterized by ^{13}C chemical shift of the carbocationic centre: 1) 1,2,2,3,4-pentamethylnaphthalenium ion; 2) hexamethylbenzenium ion; 3–7) 9,9,10-trimethylphenanthrenium ion and its derivatives; 8) 9,9,10-trimethylpyrenium ion

migrating group are equal the change of the activation parameters of rearrangement will be determined for a first approximation by the value of ΔE^{+}_{loc}. In terms of the PMO theory [567,568]

$$\Delta E^{+}_{loc} = -2\beta \cdot (1 - a_{ot})$$

where a_{ot} is the coefficient at the C_t atomic orbital in the nonbonding molecular orbital of the alternant π-carbocation. Taking $a^2_{ot} = q_t$ where q_t is the π-electron deficiency at the C_t atom we have

$$\Delta E^{+}_{loc} = -2\beta(1 - \sqrt{q_t})$$

As in the range of q_t from 0.1 to 1.0 there is an approximate linear relation between $\sqrt{q_t}$ and q_t, then, other things being equal, one can expect a linear relation between the activation parameters of the 1,2-migration and the values characterizing the electron deficiency on C_t.

Among the characteristics of charge on C_t the most adequate one seems to be the chemical shift of C_t in the NMR-^{13}C spectrum (see Sect. III.2). However, other characteristics should not be excluded. For example, a linear relation exists between the activation parameters of the 1,2-migrant shift ($R = CH_3, C_6H_5$) in arenium ions and the C=O vibrations frequencies in the IR spectra of ketones formally obtained by substituting the fragment $>C_t^+-R$ for the group $>C=O$ [307,569].

Since in the PMO theory [567,568] a_{ot} is equal to the π-bond order of the neutral π-system formally formed from the ion by the loss of R as R^+, one can expect the activation parameters of the 1,2-migrant shift to be linearly related to the length of the C=C bonds in the aromatic or unsaturated hydrocarbons. Indeed such a relation was discovered for the 1,2-shifts of the CH_3 group in the heptamethylbenzenium, 1,2,2,3,4-pentamethylnaphthalenium, 9,9,10-trimethylphenanthrenium, 9,9,10-trimethylpyrenium ions and 1,1,1,2,2-pentamethylethyl cations [210,410,555].

The influence of the electron deficiency of the C_t atom on the activation parameters of rearrangements is traced for other migrants as well. Table 50 summarizes the data

Table 50. Kinetic Parameters of 1,2-Aryl Shifts[a] in Arenium Ions

Ions		Acid system	log A	E_a kcal/mole	log k (25 °C) (k, s^{-1})	ΔG* (25 °C) kcal/mole	Ref.
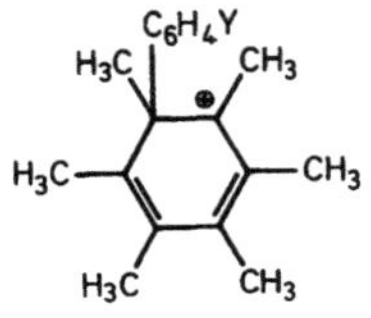	Y = H	$CF_3SO_3H—SO_2ClF$	12.5	16.8	0.19	17.2	242)
					−3.95 (−50 °C)	17.0 (−50 °C)	571)
	Y = p − CH_3	$CF_3SO_3H—SO_2ClF$			−3.07 (−50 °C)	16.0 (−50 °C)	571)
	Y = p − Cl	$CF_3SO_3H—SO_2ClF$			−4.73 (−50 °C)	17.8 (−50 °C)	571)
	Y = m − CF_3	$CF_3SO_3H—SO_2ClF$			−6.79 (−50 °C)	19.9 (−50 °C)	571)
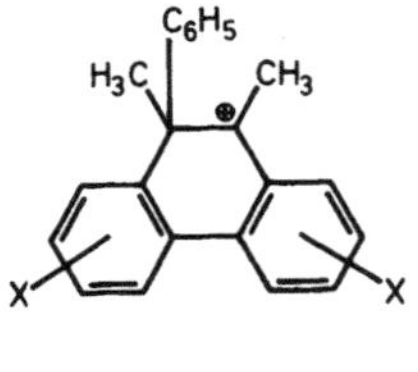	H = X	HCl—2 $AlCl_3$—CH_2Cl_2	12.1	8.9	5.56	9.7	213, 572)
		$HSO_3F—SO_2ClF$	12.2	9.0	5.60	9.8	572)
		$HSO_3F—SO_2ClF$	12.5	9.4	5.61	9.8	307)
		$HSO_3F—SO_2ClF$	11.6	8.2	5.59	9.8	307)
	X = Br (3,6)	$HSO_3F—SO_2ClF$	11.9	10.7	4.06	11.9	307)
	X = CH_3 (3,6)	$HSO_3F—SO_2ClF$	12.7	6.4	8.01	6.5	307)
	X = CF_3 (3,6) / X = Br (2,7)	$HSO_3F—SO_2ClF$	11.8	7.3	6.45	8.7	307)
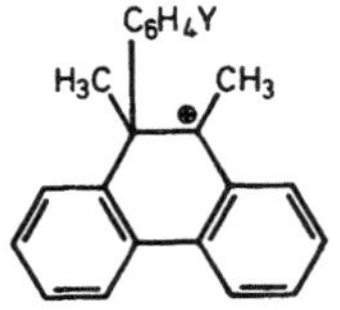	Y = p − CH_3	$HSO_3F—SO_2ClF$	11.0	6.5	6.24	8.9	213, 572)
	Y = p − CH_3O	$HSO_3F—SO_2ClF$			8.0	6.5	213, 572)
	Y = p − Cl	$HSO_3F—SO_2ClF$	12.4	9.8	5.22	10.3	213, 572)
	Y = p − F	$HSO_3F—SO_2ClF$	12.5	9.3	5.68	9.7	213, 572)
	Y = m − F	$HSO_3F—SO_2ClF$	12.6	11.5	4.17	11.8	213, 572)
	Y = p − CF_3	$HSO_3F—SO_2ClF$	12.0	11.7	3.42	12.8	213, 572)
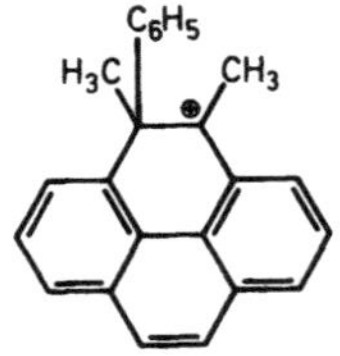		$HSO_3F—SO_2ClF$	12.4	9.4	5.51	9.9	307)

[a] Statistical factor (2) is taken into account for 1-aryl-1,2,3,4,5,6-hexamethylbenzenium ions

on the kinetic characteristics of the 1,2-shift of phenyl and substituted phenyl groups. For the phenyl group in the benzenium, phenanthrenium and pyrenium ions listed in Table 50 the following relations are fulfilled:

$$\Delta G^{\ddagger}_{C_6H_5}\,(-110\ ^\circ C) = 65.0 - 0.245\,\delta_c; \quad r = -0.993; \quad s = 0.43;$$

$$\Delta G^{\ddagger}_{C_6H_5}\,(+25\ ^\circ C) = 65.0 - 0.244\,\delta_c; \quad r = -0.995; \quad s = 0.36;$$

$$E_a = 65.3 - 0.247\,\delta_c; \quad r = -0.984; \quad s = 0.66\,.$$

Similar relations may also be valid for the 1,2-shifts of the C_2H_5, NO_2 and Cl as migrants [555; 564, 50].

The linear dependence of the kinetic parameters of the 1,2-migrant shift on the electron deficiency of the carbon atom to which the migrant is shifted are established for structurally similar arenium ions in which the migrant is shifted in the six-membered ring between the carbons bonded with the CH_3 groups. It seems unlikely that the effect of the ion skeleton on the change of the free activation energy for the 1,2-migrant shift can in all cases be described only by the effect of the electron deficiency on C_t. It is hardly possible, in particular, to ignore the disturbances brought about by the sterical strain of the initial and transition states of ions containing smaller cycles. For instance, the degenerate hydrogen shift is realized essentially slower than predicted by the above dependences in the 1-H-1,2-dimethylacenaphthylenium ion and the shift of the CH_3 group in the 1,1,2-trimethylacenaphthylenium ion [316]. Even by labelling the CH_3 shift in the 1,1,2,3,4-pentamethylcyclobutenyl cation has not been detected [573].

The charge and other factors determining the influence of the skeleton on ion rearrangement can roughly be taken into account if we choose as generalized parameter the free activation energy (or the logarithms of the rate coefficient of the 1,2-shift of a migrant selected) as a reference standard [308, 574]. In Fig. 18 the free activation energies of the 1,2-hydrogen shift are plotted against those of 1,2-shift of the CH_3 group taken as a standard. In this case the approximate linear relation

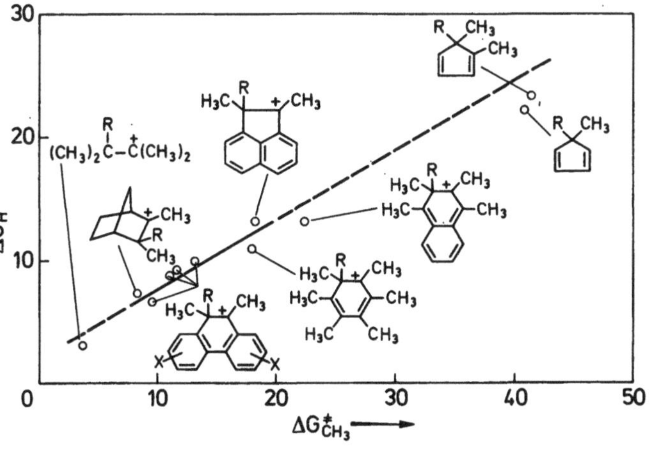

Fig. 18. Comparison of the free activation energy of the 1,2-shifts for hydrogen and CH_3 group

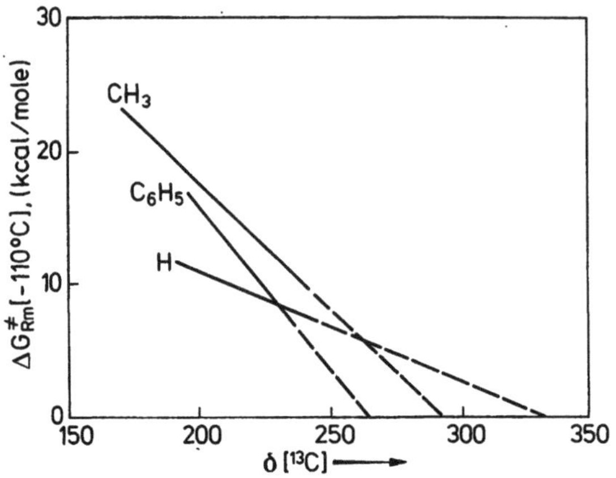

Fig. 19. Dependence of the free activation energy of the 1,2-shift for different migrants on the extent of electron deficiency characterized by the ^{13}C chemical shift of the carbocationic centre

describes a very wide range of changes in $\Delta G^{\ddagger}$ and covers aliphatic carbocations, arenium ions and neutral compounds [308, 555, 574].

A similar correlation is established in comparing the 1,2-shifts of the phenyl and the methyl groups [307], as well as chlorine atoms and the methyl group [564].

The data indicate that the degenerate rearrangements of arenium ions greatly depend on both the structure of the ion skeleton and the nature of the migrating group. Figure 19 depicts [308, 555] the dependences of the free activation energies for the 1,2-shifts of H, CH_3 and C_6H_5 on the chemical shifts of C_t. The different slopes of these dependences points to the impossibility of finding a common order for the migrating ability of the groups. When the ion skeleton changes the migrating ability of the groups may be reversed. Thus, in passing from the 1-X-1,2,3,4,5,6-hexamethylbenzenium ions to 9-X-9,10-dimethylphenanthrenium ions the relative migrating ability of chlorine and hydrogen is reversed: at −110 °C the ratio k^{Cl}/k^{H} for the above ions is $6 \cdot 10^{-6}$ and $7 \cdot 10^{3}$, respectively [308, 564].

The differences in the susceptibility of migrants to the electron deficiency changes on the C_t atom are of great interest. If the dependences presented in Fig. 19 are extrapolated to δ_c corresponding to the localized positive charge (300–330 ppm), then the straight line for the hydrogen as a migrant reaches the value of $\Delta G^{\ddagger} \approx 0$, while those for the CH_3 and C_6H_5 groups go to the region of negative values. This can be interpreted as a reversion of the relative stability for the "open" and the "bridged" structures with increasing positive charge on C_t,

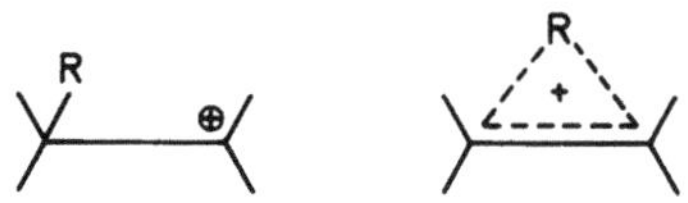

the tendency of migrants to form bridged structures increasing for the aliphatic carbocations in the order H < CH_3 < C_6H_5; this agrees with the results of ab initio calculations (basis STO-3G) for the $RCH_2CH_2^+$ ions [575–577] 57.

57 The dependence of $\Delta G^{\ddagger}_{Cl}$ on δ_c has an even greater slope [564], agreeing with the tendency of β-chloro-substituted carbocations to form bridged chloronium ions.

For similar migrants the changes of the relative ability for the 1,2-shift as the ion skeleton changes may be not so significant. An example are the $k^{C_2H_5}/k^{CH_3}$ ratios for the degenerate rearrangements of the ions:

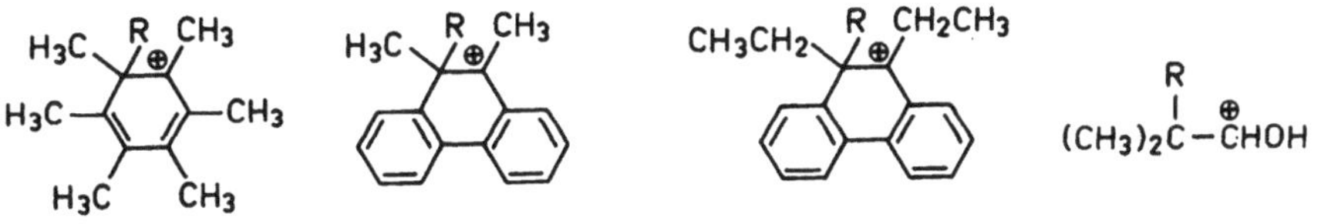

31 [215]   13.5 [212]   ~10 [215]   ~6 [578]

With rising electron deficiency at the C_t atom the $k^{C_2H_5}/k^{CH_3}$ ratio regularly decreases (for details see [215]), but only to be a few times smaller while the rate coefficients change very much (in passing from benzenium to phenanthrenium ions — by 4–5 orders of magnitude at 25 °C).

The effect of migrants on rearrangements of arenium ions is most extensively studied with the 1-X-1,2,3,4,5,6-hexamethylbenzenium ions. The now available data on the kinetics of the 1,2-shifts of the X group in these ions are summarized in Table 51.

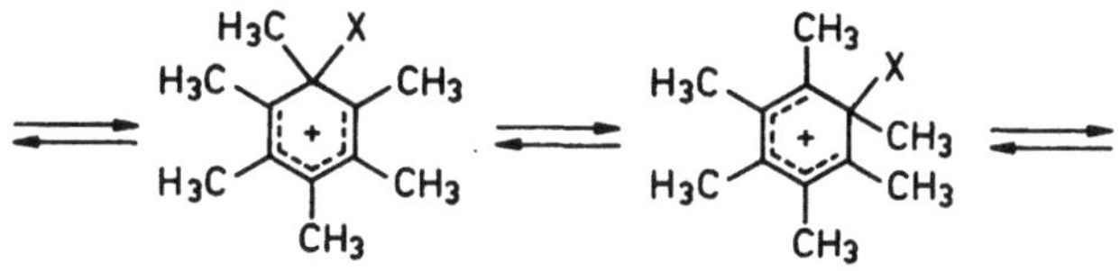

According to their migrating ability at 25 °C on the background of the hexamethylbenzenium skeleton the migrants studied are arranged in the series:

$$Br > H \geqq CH_2C_6H_5 > NO_2 > Cl > SO_3H > C_2H_5 > C_6H_5$$
$$> CH_3 > CH_2Cl, CHCl_2$$

Attention should be given to the sequence of the CH_2Y-groups (Y = CH_3, H, Cl) showing that the substituents Y characterized by —I effect facilitate, and those of +I-effect inhibit, the 1,2-shift of the CH_2Y group. This is why one does not succeed in realizing rearrangements involving the 1,2-shift in the benzenium ions of the $CHCl_2$ group (see Sect. IV.2.C). Nor has one observed the shift of $CH_2CH_2\overset{+}{N}(CH_3)_3$ in the corresponding hexamethylbenzenium ion.

For the 1-X-1,2,3,4,5,6-hexamethylbenzenium ions, in addition to the degenerate 1,2-shifts of the X group, those of the CH_3 group are possible as well, which results in formation of isomeric ions. This process complicates (e.g., for X = Cl, C_6H_5) and in some cases even makes impossible (X = OH, OCH_3 [243] etc.) the measurement of the kinetic parameters of the degenerate rearrangement.

To confirm the degenerate rearrangements of 1-X-1,2,3,4,5,6-hexamethylbenzenium ions under conditions chosen for determining kinetic parameters being accomplished by successive intraionic 1,2-shifts of the X group, just as in the rearran-

Table 51. 1,2-Shift of Different Migrants in 1-X-1,2,3,4,5,6-Hexamethylbenzenium ions[a]

X	Acid system	log A	E_a kcal/mole	log k (25 °C), k, s^{-1}	$\Delta G^{\neq}$ (25 °C) kcal/mole	Ref.
H	$HF—BF_3$ ($HF—SbF_5$)	13.3	11.3	5.01	10.6	4,275)
	$HSO_3F—SbF_5$	13.5	11.8	4.85	10.8	308)
CH_3	HSO_3Cl	12.9	18.2	—0.44	18.1	279)
	9.4 M H_2SO_4	11.7 (13.0)	15.2 (18.3)			555, 559)
	CF_3COOH	13.1	18.4	—0.38	17.9	560, 561)
C_2H_5	CF_3COOH			1.1	16.0	152, 555, 579)
$CH_2C_6H_5$	$HSO_3F—SO_2ClF$	13.9	12.4	4.81	10.9	244)
CH_2CHO	HSO_3F			<0 (—70 °C)	>11.7 (—70 °C)	155, 579)
CH_2Cl	HSO_3F (CF_3COOH)			<0 (60 °C)	>19.5 (60 °C)	152, 579)
$CHCl_2$	$CHCl_3$ (anion $Al_2Cl_7^-$)			<0 (40 °C)	>18.3 (40 °C)	155 579)
C_6H_5	$CF_3SO_3H—SO_2ClF$	12.5	16.8	0.19	17.2	242)
p-$CH_3C_6H_4$	$CF_3SO_3H—SO_2ClF$			—3.07 (—50 °C)	16.0 (—50 °C)	571)
p-ClC_6H_4	$CF_3SO_3H—SO_2ClF$			—4.73 (—50 °C)	17.8 (—50 °C)	571)
m-$CF_3C_6H_4$	$CF_3SO_3H—SO_2ClF$			—6.79 (—50 °C)	19.9 (—50 °C)	571)
Cl	HSO_3F	13.5	15.7	2.0	14.7	251, 564)
	$AlCl_3—CH_2Cl_2$	13.0	14.9	2.1	14.6	564)
Br	$HSO_3F—SO_2ClF$	13.6	10.3	6.05	9.2	371)
NO_2	$HSO_3F—SO_2$	(13.1)	16.8 (13.3)	3.35	12.9	166)
						cf. 372, 555)
	$HSO_3F—SO_2ClF$	13.5	13.2	3.80	12.2	570)
	$HSO_3F—SbF_5—SO_2ClF$	12.9	12.6	3.70	12.4	570)
	$HSO_3F—SO_2ClF$	13.5	13.2	3.82	12.2	303)
SO_3H	HSO_3F	15.4	19.2	1.32	15.6	372)
OCH_3	HSO_3F			<0 (—75 °C)	>11.7 (—75 °C)	243, 579)

[a] statistical factors are taken into account

gement of the heptamethylbenzenium ion (see above), use was made of the "labeling via saturation" (X = Br [371], NO_2 [372], SO_3H [372] [58]).

The kinetics of the heptaethylbenzenium ion rearrangement has been discussed in [156] (cf. [157]).

In the case of the 9-X-9,10-dialkylphenanthrenium ions the kinetics of the degenerate rearrangement is determined both by the broadening and coalescence of the alkyl signals and by the temperature dependent changes of the aromatic ring multiples [212, 216, 217, 252, 309, 311].

R = X = alkyl [211-215, 410]
R = alkyl, X = aryl [307, 309, 313, 572]
R = X = aryl [216]
R = aryl; X = alkyl [217]

The comparison of the results from these two techniques produces reliable data on the rate coefficient of the 1,2-shift of the X substituent. The results obtained for phenanthrenium ions are listed in Table 52.

Table 52. Kinetic Parameters of 1,2-X-shifts[a] in 9-X-9,10-Phenanthenium Ions

X	R	Acid system	log A	E_a kcal/mole	log k (25 °C) (k, s^{-1})	$\Delta G^{\neq}$ (25 °C) kcal/mole	Ref.
CH_3	CH_3	$HCl-2\,AlCl_3-CH_2Cl_2$	13.2	12.1	4.34	11.5	211)
		HSO_3F-SO_2ClF	13.3	12.2	4.36	11.5	410)
CH_3	C_2H_5	$HCl-2\,AlCl_3-CH_2Cl_2$			~4	~12	215)
C_2H_5	CH_3	$HCl-2\,AlCl_3-CH_2Cl_2$	13.2	10.6	5.43	10.1	212)
C_2H_5	C_2H_5	$HCl-2\,AlCl_3-CH_2Cl_2$	12.4	10.1	5.00	10.6	214)
C_6H_5[b]	CH_3	$HCl-2\,AlCl_3-CH_2Cl_2$	12.1	8.9	5.56	9.7	213, 572)
		HSO_3F-SO_2ClF	12.2	9.0	5.60	9.8	572)
		HSO_3F-SO_2ClF	12.5	9.4	5.61	9.8	307)
Cl	CH_3	$CF_3SO_3H-SO_2ClF-SO_2F_2$		~5.6	4 (−110 °C)	~6.3 (−110 °C)	251, 564)
NO_2	CH_3	$CF_3SO_3H-SO_2ClF-SO_2F_2$			3 (−105 °C)	7 (−105 °C)	570)

[a] Statistical factors are taken into account; [b] Data for C_6H_4Y see table 50

The phenanthrenium ions is a convenient model since it enables one, by introducing substituents into benzene rings, to vary the electron eficiency on C_t [307, 308, 410]. It was also used to estimate the influence of the orientation of the vacant orbital of C_t relative to that of the carbon-migrant bond on the activation barrier of the 1,2-shift [252, 580]. The orbital orientation was changed by introducing CH_3 groups at positions 4 and 5. The spatial distortion of skeleton (like a "spring washer") results in that two

58 In passing from HSO_3F to HSO_3F-SO_2 with a small content of HSO_3F the mechanism of successive intramolecular 1,2-shifts of the sulphonic group is replaced by the random shiftings to all the ring sites, corresponding to the intermolecular mechanism.

9-CH_3 groups in the 4,5,9,9,10-pentamethylphenanthrenium ion are differently oriented to the atomic C_{10} p-orbital. A theoretical and experimental analysis of this ion rearrangement limited the earlier suggestion on the role of orbital orientation and necessitated taking into account the mutual orientation of the carbon-migrant bond and the vacant C_t atom p-orbital not only in the original state but also on the way to the transition state.

The phenanthrenium model served for studying the rearrangement of the norbornyl ions too [408]:

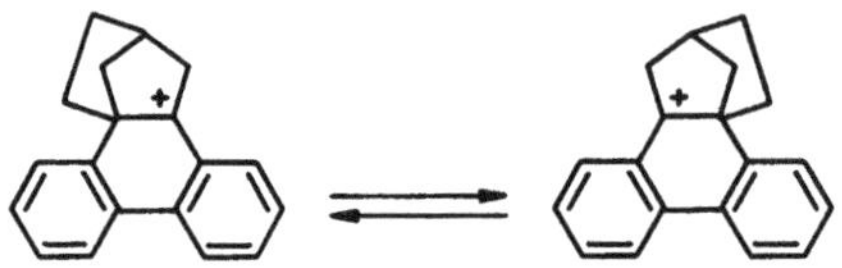

In 9-X-9,10-dimethylphenanthrenium and 1-X-1,2-3-4-5-6-hexamethylbenzenium ions the order of relative migrating ability of para- and meta-substituted phenyl groups has been conserved. The rate coefficients of degenerate rearrangements due to 1,2-shifts of Y-substituted phenyl groups in phenanthrenium ions are well described [213,572] by the equation $\log (k_y/k_H) = \varrho\sigma_y^+$ where $\varrho(-50\ ^\circ C) = -4.5$.

In hexamethylbenzenium ions the rate of the 1,2-aryl shift sharply decreases (by more than 7 orders of magnitude at $-50\ ^\circ C$), the value of the reaction constant ϱ, however, practically does not change (-4.57 ± 0.50) [571]. The order of relative migrating ability of substituted phenyl groups is also retained in rearrangements accompanying the solvolysis of β-arylalkyl systems (see [555]). The applicability of the Hammett-type equation in these cases is not surprising, since the 1,2-shifts or aryl groups can be regarded as electrophilic substitutions at the C^1 atom of aryl fragments, the carbenium centre serving as electrophilic reagent. However, the coincidence of the ϱ values for the benzenium and the phenanthrenium series throws some doubt upon the generality of the relations between the activity and the selectivity of the electrophilic agent.

The quantitative description of the migrating ability of various migrants is much more complex. The free activation energy of the 1,2-shift of the X group in ions with similar structural surrounding C_o and C_t,

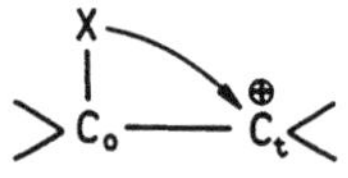

is determined by the energy losses from the lengthening of the C_o—X bond and the deformation of the valence angles upon attaining the transition state and by the energy gain from the interaction between the migrant and the carbocationic centre [581]. The variations of the first component in passing from one migrant to another can be roughly approximated by the changes in the C—X bond energy for $CH_3X(E_{CH_3X})$, and those of the second — by the changes in the affinity of CH_3X molecules to the CH_3^+ cation. Since the data on the latter affinity are scarce this parameter was replaced by the affinity to the proton (PA_{CH_3X}). In this simplified model, as the charge of the carbocationic centre (q_t^+) is usually different from unit, one can write the following relationship:

$$\Delta G^{\ddagger}_{X} = a + bE_{CH_3X} + cq_t^+ \ (d + ePA_{CH_3X})$$

where $q_t^+ = f + g\delta_c$.

For 29 ions the free activation energies of the 1,2-shift for CH_3, C_2H_5, C_6H_4Y, Cl and Br were compared with the respective values of E_{CH_3X}, PA_{CH_3X} and δ_c[59] to yield the following dependence of rearrangement on both the electronic structure of the ion skeleton (δ_c) and the migrant (E_{CH_3X}, PA_{CH_3X}):

$$\Delta G^{\ddagger}_{X}\ (-110\ ^{\circ}C) = \ + 0.313\ E_{CH_3X} - 0.00133\ PA_{CH_3X}(\delta_c - 130)\,,$$
$$r = 0.985,\ s = 0.8;$$

$$\Delta G^{\ddagger}_{X}\ (+25\ ^{\circ}C) = 0.9 + 0.329\ E_{CH_3X} - 0.00129\ PA_{CH_3X}(\delta_c - 130)$$
$$r = 0.987,\ s = 0.7\,.$$

These relationships are not valid for the free activation energy of the 1,2-hydrogen shifts which may be due to the extraordinary facility of deformation of the $H—C_o—C_t$ fragment as compared with other fragments $X—C_o—C_t$. If the force constant of the bending vibration of the $X—C_o—C_t$ fragment is used instead of E_{CH_3X} as the parameter reflecting the energy expenditure on the way to the transition state, one can obtain more general (covering X=H) relationships, but their accuracy decreases markedly [581].

For the above relationships difficulties may arise in finding the required values of PA_{CH_3X}. For $X=NO_2$ the experimental value of $PA_{CH_3NO_2}$ (178 kcal/mole) seem to refer to the addition of the proton to the oxygen, while the 1,2-shift of the NO_2 group requires the value of the proton affinity of the nitrogen. In such cases it is reasonable to estimate the value of PA by reverse calculation using the above relationships for $\Delta G^{\ddagger}_{X}$. For instance, for the 1,2-shift of the NO_2 in the 1-nitro-1,2,3,4,5,6-hexamethylbenzenium ion $\Delta G^{\ddagger}(-110\ ^{\circ}C) = 12.5$ kcal/mole [570] which at $E_{CH_3X} = 61.3$ kcal/mole [582] yields $PA_{CH_3NO_2(N)} = 104.7$ kcal/mole. With this value, the calculated value $\Delta G^{\ddagger}$ (—110 °C) = 7.5 kcal/mole for the 1,2-shift of the NO_2 group in the 9-nitro-9,10-dimethylphenanthrenium ion is in good agreement with the experimental one (7.0 kcal/mole [570]).

To estimate more reliably the applicability of the described relationships to the quantitative description of 1,2-shifts of various migrants it is necessary to do further experimental research on degenerate rearrangements of arenium ions and other types of carbocations.

The degenerate conversions revealed the regularities connecting the structure of carbocations with their aptitude to rearrange (see [555,579]). The applicability of these regularities in describing the nondegenerate 1,2-shift of substituents in arenium ions is discussed in the next section.

---

59 For ions with $X=CH_3$.

Quantum chemical analysis of the correlations between the structure of arenium ions and their aptitude to rearrangements (see [583]): Quantum chemical calculations of activation parameters for 1,2-shifts of R in 1-H-1-R-benzenium ions give so far rather approximate results. Indicative in this respect is the discrepancy of results from different methods in estimating the relative stability of the unsubstituted benzenium ion and the π-complex in which the added proton is located over one of the ring C—C bonds. Calculations on the CNDO/2 approximation [97,99] point to a greater stability of the π-structure with a three-centre bond whereas those by the CNDO/2-FK and MINDO/2 methods [100,584] and non-empirical calculations with the STO-3G and 4-31G bases [109] yield an opposite result which is in qualitative agreement with experimental data. The activation barrier of the 1,2-hydrogen shift in the benzenium ion, according to ab initio calculations must be 20.6 kcal/mole (basis 4-31G) or even 27.7 kcal/mole (STO-3G basis) [109], while the experimentally measured value for a superacidic media is 10 kcal/mole [139,282]. The MINDO/2′ calculation leads to the conclusion that the π-complex is an unstable intermediate of the 1,2-hydrogen shift in the benzenium ion [108] (cf. [99]), the difference between the calculated energies of the σ- and π-complexes amounting to about 8 kcal/mole which is rather close to the experimental value of the activation energy of 1,2-hydrogen shift in the benzenium ion.

Nonempirical calculation of the 1-H-1-fluorobenzenium ion and the π-complex in which the fluorine atom is involved in a three-centre bond with the STO-3 G basis indicates a higher stability of the π-structure while with the 4-31 G basis — that of the σ-structure [114].

The 1,2-nitro-group shift is discussed in [585].

C Kinetics of Isomerizations Involving Arenium Ions

After characterizing the rates of the 1,2-shift of hydrogen and different groups it is reasonable to return to isomerizations of substituted benzenium ions. Isomerization of methylated benzenes: In HF—SbF_5 at 25 °C o- and p-xylenes are present as ions (*105*) and (*106*) which are further practically completely converted to a conjugate acid of m-xylene, i.e. ion (*107*).

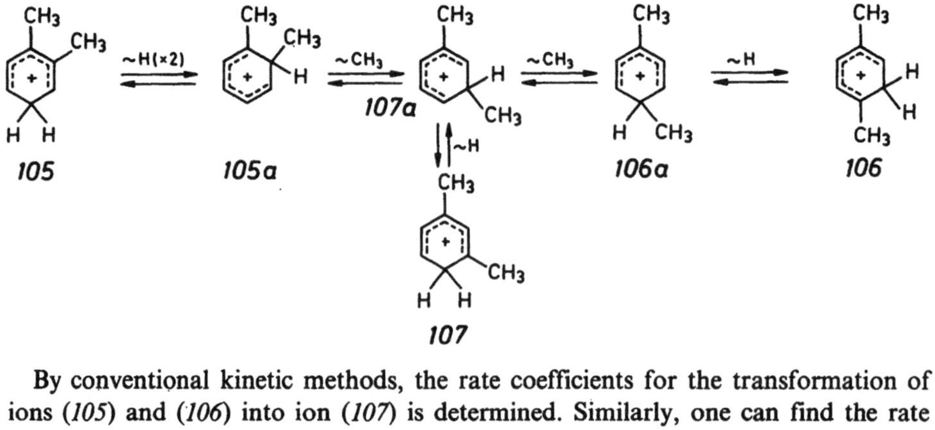

By conventional kinetic methods, the rate coefficients for the transformation of ions (*105*) and (*106*) into ion (*107*) is determined. Similarly, one can find the rate coefficients for the transformation of the trimethylbenzenium ion (*108*) into ion (*109*) and then into ion (*6*):

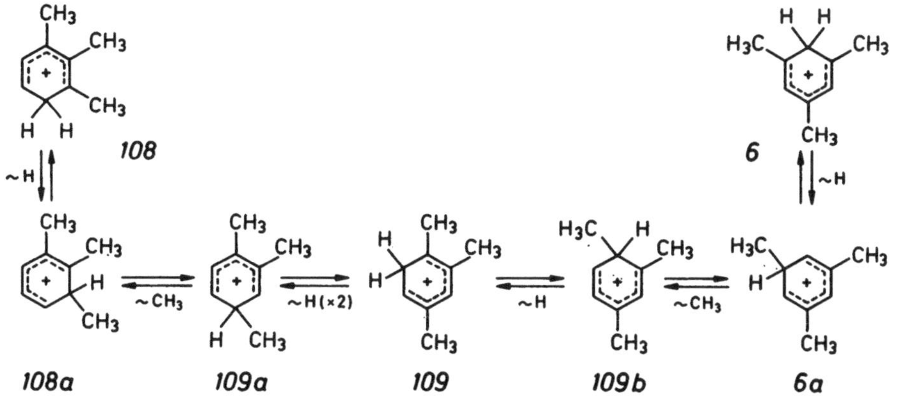

and for that of the tetramethylbenzenium ions (*110*) and (*103*) into ion (*111*):

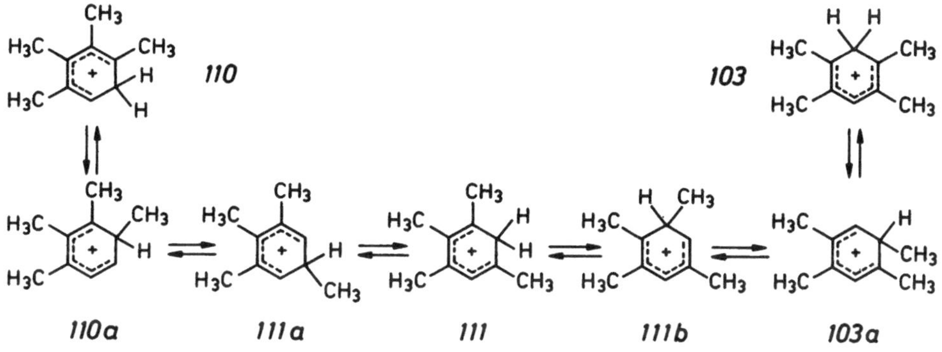

The rate coefficients for most of the mentioned transformations are listed in Table 53 [512].

The data obtained from the degenerate rearrangements of benzenium ions testify that the rate coefficients of intraionic 1,2-hydrogen shifts are usually by several

Table 53. Kinetic and Thermodynamic Parameters of Methylbenzenium Ion Rearrangements (25 °C) [512]

Reaction	10^5 k (s^{-1})	ΔG* kcal/mole	Stage of 1,2-CH_3-shift	10^5 k_{CH_3} (s^{-1})	$\Delta G^{\ddagger}_{CH_3}$ kcal/mole	ΔG_{CH_3} kcal/mole
(105)→(107)	80	21.7	(105a)→(107a)	13000	18.7	+2.2
(106)→(107)	335	20.8	(106a)→(107a)	13000	18.7	+2.7
(108)→(109)	8.9	23.0	(108a)→(109a)	17000	18.5	+1.9
(109)→(6)	1.39	24.1	(109b)→(6a)	65000	17.7	−2.2
(110)→(111)	155	21.3	(110a)→(111a)	65	21.8	+6.5

orders of magnitude higher than those for CH_3 groups. Consequently in the transformations discussed the rate-determining step is that of the 1,2-shift of a CH_3 group. Their rate coefficients can be determined from the measured rate coefficient of reaction by estimating the equilibrium fractions of methylbenzenium ions in which the 1,2-CH_3-group shift occurs.

Assuming the effect of methyl groups on the ring carbon basicity of methylbenzenes to be additive and using measured basicities D. Brouwer [512)] estimated the increments of CH_3 located at positions ipso, ortho, meta and para to the site of protonation[60]; he then calculated the equilibriums of methylbenzenium ions and thereby evaluated the kinetic parameters of the 1,2-CH_3-group shift for isomerization of methylbenzenes (see Table 53). He established that the free activation energy of the 1,2-CH_3-group-shift ($\Delta G^{\ddagger}_{CH_3}$) changes parallel with the change in the free energy of this step (ΔG_{CH_3}), i.e., it is determined by the difference in thermodynamic stabilities of the rearranging ion and the formed one. A change in the free energy of the 1,2-shift by 1 kcal/mole causes a change in that of activation by about 0.5 kcal/mole.

This relation is explained by V. Koptyug and co-workers [296, 302, 511, 514, 515, 531, 586)] who founded the quantitative theory of isomeric conversions for aromatic compounds. The transfer of substituents in the aromatic ring usually proceed with acid catalysts and permit the derivatives of aromatic compounds obtained from coal and oil processing as well as by electrophilic substitutions to be converted into isomers hardly accessible by other means [13, 14, 17, 587, 588)]. Numerous investigations have clearly elucidated the mechanism of these reactions. The substituent transfer in the aromatic ring turned out to be feasible both intra- and intermolecularly. In either case the reaction starts by adding the proton to the ring carbon bonded to an R substituent, to form 1-H-1-R-arenium ion. The 1,2-shift of the R substituent and the subsequent elimination of the proton correspond to the *intramolecular* mechanism of isomerization (route 1) while the elimination of R as R^+ (route 2) and transferring it to a neutral aromatic molecule (route 3) in combination with reverse processes provide the *intermolecular* course of the reaction

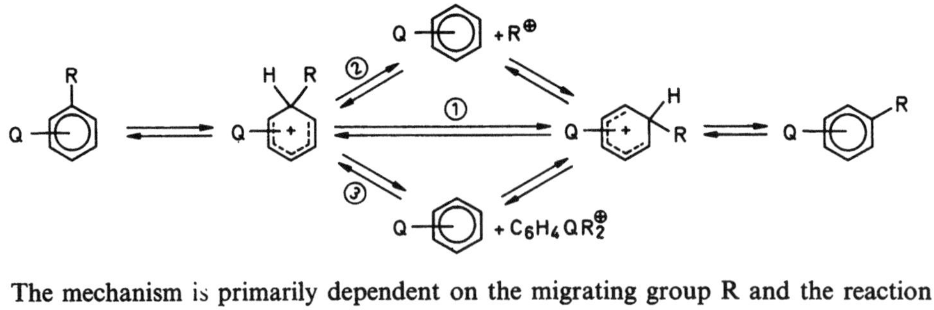

The mechanism is primarily dependent on the migrating group R and the reaction condition.

Intramolecular isomerization of a substituted benzene A: In general the compound A being isomerized is protonated to yield arenium ions AH_i^+ differing in the site (i) of proton attachment. Of these only the ion AH_r^+ with the proton added to the

60 These increments differ from the respective $\rho\sigma^+$ values calculated by the way cited in Sect. IV.1.A.

R-substituted ring carbon can be rearranged by the 1,2-shift of this substituent into the ion BH_r^+; due to the transfer and removal of the proton, the equilibrium is established with the other ions of the BH_i^+ set and the corresponding neutral compound B.

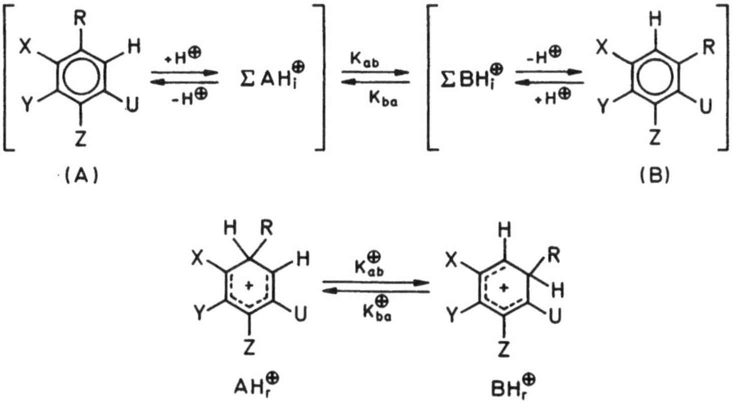

The measured isomerization rate coefficients k_{ab} and the respective free activation energy $(\Delta G^{\neq})_{ab}$ refer to the conversion of both particles A and AH_i^+ (though in fact it is only the ion AH_r^+ that is rearranged) into that of B and BH_i^+. Since the intra- and intermolecular proton transfers are realized at high rates (see the preceding sections), one can assume the rate-determining step of the isomerization with strong acid catalysts to be that of the 1,2-shift of R in the AH_r^+ ion. In this case the rate of the process as a whole is determined by the degree of protonation of A under reaction conditions (δ_a), by the equilibrium fraction (g_{ar}) of the ion AH_r^+ among the AH_i^+ ions differing in the site of protonation, and by the kinetic parameters of the 1,2-shift of R in the AH_r^+ ion. Each of these factors contributes to the k_{ab} rate coefficient being measured and to the respective value of $(\Delta G^{\neq})_{ab}$. Represent the apparent free activation energy for the isomerization as the sum of the contributions:

$$(\Delta G^{\neq})_{ab} = \Sigma N_i$$

The degree of protonation and the equilibrium fraction of the rearranging ion are taken into account by the term N_1:

$$N_1 = -2.3RT \log \delta_a g_{ar}$$

where

$$\delta_a = \frac{\Sigma[AH_i^+]}{[A] + \Sigma[AH_i^+]} \quad \text{and} \quad g_{ar} = \frac{[AH_r^+]}{\Sigma[AH_i^+]}$$

$\log g_{ar}$ can be calculated according to the $\varrho\sigma^+$ approach described in Sect. IV.1.A.

The kinetic parameters of the 1,2-R-shift in the AH_r^+ ion, if steric factors are neglected, must depend, as shown in Sect. IV.2.B, on the charge characteristics of the ion and on the difference between the thermodynamic stability of the ion being rearranged, and that of the formed one. According to R. Marcus [589, 591] the free activation energy of the elementary stage is related to the difference between the free energies of parent particles and those of formed ones by the following equation[61]

$$\Delta G^{\neq} = \Lambda\left(1 + \frac{\Delta G^{\circ}}{4\Lambda}\right)^2$$

The parameter Λ reflects the "intrinsic reaction barrier". It coincides with the free activation energy for degenerate conversions ($\Lambda = \Delta G^{\neq}$ at $\Delta G^{\circ} = 0$). If the quadratic term in the Marcus equation is neglected it takes the form

$$\Delta G^{\neq} = \Lambda + 0.5\,\Delta G^{\circ}\,,$$

bringing out more clearly (at $\Lambda \approx$ const) the linear relation found out by Brouwer between the free activation energy of the intramolecular 1,2-CH_3-shift and the difference in the free energies between the rearranging and the forming ions.

As shown in Sect. IV.2.B the activation barrier of the degenerate 1,2-shift of R in structurally similar ions is determined by the electron deficiency (by the value of positive charge q) on the atom to which the R group is transferred, i.e.,

$$\Lambda^R = C_1 + C_2 q$$

According to R. Marcus [590)] the intrinsic barrier of a nondegenerate process (Λ_{ij}) can be represented as the arithmetical mean of the respective degenerate ones:

$$\Lambda_{ij} = 0.5(\Lambda_{ii} + \Lambda_{jj})\,.$$

Then the term N_2 considering the effect of the charge on the intrinsic barrier of $AH_r^+ \rightarrow BH_r^+$ interconversion can be written as

$$N_2 = \Lambda_{ab}^R = C_1^R + 0.5C_2^R(q_a + q_b)\,,$$

where q_a and q_b are positive charges in the ortho positions of AH_r^+ and BH_r^+ ions to which the group R is shifted, and C_1^R and C_2^R are coefficients depending on the group R estimated from the equation described in Sect. IV.2.B.

A third term accounts for the difference in the thermodynamic stability of the rearranging and forming ions:

$$N_3 = 0.5(\Delta G^{\circ})_{ab}^+ = -0.5(2.3RT\log K_{ba}^+)$$

where K_{ba}^+ is a $BH_r^+ - AH_r^+$ equilibrium constant calculated by the $\varrho\sigma^+$-approach if the constant of (or the respective thermodynamic characteristics for) the equilibrium between neutral compounds A and B is known.

61 It is not clear how general this equation is and whether it is applicable to reactions involving the transfer of atom groups; however, in our case the introduction of the "intrinsic reaction barrier" is essential rather than the form of this dependence.

Finally, the symmetry of the forming and the rearranging ions is introduced in the term N_4:

$$N_4 = -0.5(2.3RT \log s_{ar}s_{br})$$

where s_{ar} is the statistical factor equal to the number of ortho carbon atoms in the AH_r^+ ion to which the R substituent is shifted to form the BH_r^+ ion; $s_{ar} = 2$ at X = H and Y = V; in the other cases $s_{ar} = 1$ (similarly s_{br} refers to the conversion of BH_r^+ into AH_r^+).

Predicting activation barriers and rate coefficients of isomerization reactions: For the degenerate rearrangements of arenium ions realized by the 1,2-shift of the CH_3 group, according to [410]:

$$\Delta G^{\neq}\ (25\ °C, \text{cal/mole}) = 53700 - 181\delta_c$$

where δ_c is the chemical shift of the carbon atom to which the CH_3 group is shifted. Assuming the change in δ_c for sp^2-hybridized carbon atoms to be about 180 ppm/charge unit [299] we find the $C_2^{CH_3}$ coefficient for the N_2 term to be $-3.3 \cdot 10^4$. Further, knowing that for the heptamethylbenzenium ion $\Delta G^{\neq}$ (25 °C) = 18100 cal/mole [279] and the calculated[62] value of π-electron deficiency $q_{ortho} = 0.248$ we find for N_2 the coefficient $C_1^{CH_3} \approx 2.6 \cdot 10^4$. The obtained values of $C_1^{CH_3}$ and $C_2^{CH_3}$ refer to 25 °C. For a first approximation, however, these, just as the values of $\Delta G^{\neq}$ can be considered independent of temperature.

According to the data from the preceding sections the constants of equilibrium between isomeric carbenium ions and the rates of their rearrangements are weakly sensitive to changes in the character of the acid medium. Therefore, the only parameter greatly changing the value of $\Delta G^{\neq}_{ab}$ with changing medium acidity is the protonation of the parent compound (δ_a). The discussion will be confined to isomerization under complete protonation of parent compounds ($\delta_a = 1$).
Isomerization of p-xylene into m-isomer:

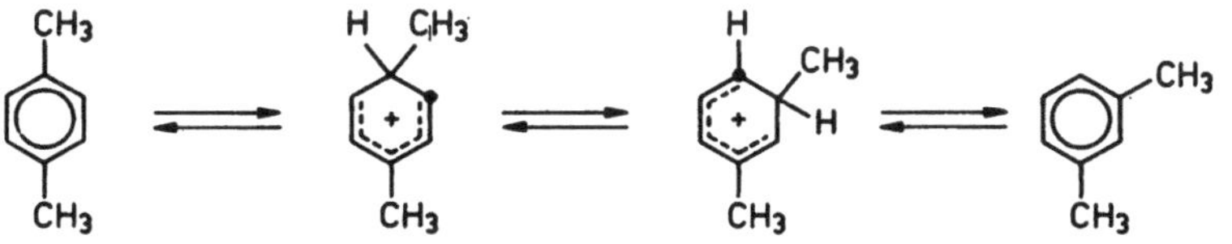

Using ϱ (25 °C) = −10.2 and the σ^+-constants for the CH_3 group listed in Table 44 and considering that for statistical reasons the protonation at an unsubstituted position of p-xylene is twice as high as that at a substituted one we find:

$$N_1 = -2.3RT[\varrho(\sigma_i^+ + \sigma_p^+ - \sigma_o^+ - \sigma_m^+) - \log 2] = 1980\ \text{cal/mole}.$$

Calculation by the CNDO/2 method for the closed-circle-marked positions of the rearranging and the forming ions yields the values $q_a^\pi = 0.263$ and $q_b^\pi = 0.269$. This

62 Cited here and further are the values of q^π calculated by the CNDO/2 method.

makes it possible to determine the term N_2:

$$N_2 = 2.6 \cdot 10^4 - 3.3 \cdot 10^4 \cdot 0.266 = 17220 \text{ cal/mole}$$

From the equilibrium constant between the m- and the p-xylenes (K_{ba}) at 25 °C equal to 2.50 [512] we find N_3:

$$N_3 = -1.15RT[\varrho(\sigma_m^+ - \sigma_p^+) + \log K_{ba}] = 1420 \text{ cal/mole}$$

Since the parent ion can rearrange in two ways to equivalent ortho positions and the reverse conversion of the forming ion only in one way, we have

$$N_4 = -1.15RT \log 2 = -210 \text{ cal/mole}$$

Thus, from the calculation performed it is to be expected that for the isomerization of p-xylene into m-xylene under complete protonation at 25 °C the apparent free activation energy should be $\Delta G^{\neq} = \Sigma N_i = 20.4$ kcal/mole which corresponds to $\log k\ (s^{-1}) = -2.2$. Experimental measurements in $HF-SbF_5$ at 25 °C give $\log k\ (s^{-1}) = -2.47$.

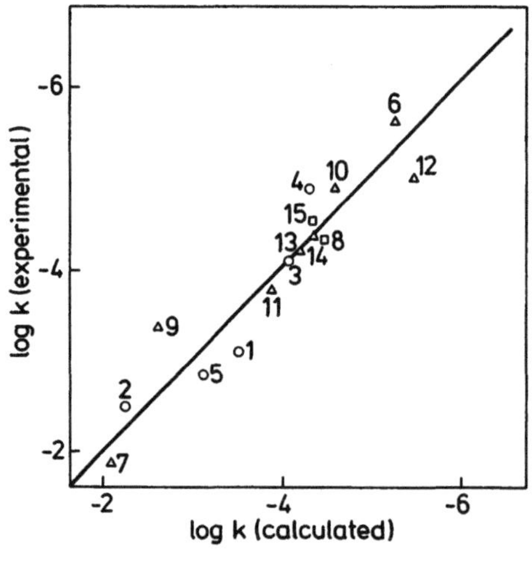

Fig. 20. Comparison between the experimental and calculated rate constants for isomerization of substituted benzenes under complete protonation:
1) $1,2\text{-}(CH_3)_2 \rightarrow 1,3\text{-}(CH_3)_2$; 2) $1,4\text{-}(CH_3)_2 \rightarrow 1,3\text{-}(CH_3)_2$; 3) $1,2,3\text{-}(CH_3)_3 \rightarrow 1,2,4\text{-}(CH_3)_3$; 4) $1,2,4\text{-}(CH_3)_3 \rightarrow 1,3,5\text{-}(CH_3)_3$; 5) $1,2,3,4\text{-}(CH_3)_4 \rightarrow 1,2,3,5\text{-}(CH_3)_4$; 6) $1\text{-F-}2\text{-}CH_3 \rightarrow 1\text{-F-}3\text{-}CH_3$; 7) $1\text{-F-}4\text{-}CH_3 \rightarrow 1\text{-F-}3\text{-}CH_3$; 8) $4\text{-F-}1,2\text{-}(CH_3)_2 \rightarrow 4\text{-F-}1,3\text{-}(CH_3)_2$; 9) $4\text{-F-}1,3\text{-}(CH_3)_2 \rightarrow 2\text{-F-}1,4\text{-}(CH_3)_2$; 10) $2\text{-F-}1,4\text{-}(CH_3)_2 \rightarrow 4\text{-F-}1,3\text{-}(CH_3)_2$; 11) $4\text{-F-}1,3\text{-}(CH_3)_2 \rightarrow 3\text{-F-}1,2\text{-}(CH_3)_2$; 12) $3\text{-F-}1,2\text{-}(CH_3)_2 \rightarrow 4\text{-F-}1,3\text{-}(CH_3)_2$; 13) $2\text{-F-}1,4\text{-}(CH_3)_2 \rightarrow 5\text{-F-}1,3\text{-}(CH_3)_2$; 14) $1\text{-OH-}4\text{-}CH_3 \rightarrow 1\text{-OH-}3\text{-}CH_3$; 15) $1\text{-OH-}3,4\text{-}(CH_3)_2 \rightarrow 1\text{-OH-}3,5\text{-}(CH_3)_2$.
Points 1–12 are for 25 °C; point 13 is for 62 °C; point 14 is for 81 °C; point 15 is for 100 °C. The straight line is drawn at an angle of 45°

On the whole the coincidence of predicted and measured kinetic characteristics of isomeric benzene derivative conversions realized through intramolecular 1,2-CH_3-shift can be judged by the data presented in Fig. 20. They cover the isomerization of methylated benzenes [511) 63] and fluoro-substituted methylbenzenes [514, 531)] in HF—SbF_5 as well as methylated phenols [296)] in HF—BF_3. The suggested approach permits the isomerization rate coefficients to be predicted with an accuracy of about 0.5 logarithmic units.

This approach to the quantitative description of isomeric conversions with acid catalysts reveals the role of each of the factors affecting the kinetic parameters and allows one to understand the reasons for the great differences in isomerization rates of different compounds.

Consider, e.g., the slower isomerization of methylated phenols as compared with methylbenzenes [296)]. The equilibrium between individual xylenes with 0.5 mole of $AlCl_3$ in presence of hydrogen chloride is established at 50 °C within 20–25 hours [592)]. The xylenes are isomerized still more readily (due to a higher degree of protonation) in the stronger acid HF—SbF_5 [512)]. Cresols are isomerized with 2 moles of $AlCl_3$ at 120–130 °C [13, 593)]. The reason for the low isomerization rate for methylphenols becomes clear when the values of free activation energies are compared (Table 54). The higher values of the apparent free activation energy for methylphenols as compared to methylbenzenes are mainly due to N_3: the thermodynamic disadvantage of the rearrangement of the ion with the para-hydroxy group into that with the meta-hydroxy group being unable to effectively delocalize the positive charge of the ion. As a result, the energy barrier for the isomerization of p-cresol and 3,4-xylenol proves to be about 5 kcal/mole greater than for the respective methylbenzenes.

The quantitative description [296, 511, 514, 515, 531)] of isomerizations of the benzene derivatives can be extended to other aromatic systems as well. This was demonstrated by isomerizing methyl-substituted naphthalenes [586)]. As it is difficult to determine a complete set of $\sigma^+_{i\text{-}CH_3}$ constants for the naphthalene ring, the increments of the effect of i-CH_3 groups on the relative stability of naphthalenium ions were estimated from experimental data referring to isotopic hydrogen exchange reactions. The predicted values of the free activation energy and the respective experimental values (HF—BF_3, 28 °C [594)]) for mono- and dimethylnaphthalenes under complete protonation are presented in Table 55.

Table 54. Calculated Components of Free Activation Energies for Isomerization of Methylbenzenes and Methylphenols (δ_a = 1; 100 °C) [296)]

Reaction	N_1	N_2	N_3	N_4	$\Delta G^* = \Sigma N_i$ kcal/mole
1,4-$(CH_3)_2C_6H_4$ → 1,3-$(CH_3)_2C_6H_4$	2.11	17.22	1.36	−0.26	20.4
4-$CH_3C_6H_4OH$ → 3-$CH_3C_6H_4OH$	1.35	17.42	7.10	−0.26	25.6
1,2,4-$(CH_3)_3C_6H_3$ → 1,3,5-$(CH_3)_3C_6H_3$	2.62	18.20	2.69	−0.26	23.3
3,4-$(CH_3)_2C_6H_3OH$ → 3,5-$(CH_3)_2C_6H_3OH$	1.84	18.33	8.34	0	28.5

63 In a similar manner the isomerization of toluene-1-^{14}C [600,601)] has been analysed.

Table 55. Estimated and Experimental Values of Free Activation Energy for Isomeric Transformations of Mono- and Dimethylnaphthalenes ($\delta_a = 1$)[586]

Reaction	N_1	N_2	N_3	Estimated ΔG^{*} (0 °C) $= \Sigma N_i$ kcal/mole	Experimental ΔG^{*} (28 °C) kcal/mole[a]
$1\text{-}CH_3 \rightarrow 2\text{-}CH_3C_{10}H_7$	6.1	14.9	+1.0	22.0	21.8
$2\text{-}CH_3 \rightarrow 1\text{-}CH_3C_{10}H_7$	9.1	14.9	−1.0	23.0	23.0
$1,4\text{-}(CH_3)_2 \rightarrow 1,3\text{-}(CH_3)_2C_{10}H_6$	1.3	15.3	+1.8	18.4	18.3 (−12 °C)
$1,3\text{-}(CH_3)_2 \rightarrow 1,4\text{-}(CH_3)_2C_{10}H_6$	12.3	15.3	−1.8	25.8	—
$1,5\text{-}(CH_3)_2 \rightarrow 1,6\text{-}(CH_3)_2C_{10}H_6$	5.9	15.0	+0.9	21.8	22.7
$1,6\text{-}(CH_3)_2 \rightarrow 1,5\text{-}(CH_3)_2C_{10}H_6$	9.7	15.0	−0.9	23.8	24.5
$1,6\text{-}(CH_3)_2 \rightarrow 2,6\text{-}(CH_3)_2C_{10}H_6$	7.2	15.2	0.0	22.4	23.0
$2,6\text{-}(CH_3)_2 \rightarrow 1,6\text{-}(CH_3)_2C_{10}H_6$	7.1	15.2	0.0	22.3	22.6
$1,7\text{-}(CH_3)_2 \rightarrow 2,7\text{-}(CH_3)_2C_{10}H_6$	6.2	15.3	+1.2	22.7	22.7
$2,7\text{-}(CH_3)_2 \rightarrow 1,7\text{-}(CH_3)_2C_{10}H_6$	9.6	15.3	−1.2	23.7	24.0
$2,3\text{-}(CH_3)_2 \rightarrow 1,3\text{-}(CH_3)_2C_{10}H_6$	8.9	15.8	−1.2	23.5	22.1
$1,3\text{-}(CH_3)_2 \rightarrow 2,3\text{-}(CH_3)_2C_{10}H_6$	9.4	15.8	+1.2	26.4	—

[a] Calculated according to $\Delta G^{*} = 2.3\,RT\left(\log\frac{\varkappa T}{h} - \log k\right)$ using experimental values[594] of rate coefficient (k)

As follows from these data the main factor affecting ΔG^{*} for isomerization of the mono- and dimethylnaphthalenes is the term N_1 reflecting the difference in the free energies of the rearranging AH_r^+ ion and of the most stable isomer of the ions AH_i^+ differing in the site of proton addition. Appreciable, but smaller changes to the values of ΔG^{*} are introduced by the difference in the free energies of the rearranging and forming ions (term N_3). As shown by calculations, this term accounts for the experimental difficulty in the β-β′-transfers of the methyl group[594, 595]. Thus, for the degenerate conversion of $2\text{-}CH_3C_{10}H_7$ into $3\text{-}CH_3C_{10}H_7$ the calculation yields $N_1 = 9.1$, $N_2 = 22.6$, $N_3 = 0$ and ΔG^{*} (0 °C) ≈ 32 kcal/mole[586]. The high value of N_2 is due to the low value of positive charge ($q_a = q_b = 0.103$) at that position of the AH_r^+ ion to which the CH_3 group is shifted. The above value of N_2 is very close to the experimentally estimated value of the free activation energy for the degenerate rearrangement of the 1,2,2,3,4-pentamethylnaphthalenium ion — ΔG^{*} (−125 °C) = 22.7 kcal/mole[210].

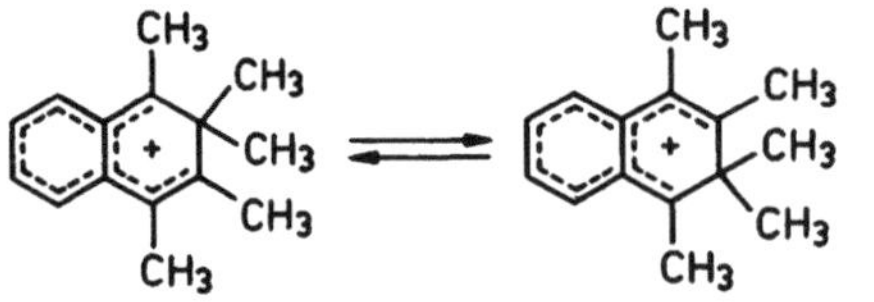

Sterical factors can be taken into account in analyzing the isomerizations of substituted aromatic compounds too; this can be demonstrated, e.g., by considering methylated naphthalenes with peri-methyl groups. Isomerization of peri-substituted naphthalenes is characterized by two features[13]: by an anomalous facility of the reaction and by an apparent irreversibility of the process. These peculiarities are explained by the steric strain release in (*112*) and in the peri-substituent (Y) shift to the neighbouring β-position:

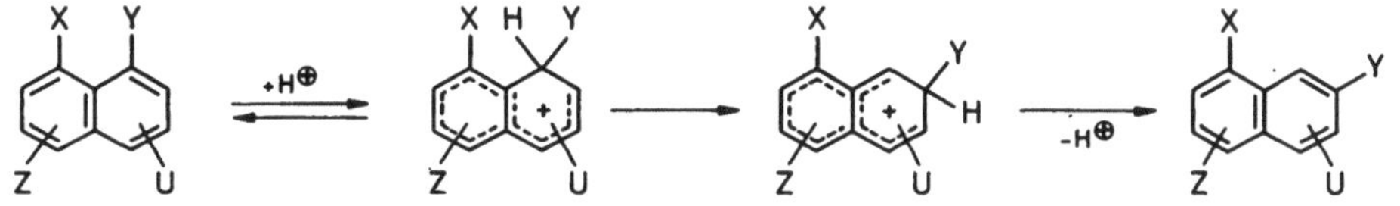

112

The Japanese investigators [596] who have studied the isomerization of peri-substituted polymethylnaphthalenes in CF_3COOH have failed to explain many peculiar effects of the number and position of methyl groups on the kinetic parameters.

This problem is discussed in terms described above [302]. A decrease in the steric strain upon formation of (*112*) naturally leads to an increase of its equilibrium fraction among isomeric ions, differing in the site of proton attachment. According to available data [597, 598], the steric strain of 1,8-dimethylnaphthalene resulting from the Van der Waals interaction of peri-CH_3 groups amounts to 6 kcal/mole. The addition of the proton to atom C_1 or C_8 essentially decreases the steric interaction of methyl groups, but it can hardly be completely removed. The extent of reducing the steric strain can be estimated when peri-substituted polymethylnaphthalene are protonated, as the NMR spectra show, to form isomeric ions, one with the proton added to a peri-substituted carbon atom and the other to some other site. This occurs, e.g., in the protonation of 1,4,5-trimethylnaphthalene [302]. In $HSO_3F—SO_2ClF$ at —80 °C this compound yields ions protonated at the 4- and 8-positions in the equilibrium ratio of 4:1.

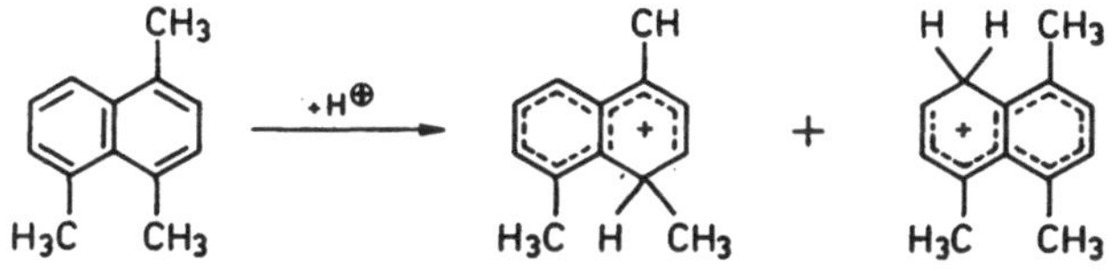

113a *113b*

Using the electronic effect of CH_3 groups in different positions of the naphthalene ring on the basicity one can show, that ion (*113b*) should be 3.1 kcal/mol (ΔΔG) more stable than ion (*113a*). The observed equilibrium gives $\Delta\Delta G = —0.7$ kcal/mole. Thus, the change of the free energy due to the steric interaction of CH_3 groups amounts to about 3.8 kcal/mole. A knowledge of this value in addition to the increments of electronic effect of methyl groups in the naphthalene ring makes it possible to estimate the equilibrium fraction of (*112*)-type ions among other ions differing in the site of proton attachment which is necessary for calculation of N_1.

The calculated values of N_i terms [302] for the isomerization of peri-substituted polymethylnaphthalenes under complete protonation ($\delta_a = 1$) are presented in Table 56 together with the experimental values of $\Delta G^{\neq}$ (at $\delta_a = 1$) for the conversions of 1,8-di-, 1,4,5-tri- and 1,2,3,4,5-pentamethylnaphthalenes. The differences in the relative ability to isomerize for various peri-substituted polymethylnaphthalenes

Table 56. Estimated and experimental values of free activation energy for isomeric transformations of peri-substituted polymethylnaphthalenes [302]

Reaction	N_1	N_2	N_3	$\Delta G^{\neq}$ (0 °C) ΣN_i, $\delta_a = 1$	$\Delta G^{\neq}_{exp}$ (0 °C) $\delta_a = 1$	$-2.3\,RT$ lg δ_a	$\Delta G^{\neq}_{calc}$ (60 °C) CF_3COOH	$\Delta G^{\neq}_{exp}$ (60 °C)[a] CF_3COOH
$1\text{-}CH_3 \rightarrow 2\text{-}CH_3C_{10}H_7$	6.1	14.9	+1.1	22.1	21.8	14.9	37.0	
$1,8\text{-}(CH_3)_2 \rightarrow 1,7\text{-}(CH_3)_2C_{10}H_6$	2.6	15.0	−0.2	17.4	16.2	13.6	31.0	29.6
$1,4,5\text{-}(CH_3)_3 \rightarrow 1,3,5\text{-}(CH_3)_3C_{10}H_5$	0.1	15.3	+1.2	16.6	16.5	12.7	29.3	
$1,4,5,8\text{-}(CH_3)_4 \rightarrow 1,4,5,7\text{-}(CH_3)_4C_{10}H_4$	0.0	15.6	−0.2	15.4		10.9	26.3	25.0
$1,3,5,8\text{-}(CH_3)_4 \rightarrow 2,3,5,8\text{-}(CH_3)_4C_{10}H_4$	5.6	16.0	−0.3	21.3		8.2	29.5	28.2
$1,3,5,8\text{-}(CH_3)_4 \rightarrow 1,3,5,7\text{-}(CH_3)_4C_{10}H_4$	3.1	15.6	+0.2	18.8		8.2	27.0	26.6
$1,3,6,8\text{-}(CH_3)_4 \rightarrow 1,3,6,7\text{-}(CH_3)_4C_{10}H_4$	6.5	16.1	−1.3	21.3		7.1	28.4	28.2
$1,2,3,4,5\text{-}(CH_3)_5 \rightarrow 1,2,3,4,6\text{-}(CH_3)_5C_{10}H_3$	9.0	15.7	−0.9	23.8	23.9 (60 °C)	3.3	27.1	27.2 (70 °C)

[a] Calculated according to $\Delta G^{\neq} = 2.3\,RT\left(\log \frac{\varkappa T}{h} - \log k\right)$ using experimental values [596] of rate coefficients (k)

under complete protonation are mainly due to N_1 (see above). Thus, a partial release of steric strain by adding the proton to one of the peri-substituted carbon atoms of 1,8-dimethylnaphthalene increases considerably the equilibrium fraction of the ion responsible for the isomerization. Such a situation, however, does not remain unchanged when CH_3 groups are accumulated. For example, in 1,3,5,8- and 1,3,6,8-tetramethylnaphthalenes the additional methyl groups increase the basicity of unsubstituted positions. Inspite of reducing the steric strain under complete protonation the isomerization rates must be comparable to that of 1-methylnaphthalene, and the conversion of 1,2,3,4,5-pentamethylnaphthalene into 1,2,3,4,6-isomer must proceed less readily than that of 1-methylnaphthalene. This is experimentally confirmed.

In weak acids when the isomerized compounds fail to be completely protonated the analysis is complicated by the difference in the protonation of parent compounds; here, the reaction conditions may bring about the reversion of the aptitude to isomeric conversions. Thus, in CF_3COOH at 60 °C the rate of conversion of 1,2,3,4,5-pentamethylnaphthalene into 1,2,3,4,6-isomer is by about 10^4 times higher than for the conversion of 1,4- into 1,3-dimethylnaphthalene [596] whereas under complete protonation the picture is exactly opposite.

To estimate the kinetic parameters of isomerization for aromatic compounds under a small extent of protonation and an excess of acidic agent one can assume for a first approximation that

$$-2.3\ RT \log \delta_a = A + B \log K_a^{\Sigma}$$

where K_a^{Σ} is the basicity constant of the isomerized compound (for all sites of protonation) calculated from the additive substituent effect. Assuming, further, that the difference in the free activation energy of isomeric conversion for complete protonation and for a weak acid is equal to $-2.3\ RT \log \delta_a$ one can determine from the activation parameters for two compounds the A and B coefficients and calculate the activation parameters of isomerization for the other compounds. Thus, for methylated naphthalenes (see Table 56) in CF_3COOH at 60 °C the values of $\Delta G^{\ne}$ for conversion of 1,8-dimethyl- and 1,2,3,4,5-pentamethylnaphthalenes under complete protonation (CF_3SO_3H, $H_0 = -13.0$ [133]) and in CF_3COOH ($H_0 = -3.0$ [134]) were used [303] to obtain $A = 12.0$ and $B' = -1.7$ which were then used to calculate the apparent free activation energy for the other methylnaphthalenes in CF_3COOH. The values of $\Delta G^{\ne}$ obtained in this way agree sufficiently well with experimental data (see the last two columns of Table 56).

As the naphthalene molecule accumulates methyl groups and as, due to this, the basicity increases the values of $-2.3\ RT \log \delta_a$ decreases and so does the difference in the isomerization rates in strong and weak acids. Thus, for the isomerization of 1-methylnaphthalene the free activation energy, according to calculations, increases from 22.1 kcal/mole under complete protonation to 37 kcal/mole in CF_3COOH (due to the small extent of protonation) making this reaction practically impossible. At the same time for isomerization of 1,2,3,4,5-pentamethylnaphthalene the difference in the apparent free activation energy for the two types of experiments is only about 3 kcal/mole.

The above analysis of aromatic compounds with migration of CH_3-group demonstrates the wide use of the information accumulated in the study of arenium ions to describe reactions with their participation.

Dienone-phenol rearrangement: Consider, e.g. the data obtained from a detailed study of acidcatalyzed conversion of 4,4-dimethylcyclohexa-2,5-dienone into 3,4-dimethylphenol:

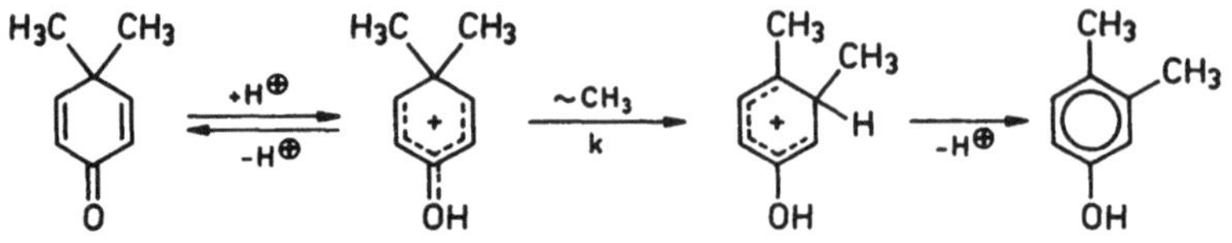

114 115

Deuteration of the methyl group hinders ist shift [599]. Introduction of deuterium at the 3(5)-position of the parent dienone does not decelerate the reaction in 55.2 and 97.3% H_2SO_4 [425,602]. Hence, the last stage (proton removal) does not affect the observed reaction rate. In this case the dependence of the reaction rate on the medium acidity is to be determined by the equation

$$k_{obs.} = \frac{[BH^+]}{[B] + [BH^+]} \cdot \frac{f_{BH^+}}{f_{\ddagger}} \cdot k_0$$

where [B] and $[BH^+]$ are the concentrations of the free and the O-protonated dienone; f_{BH^+} and $f_{\ddagger}$ are the activity coefficients of the protonated form of dienone and of the transition state for 1,2-CH_3-shift.

The measurements made in H_2SO_4 of different concentrations at 25 °C [228,602] have shown that the rate coefficient of the 1,2-CH_3-shift at the 1,1-dimethyl-4-hydroxybenzenium ion

$$k = \frac{f_{BH^+}}{f_{\ddagger}} k_0$$

grows with growing acid concentration reaching a definite limit and then (in 90–100% H_2SO_4) remains constant. The dependence of the ratio $f_{BH^+}/f_{\ddagger}$ on the medium acidity is most probably due to the fact that the hydroxybenzenium ion (*114*) is solvated by a larger number of water molecules than the transition state on the way to ion (*115*) [228,602][64]. Ion (*115*) must be a weaker acid than ion (*114*) and so its hydroxy group forms weaker hydrogen bonds with water molecules. With rising acidity the activity of water molecules falls and f_{BH^+} grows faster than $f_{\ddagger}$; with water molecules of very low activity the solvation factor becomes inessential and the $f_{BH^+}/f_{\ddagger}$ ratio may be considered constant.

The rearrangement of 4,4-dimethyl-, 3,4,4-trimethyl- and 3,4,4-triethylcyclohexa-2,5-dienones in aqueous H_2SO_4 at 25 °C showed a similar growth [226] for the isomerization rate coefficient ions with rising medium acidity, a linear dependence of the logarithm of the rate coefficient on the acidity function (H_0 of H_A) being observed within the studied range of acidity. More recently the kinetics were

64 For the solvation factors in acid-catalyzed conversions of carbonyl compounds see [603].

measured for a number of other alkyl-substituted 4,4-dimethylcyclohexa-2,5-dienones [230,232,604].

The basicity of dienones and the kinetic parameters of rearrangement for hydroxybenzenium ions from the data obtained in media of various acidity are discussed in detail in [229].

These data show that for determining the kinetic parameters of the 1,2-substituent shifts sufficiently strong acids are desirable when the effect of solvation becomes insignificant and the rate coefficient of the 1,2-shift is practically independent of the medium acidity. Thus far only a few measurements meet this requirement.

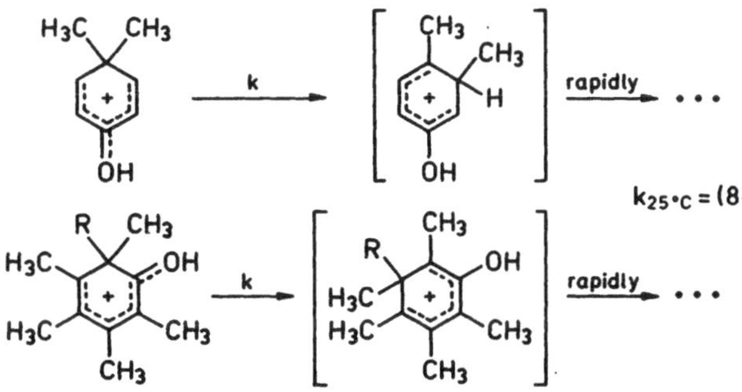

$k_{25°C} = (8-9) \cdot 10^{-4} s^{-1}$ (conc. H_2SO_4) [228,425]

$R = CH_3$, $k_{28.4°C} = 1.04 \cdot 10^{-5} s^{-1}$ ($CF_3COOH-H_2SO_4$) [330], cf. [332][65]

$R = CH_2C_6H_5$, $k_{-50°C} = 2 \cdot 10^{-4} s^{-1}$ (HSO_3F-SO_2ClF) [244]

$R = C_2H_5$, $k_{28.4°C} = 1.86 \cdot 10^{-4} s^{-1}$ ($CF_3COOH-H_2SO_4$) [330]

$R = C_6H_5$, $k_{28.5°C} = 5.43 \cdot 10^{-5} s^{-1}$ ($CF_3COOH-H_2SO_4$) [330]

The relative migration ability of CH_3, C_2H_5 and C_6H_5 groups is 1:36:10 (a detailed analysis of rearrangement kinetics and the temperature dependence of rate coefficients see [330]).

Investigations were undertaken to uncover the effect of structural factors on the rate of the rate-determining step for the dienone-phenol rearrangement. The logarithm of the rate coefficient for the 1,2-shift of the CH_3 group in O-protonated alkyl- and halogen-substituted 4,4-dimethylcyclohexa-2,5-dienones is connected by a rough linear relation [605] with the difference of the sums of σ^+-constants of substituents (Q) of the forming (BH^+) and the rearranging (AH^+) hydroxybenzenium ions (80% H_2SO_4, 25 °C):

$$\log k = -3.97 - 3.96\,[\Sigma\sigma^+(BH^+) - \Sigma\sigma^+(AH^+)]$$

This correlation can be easily explained in terms of the mentioned above approach to the quantitative description of rearrangements involving arenium ions [515]. For a

65 The rate coefficient does not change in passing to CF_3SO_3H.

rearrangement of dienones with two equal groups R at the ring sp^3-hybridized carbon atom:

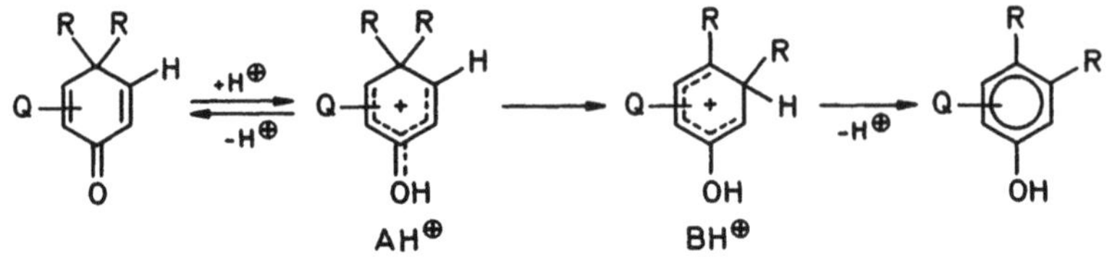

the free activation energy under complete protonation is determined by two components:

$$\Delta G_R^{\ddagger} = \Lambda_R + 0.5 \cdot \Delta G_{BH^+,AH^+}$$

where Λ_R is the intrinsic barrier of the 1,2-shift of R dependent on the positive charges, and $\Delta G_{BH^+,AH^+}$ is the difference of the free energies between the forming and the rearranging ions.

When the substituents Q possess less pronounced electronic effects than the hydroxy group the changes in the charges with varying Q can be neglected and $\Lambda_R \approx$ const; $\Delta G_{BH^+,AH^+}$, as shown earlier, can be estimated from the $\varrho\sigma^+$-description which eventually yields a relationship of the above type:

$$\Delta G_R^{\ddagger} = A + B\,[\Sigma\sigma^+(BH^+) - \Sigma\sigma^+(AH^+)]$$

If there are different substituents at the ring sp^3-hybridized carbon of dienone, the preference of the shift for both of them is determined by their aptitude to migration and by the ability of the remaining group to delocalize the positive charge of the forming ion [606]. Seemingly, due to the latter circumstance the 1-dichloromethyl-1-methyl-4- and -2-hydroxybenzenium ions and their derivatives are not apt to isomerize by a shift of the CH_3 group from the C_1 atom [224,225,326,426,607]. Such a shift must overcome a high energy barrier because the forming ion contains the electron-withdrawing $CHCl_2$ group at an electron-deficient position.

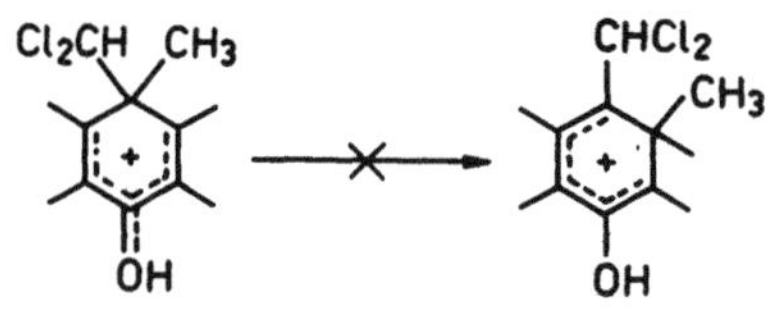

The $CHCl_2$ group seem to possess a very low migrating ability, like the CCl_3 group [224,326,608].

On the contrary, the 1-methoxy-1-methyl-4-hydroxybenzenium ion is rearranged by the 1,2-shift of the CH_3 group and, what is more, at a rate about 30 times (at 25 °C, $H_0 = -2$) higher than the rearrangement of ion (*114*) [231] reflecting the involvement of the CH_3O group in delocalizing the positive charge on formation of the transition state.

Preferable rearrangement of 4-R-4-methylcyclohexa-2,5-dienones:

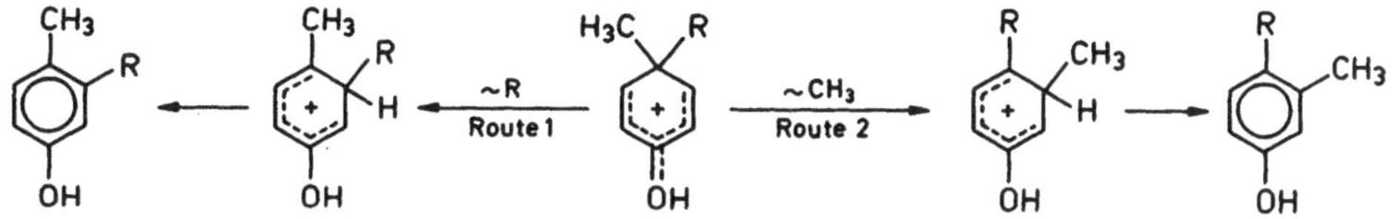

116

$R = C_2H_5$, $n\text{-}C_3H_7$ [230] $R = OCOCH_3$ [611] $R = C_6H_5$ [606] $R = Cl$ [610] $R = OH$ [609] $R = OCH_3$ [231]

The R groups in ions (*116*) have the following relative migrating ability

$CH_3 : C_2H_5 : n\text{-}C_3H_7 : sec\text{-}C_4H_9 = 1:50:40:7000$ [230, 612]
(aqueous H_2SO_4, 25 °C)

$CH_3 : COOC_2H_5 : C_2H_5 : C_6H_5 = 1:14:55:260$ [606]
(CF_3COOH, 38.5 °C)

Migrating ability of the carbethoxy group (see also [613, 614]), carrying on the carbonyl carbon atom a partial positive charge, seems to be due to the readily polarizable π-orbital which interacts with the carbocationic centre when the transition states of the 1,2-shift is formed. The carbethoxy group is analogous to the nitro group which is characterized by the facility of 1,2-shifts in methyl-substituted arenium ions. For ion (*116*) $R = NO_2$ the intramolecular 1,2-shift seems to be realized less readily than the rupture of the $C—NO_2$ bond to form 4-methyl-2-nitrophenol [615–618].

Consider the possibility of predicting [515] the relation of rearrangement of O-protonated cyclohexadienones with two different substituents at the ring sp^3-hybridized carbon atom, e.g., (*116*). The competitive directions of rearrangement differ both in the intrinsic barriers and in the thermodynamic parameters of the 1,2-shift. The estimation of the intrinsic barriers have already been discussed in this and preceding sections. So at this point only the difference in the free energies of the forming and the rearranging ions is calculated. This difference between two isomeric ions one of which carriers two substituents at the ring sp^3-hybridized carbon atom:

for a first approximation is equal to the difference of the free energies of the ions formed from the previous pair by substituting R_2 located at the ring carbon by a hydrogen:

Since the last two ions differ in the site of proton the value of ΔG can be calculated for them by the $\varrho\sigma^+$-approach using $\varrho = -3100/T$ (see above).

For the rearrangements of the O-protonated 4-chloro-4-methyl- and 4-phenyl-4-methylcyclohexa-2,5-dienones (*116*; R = Cl, C_6H_5) the estimation of the thermodynamic term of routes 1 and 2 yields:

$$\text{for R = Cl} \quad 0.5 \cdot \Delta G_1 = 4.7 \quad \text{and} \quad 0.5\,\Delta G_2 = 4.2\ \text{kcal/mole}$$

$$\text{for R} = C_6H_5 \quad 0.5 \cdot \Delta G_1 = 4.7 \quad \text{and} \quad 0.5\,\Delta G_2 = 4.6\ \text{kcal/mole}$$

Thus the thermodynamic terms of the free activation energy in either case are somewhat lower for route 2 than route 1 (by 0.5 and 0.1 kcal/mole). The intrinsic barriers (Λ_R) of the 1,2-shift of both the phenyl and the chlorine atom in benzenium ions are usually lower than for the methyl group. Thus, e.g., the free activation energy of the 1,2-shift in 1-R-1,2,3,4,5,6-hexamethylbenzenium ions for R = Cl, C_6H_5 and CH_3 at 25 °C is, respectively, 14.7 [299)], 17.2 [242)] and 18.1 kcal/mole [279)]. Neglecting, for a first approximation, the difference in the charge of hydroxy-substituted and methylated benzenium ions[66] and assuming that in ion (*116*) the intrinsic barrier (Λ_R) for the 1,2-shift of chlorine is 3.4 kcal/mole lower and of phenyl 0.9 kcal/mole lower than for that of the CH_3 group, one can, by summing up the data on Λ_R and ΔG, conclude that the shift of a group other than methyl is predominant for either ion (route 1), just as experimentally observed [606,610)].

A more detailed analysis should take into account the dependence of the intrinsic barrier of the 1,2-shift of migrants on the charge characteristics of the ions.

3 Hydrogen Exchange Reaction of Arenium Ions and Their Precursors

The protonation of aromatic compounds at unsubstituted carbon atom and the reverse process of transferring a proton of the rings CH_2 group of the arenium ion to nucleophilic particles of the medium simulate the acid catalyzed hydrogen isotope exchange of aromatic compounds (see reviews [7,619)]). The two processes generate, by interaction of aromatic compounds with deuteroacids, arenium ions in which the ring hydrogen atoms are replaced by deuterium (see, e.g., [27,50,65,444)]).

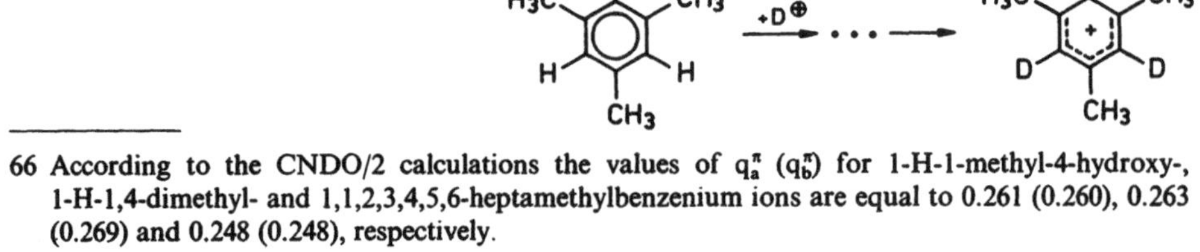

66 According to the CNDO/2 calculations the values of q_a^π (q_b^π) for 1-H-1-methyl-4-hydroxy-, 1-H-1,4-dimethyl- and 1,1,2,3,4,5,6-heptamethylbenzenium ions are equal to 0.261 (0.260), 0.263 (0.269) and 0.248 (0.248), respectively.

In compounds containing non-equivalent ring hydrogen atoms the most basic positions are primarily substituted (see, e.g., the data on 1,3-dimethoxybenzene [346]), but with a sufficiently long time of reaction and with stronger acids it is possible to exchange the less basic positions as well [7].

Alkylarenium ions possess one more centre for the hydrogen isotope exchange: the α-CH_n (n = 1, 2, 3) groups located at the positions para and ortho to the ring sp^3-hybridized carbon since for such ions the conjugate bases are not only aromatic compounds (*118*, R_2 = H), but also alkylidencyclohexadienes (*119*) and (*120*).

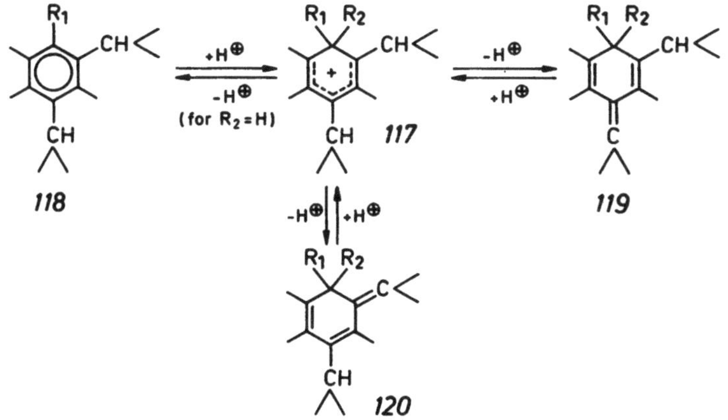

When substituents R_1 and R_2 in ion (*117*) are not hydrogen, alkylidencyclohexadienes of types (*119*) and (*120*) can be obtained by interaction of an ion salt with a base. From 4-alkylated ions the para-semibenzoid isomers (*119*) are preferably formed, e.g. 1-R-1,2,3,5,6-pentamethyl-4-methylenecyclohexa-2,5-dienes from the salts of 1-R-1,2,3,4,5,6-hexamethylbenzenium ions:

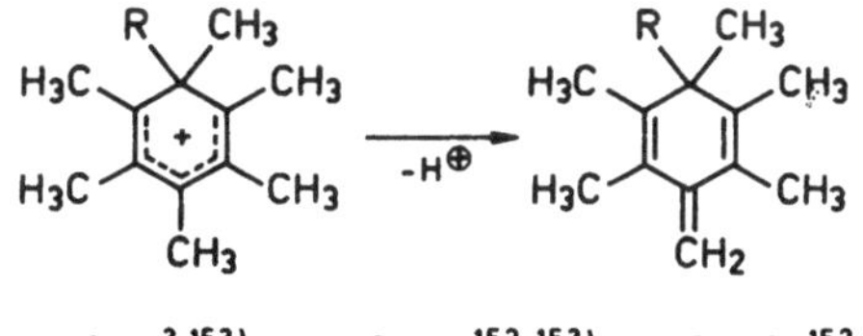

R = CH_3 [3,153] C_2H_5 [152,153] CH_2Cl [152,153]

and 1,2,3,5,6-pentaethyl-4-ethylidencyclohexa-2,5-diene from a salt of the heptaethylbenzenium ion [3,153]. Similar conversions are described for alkyl-substituted anthracenium [153] and phenanthrenium ions [153,212,214].

A rather high basicity of alkylidencyclohexadienes [3,207] and the inhibition of intermolecular proton transfer in very strong acids allows one to expect that favourable conditions for hydrogen isotope exchange in α-position of alkyl groups of arenium ions are offered in acids of moderate strength. A convenient medium for investigating this reaction proved to be deuterotrifluoroacetic acid [153,620,621] (for CF_3COOH $H_0 = -3.03$ [622]).

Dissolution of 9,9-dimethyl-10-methylene-9,10-dihydroanthracene (*121*) in CF_3COOD yields ion (*122*) in which at 70 °C all protons of the 10-methyl group are rapidly substituted by deuterium [153], whereas no deuterium is introduced into the methyl groups located at the sp^3-hybridized carbon,

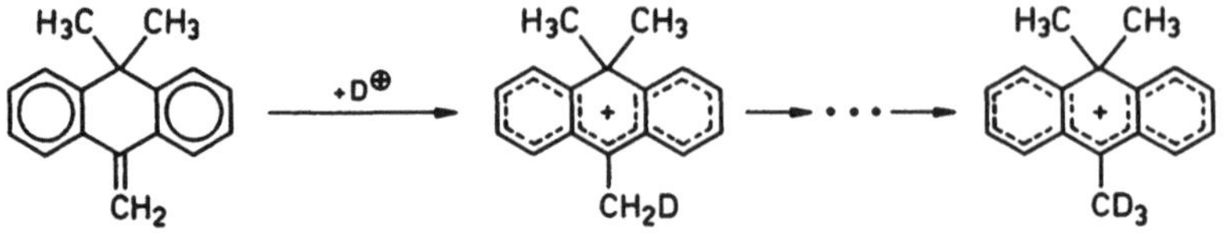

121 *122*

By contrast, in 1-R-1,2,3,4,5,6-hexamethylbenzenium ions (R = CH_3, C_2H_5, CH_2Cl) generated by dissolving 1-R-1,2,3,5,6-pentamethyl-4-methylenecyclohexa-2,5-dienes in CF_3COOD the hydrogen of all ring CH_3 groups are involved in the isotope exchange reaction [153,155,620]. This is a result of degenerate rearrangements due to 1,2-shifts of substituents R bringing all the ring CH_3 groups into exchangeable positions. The groups R = C_2H_5 and CH_2Cl being on each step of degenerate rearrangement at the sp^3-hybridized carbon atoms are not involved in the hydrogen-deuterium exchange. For R = CH_3 all the seven methyl groups of the heptamethylbenzenium ion are converted into CD_3 groups. Since in aqueous acids the heptamethylbenzenium ion readily loses one methyl group it is a convenient intermediate for perdeuterohexamethylbenzene [620].

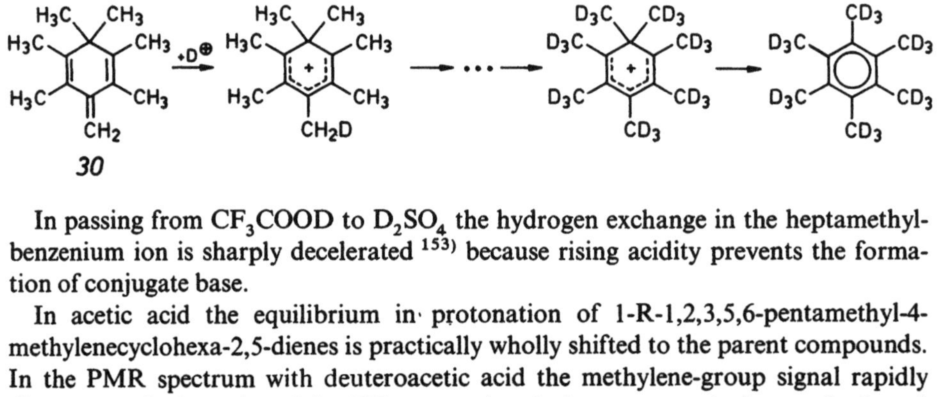

30

In passing from CF_3COOD to D_2SO_4 the hydrogen exchange in the heptamethylbenzenium ion is sharply decelerated [153] because rising acidity prevents the formation of conjugate base.

In acetic acid the equilibrium in protonation of 1-R-1,2,3,5,6-pentamethyl-4-methylenecyclohexa-2,5-dienes is practically wholly shifted to the parent compounds. In the PMR spectrum with deuteroacetic acid the methylene-group signal rapidly disappears, the intensity of the CH_3-group signals decreases much slower, the 3- and 5-CH_3 groups vanishing somewhat faster than the other ones [153]. Accordingly the hydrogen isotope exchange of methylencyclohexadiene in CH_3COOD proceeds with the intermediate formation of benzenium ions undergoing, despite a short lifetime, a degenerate rearrangement by the 1,2-shift of R. A similar picture is observed for 1,1,2,3,5,6-hexaethyl-4-ethylidencyclohexa-2,5-diene (exchange of CH_2 fragments of ethyl groups) and 9,9-dimethyl-10-methylene-9,10-dihydrophenanthrene. With 9,9-dimethyl-10-methylene-9,10-dihydroanthracene (*121*) only the hydrogens of methylene group are involved in the exchange since the 9,9,10-trimethylanthracenium ion undergoes no degenerate rearrangement leading to randomization of methyl groups. For the latter two hydrocarbons the isotope hydrogen exchange reaction in CH_3COOD proceeds at a far lower rate than for the triene (*30*) in full accord with their lower basicity [207].

Also the hexamethylcyclohexa-2,5- and -2,4-dienones [153] yield upon protonation hydroxyhexamethylbenzenium ions. In the dienone (*71*) in 58% D_2SO_4 at 70 °C all the methyl groups have been enriched by deuterium; this reflects the conversion of ion (*98*) into the conjugate base (*123*) and the randomization of all methyl groups in ion (*98*) as a result of their 1,2-shifts

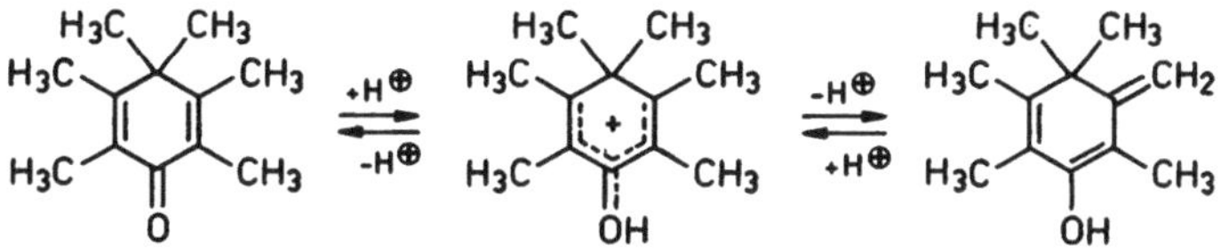

71 *98* *123*

With deuteroacetic acid the 3- and 5-CH_3 groups of the dienone (*71*) and the 3-CH_3 group of the isomeric 2,3,4,5,6,6-hexamethylcyclohexa-2,4-dienone are primarily deuterated. In concentrated D_2SO_4 the proton removal from methyl groups of hydroxymethylbenzenium ions decreases and at 70 °C hydrogen exchange is almost completely suppressed.

These data on the hydrogen exchange in the alkyl groups of arenium ions and their precursors cover the examples when both substituents located at the ring sp^3-hybridized carbon are not hydrogen. A similar exchange scheme, however, is also possible for ions formed upon protonation of alkyl-substituted aromatic compounds.

The concentration of non-aromatic conjugate bases (*119*, *120*, R_2 = H) which are in equilibrium with ions (*117*, R_2 = H) must naturally be much lower than the equilibrium concentration of aromatic compound (*118*) since the difference in the basicity of these structures amounts to several orders of magnitude [4,120,207]. But there are any doubts in the formation of compounds (*119*) and (*120*) with R_1 or R_2 = H from arenium ions because such unstable "semibenzenes" with a hydrogen atom at the ring sp^3-hybridized carbon are successfully obtained in other ways; examples are the trienes (*124*) and (*125*) [623,624], conjugate bases of 4- and 2-methylbenzenium ions (for other examples see [625]).

124 *125*

Indeed, methyl-substituted aromatic hydrocarbons [621] (mesitylene, hexamethylbenzene, 9,10-dimethylphenanthrene) exchange the hydrogen of the CH_3 group for deuterium when heated for a long time in CF_3COOD. The exchange is readily performed for the 7-CH_3 group of 7,12-dimethylbenz[a] anthracene and for the 12b-CH_2 group of 3-methylcholanthrene in $CDCl_3$—CF_3COOD—D_2SO_4 [317]. In hexaethylbenzene, according to the supposed exchange mechanism, the deuterium entered the molecule via the α-position of the ethyl groups [621]

The methyl groups of hexamethylbenzene are enriched by deuterium when kept in 97% D_2SO_4 [626]. But the exchange is accompanied by irreversible oxidative conversions of hexamethylbenzene and it is not precluded that the deuteration uses a different mechanism. In $DF—BF_3$ ([7], p. 224) and $DCl—AlCl_3$ [421] hexamethylbenzene undergoes no exchange while hexaethylbenzene, when heated with D_2SO_4 in a mixture of CF_3COOD and CCl_4, is enriched by deuterium at the β- and not α-positions of the ethyl groups [627].

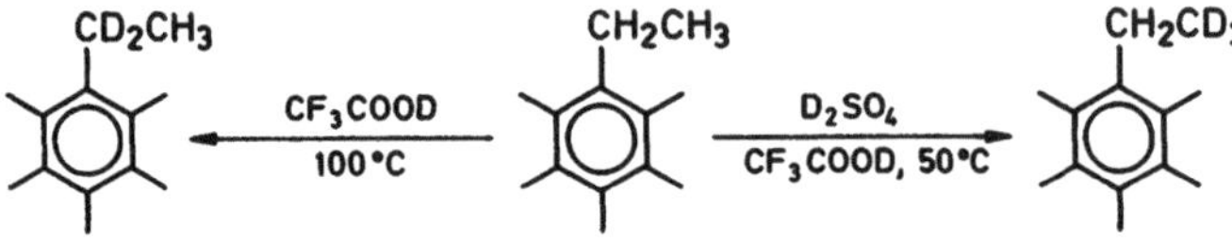

The change of the exchange direction with D_2SO_4 seems to be due to its oxidizing properties contributing to the formation of benzyl-type cations in which an exchange of β-hydrogen atoms in alkyl groups is typical (see [628]). Interestingly, for the polymethylene cycles condensed with the benzene ring the exchange in the β-CH_2 groups [628,629] 67 is predominant not only in D_2SO_4, but also in CF_3COOD, although there is a partial exchange of the α-CH_2 groups as well [621].

Hydrogen exchange of CH_3 groups in acid media was observed for some methyl-substituted heterocyclic systems, e.g., pyrimidine [630], as well as for methyl-substituted cationic systems of tropylium and pyrylium [631,632]. Hydrogen is exchanged in the chain positions of alkylbenzenes in $CH_3COOD—D_2O$ with chloroplatinates of alkaline metals as catalysts [633,634].

The hydrogen isotope exchange in the alkyl groups of arenium ions reveals their degenerate rearrangements by observing deuterated groups at positions where alkyl groups cannot directly be deuterated [153,155,620]. When the precursors of arenium ions of methylenecyclohexadiene and cyclohexadienone types are selectively deuterated in weak acids their protonation in strong acids (where the deuterium exchange reaction is suppressed) makes it possible to generate selectively deuterated arenium ions and to observe subsequent conversions by the NMR spectroscopy [153,155].

4 Removal or Modification of the Substituent Located at the sp³-Hybridized Ring Carbon Atom

One chemical transformation of arenium ions is the removal of the substituent located at the ring sp^3-hybridized atom leading to the regeneration of the aromatic system:

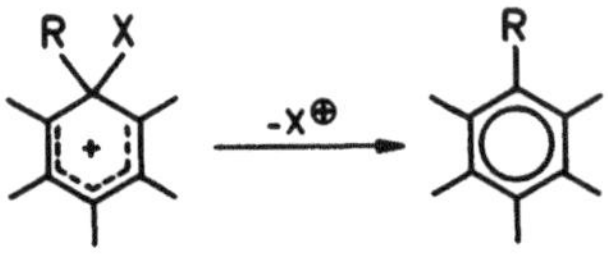

67 In the presence of the air oxygen in CF_3COOD there may occur oxidizing processes as well, e.g., formation of radical cations of polycyclic hydrocarbons has been noted in CF_3COOH [191].

The most mobile cationic substituent at the sp^3-hybridized ring carbon is hydrogen. When the medium acidity is decreased by adding water, alcohols or other bases most arenium ions with X = H are completely converted into the aromatic compounds. Strong electron-releasing substituents and bulky groups at the 2- and 6-positions essentially decrease the facility of the heterolytic rupture of the C—H bond and stabilize arenium salts with X = H in nonacid media (see the data for triaminobenzenium ions with X = H and R = Br). The number of such examples, however, is very limited. Mostly, even in strong acids the temperature dependence of NMR spectra or the hydrogen exchange reaction reveals rather a rapid reversible transfer of the proton from arenium ions to the nucleophiles of the medium.

The reversibility of C-protonation of aromatic compounds explains the possibility of replacing the hydrogen located at the sp^3-hybridized ring carbon of the arenium ion by another strong electrophilic substituent.

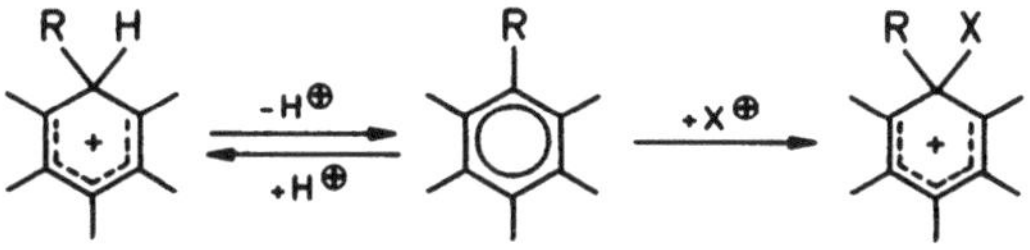

The following conversions serve as examples (for more detail see Sect. II):

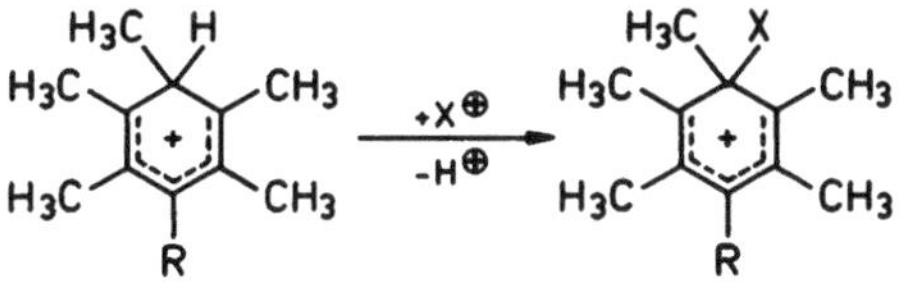

R = OH, CH_3; X = NO_2, SO_3H, Cl, Br

The reactions between salts of benzenium ions and alkylating or acylating agents lead, upon neutralization, to substituted benzenes [28,49,51,54]. The salts of alkylbenzenium ions do not react [54] with carbon dioxide as distinct from free hydrocarbons which interact with CO_2-aluminium halides to yield arylcarbonic acids and diarylketones. Hexafluoroantimonates of benzenium ions react with carbon oxide and yield aromatic aldehydes [28]:

$$ArH_2^+ \cdot SbF_6^- \xrightarrow{CO} [ArHCHO^+ \cdot SbF_6^-] \rightarrow ArCHO .$$

The heterolysis of the C—H bond may be conjugate with the addition of the leaving proton to the reagent if the latter does not yet carry a positive charge [28].

The reversibility of some electrophilic substitutions points to the possibility of removing a substituent other than hydrogen from the sp^3-hybridized ring carbon.

For example the dealkylation of alkylaromatic compounds with acid catalysts [13] is attributed to the heterolytic rupture of the C_{sp^3}-Alk bond in ions of the type

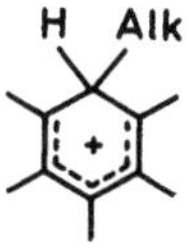

The leaving ability of alkyl groups increases with growing stability of the Alk^+ cation:

$$CH_3 < CH_3CH_2 < \text{sec.-alkyl} < \text{tert-alkyl}$$

E.g., tert-butylbenzene is protonated in $HSO_3F—SbF_5$ at −78 °C to form a 4-tert-butylbenzenium ion [65,281,282]; with rising temperature the PMR signals of this ion rapidly disappear and are replaced by signals of an unsubstituted benzenium ion and tert-butyl cation.

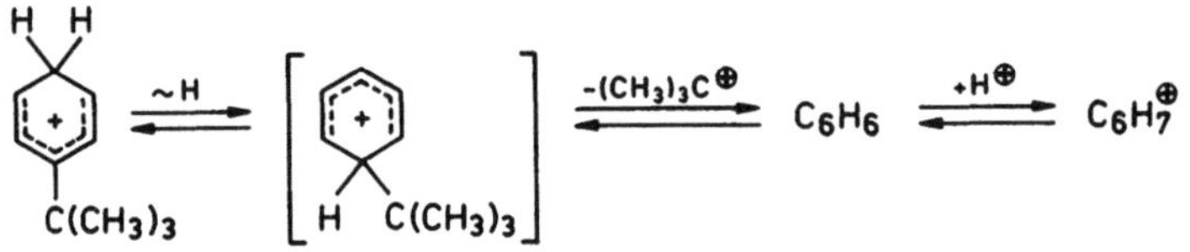

Substituted tert-butylbenzenes behave likewise [281].

The complete dealkylation of tert-butylbenzenes under these conditions is due to benzene protonation. In the weaker acid system $HF—TaF_5$, protonated 1,3-di- and 1,3,5-tri-tert-butylbenzenes are formed from tert-butylbenzene, but the $(CH_3)_3C^+$ cation does not appear [398].

In 4-isobutyl-, 4-sec-butyl- and other 4-alkylbenzenium ions the formation of benzyl-ions competes with the dealkylation. For example:

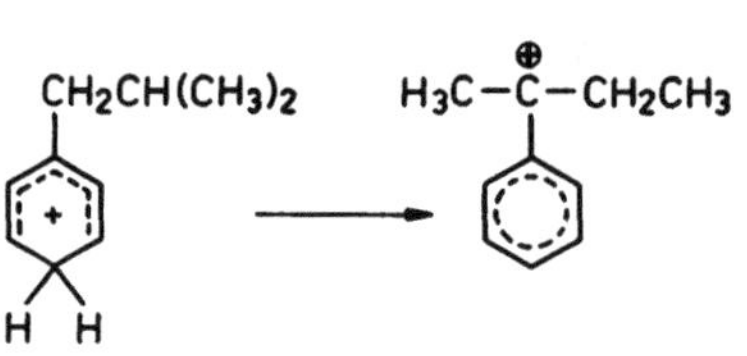

The ratio of the products depends on the reaction conditions [65].

The stability of benzyl-type cations accounts for the high leaving ability of various aralkyl groups. The following conversions are observed in strong acids ($HCl—AlCl_3$, $HF—SbF_5$, $HSO_3F—SbF_5$): [65,635–639]

$$(C_6H_5)_4C \xrightarrow{+H^+} (C_6H_5)_3C^+ + C_6H_6$$

$$(C_6H_5)_3CCH_3 \xrightarrow{+H^+} (C_6H_5)_2\overset{+}{C}CH_3 + C_6H_6$$

$$(C_6H_5)_3CH \xrightarrow{+H^+} (C_6H_5)_2\overset{+}{C}H + C_6H_6$$

$$(C_6H_5)_2C(CH_3)_2 \xrightarrow{+H^+} C_6H_5\overset{+}{C}(CH_3)_2 + C_6H_6$$

The cleavage of α-phenylindenes to form indenylic cations is also described [640]. Conversions of this type are used to generate anthracenium ions from 9-phenyl-9,10--dihydroanthracene and its derivatives (see Sect. II).

The relative leaving abilities of substituents located at sp^3-hybridized ring carbon of arenium ions are the more significant in view of the interest in electrophilic reactions with completely substituted aromatic compounds. Ch. Perrin [641] guided by the principle of microscopic reversibility in considering electrophilic substitution of hydrogen in aromatic compounds concluded that one should differentiate between two types of arenium cleavage — without nucleophilic assistance (S_N^1-reaction) and with it (S_N^2-reaction). After analysing the kinetic hydrogen isotope effect in electrophilic substitution, the replacement of substituents X by substituents Y in the aromatic nucleus, and the concurrent migration tendency of various groups Ch. Perrin arranged the X substituents according to the ease of removing them the sp^3-hybridized carbon as X^+ in S_N^1-reactions:

$$NO_2 < (CH_3)_2CH \sim SO_3^- < (CH_3)_3C \sim ArN_2 < ArC(OH)H$$
$$< NO < CO_2^- < B(OH)_3;$$

in S_N^2-reactions:

$$CH_3 < Cl < Br < D \sim RCO < H \sim I < HgZ < Si(CH_3)_3$$

The order of relative leaving ability of X groups in S_N^1 processes is parallel to the "stability" of the X^+ electrophile in solution; in other words, the lower the electrophilic activity of X^+, the easier the X group is cleaved in the S_N^1-reaction. Comparing the relative leaving ability groups of two different series is complicated by the medium nucleophilicity in the S_N^2 processes (cf. the relative leaving ability of H and Br from the 1-bromo-3,5-di-tert-butyl-4-hydroxybenzenium ion in the presence of various nucleophiles [642]). In CH_3COOH — $(CH_3CO)_2O$ — HCl the relative leaving ability of chlorine, bromine and the nitro group increases in the order: $Cl < NO_2 < Br$ [641].

Convenient models for studying the relative leaving ability of various groups are completely substituted benzenium ions.

For example, the facility of acidic cleavage of 1-R-1,2,3,4,5,6-hexamethylbenzenium ions increases in the following order (R) [209][68]:

$$CH_2CH_2X \ll CH_3 < CH_2CH_3 \ll CH_2CH{=}CHCH_2X$$

where

$$X = \overset{+}{N}(CH_3)_3 \quad \text{and} \quad \overset{+}{P}(C_6H_5)_3$$

The heptaethylbenzenium ion is rather readily converted into hexaethylbenzene [153, 157]. The heptamethylbenzenium ion is stable in strong protonic acids, but with rising nucleophilicity of the medium, e.g., in aqueous hydrochloric acid, demethylation (slowly at about 25 °C and rapidly when heated) leads to hexamethylbenzene and methanol [3]. Similarly, phenylhexamethylbenzenium ions with two CH_3 groups at the sp^3-hybridized ring carbon in aqueous acids at elevated temperature yield

68 The cleavage can be promoted by bringing the I^- anion into the solution.

2,3,4,5,6-pentamethylbiphenyl [208]. By contrast, attempts to obtain pentamethylphenol by heating hexamethylcyclohexa-2,5-dienone with acids met with failure.

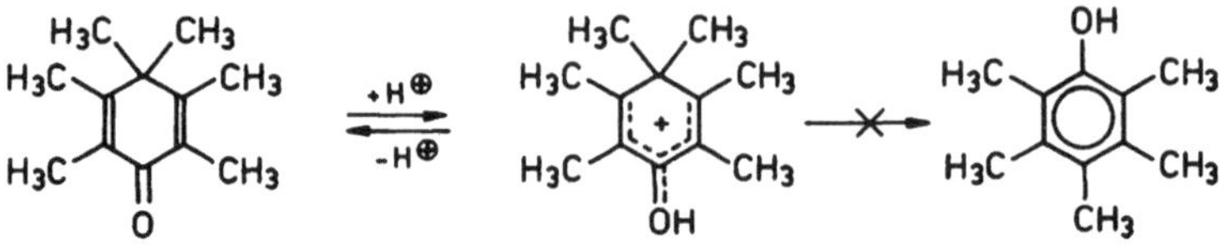

The smooth removal of alkyl groups to form phenols from substituted cyclohexadienones can be achieved by heating with alkalies [643]. Alkyl groups other than CH_3 are split off by acids from alkyl-substituted cyclohexadienones. Examples are cyclohexadienones having the following sp^3-node: $C(C_2H_5)_2$ [644], $C(Br)CHBr_2$ [645], C(Br)CH(OR)R [645,646].

Accordingly the reversibility of sulphonation and bromination of aromatic compounds (cf. also the Perrin's order) the quenching (*22b*)- and (*22d*)-ion salts with methanol or methanol/diethylamine leads to hexamethylbenzene [171,172,259,371]:

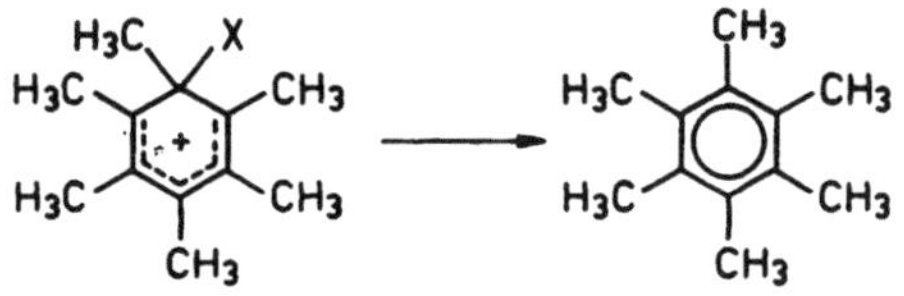

22

b X = SO_3^- (or SO_3H); *d* X = Br

C-protonated sulphonic acids of durene, naphthol and their ethers are also desulphonated [175,202,206].

The elimination of X-substituents from arenium ions with regeneration of the aromatic structure, as well as the formation of arenium ions by interaction of aromatic compounds with electrophiles (see Sect. IV.6) may turn out to be more complicated than is now generally believed.

The reaction of 1-halogeno-2,4,6-tris(dialkylamino)benzenium ions (*29a*) with triethylamine leads [197] to derivatives of biphenyl (*126*). The conversions are elucidated by the following scheme:

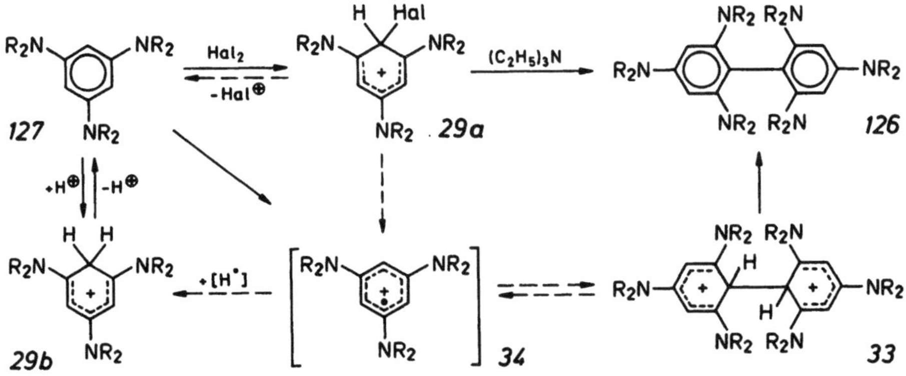

Triaminobenzenes (*127*) are oxidized by silver salts, CBr_4 or halogens [266,267] (cf. [197]) to yield salts of the dications (*33*) which seem to be dimerization products of radical-cations (*34*). With bases, the dications (*33*) react to hexaaminobiphenyls (126). They may be formed from the salts of triaminobenzenium ions (*29*) either by heterolytic cleavage with subsequent oxidation of triaminobenzene or by the homolytic rupture of the C-Hal bond leading directly to radical cation (*34*). When heated or kept in polar solvents, dications (*33*) are converted into triaminobenzenium ions (*29b*) [267]. This may be due to the dissociation of dications into radical cations (*34*) which cleave hydrogen atoms from the solvent.

The salts of ions (*29a*) are converted into sym-tris(dialkylamino)benzenes (*127*) by reducing agents ($NaHSO_3$, C_2H_5SNa). The strong bases (CH_3ONa) remove the proton to form the bromosubstituted aromatic compound [197,199].

The 4-hydroxy-1-X-1,2,3,5,6-pentamethylbenzenium ions (*31*) with electronegative groups X, with rising temperature of acid solutions, are converted into the 4-hydroxy-2,3,5,6-tetramethylbenzyl cation (*128*) [175,220–223]. The substituents X allow the completed in HSO_3F within 5–10 min:

$$X = OCOCH_3\ (-10\ °C),\ SO_3H\ (-10\ °C),\ OH\ (0\ °C),\ OCH_3\ (0\ °C)\,,$$

$$NO_2\ (10\ °C),\ F\ (80\ °C),\ Cl\ (>80\ °C),\ Br\ (>100\ °C)\,.$$

Similar reaction are described for 4-methoxy-1-X-1,2,3,5,6-pentamethylbenzenium ions with $X = NO_2, SO_3H$ [175].

The reaction mechanism is not clear. The most likely routes are:

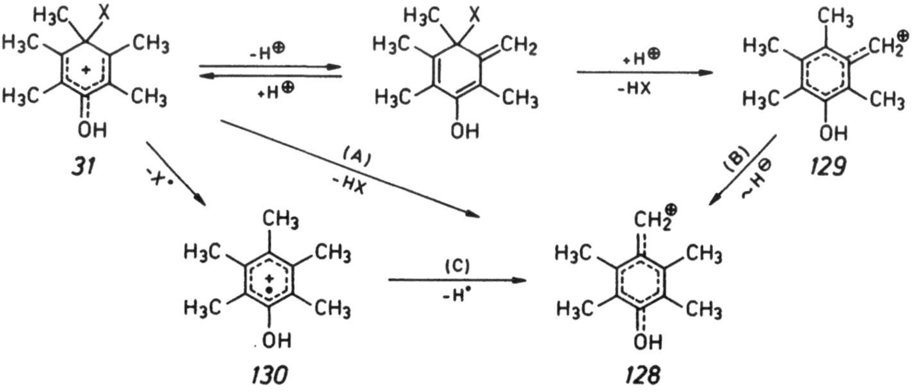

The simultaneous elimination of X and the hydrogen of the 1-CH_3 group is route A; the cleavage of a proton from the 2(6)-CH_3 group and the heterolytic rupture of the C—X bond (or the simultaneous elimination of HX) to form a 3-hydroxybenzyl cation (*129*) rearranging via hydride transfer into a more stable 4-hydroxybenzyl cation is route B; the homolytic rupture of the C—X bond with the subsequent removal of the hydrogen from the radical cation (*13*) is route C.

In neutral and acid media 4-methyl-4-nitrocyclohexa-2,5-dienone undergoes a dienone-phenol rearrangement transfering the NO_2 group to the 2-position [615–618].

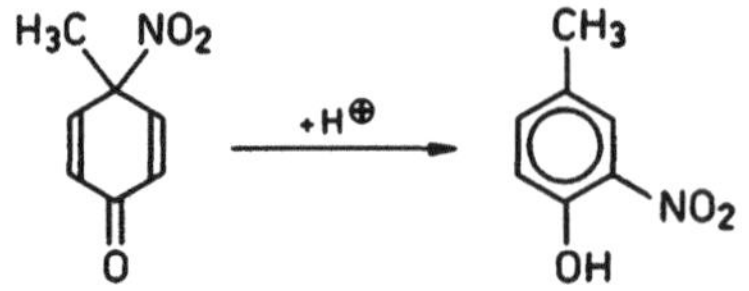

In this scission of the C—NO_2 bond, the homolytic process may be realized in neutral and weakly acid media [617] and the heterolytic one in strong acids [618].

In consideration of ion (*31*) conversions one should also take into account the possibility of the acid-catalyzed heterolytic removal of the X group as X^- anion with the formation of antiaromatic dication (*131*) which further loses a proton [69]:

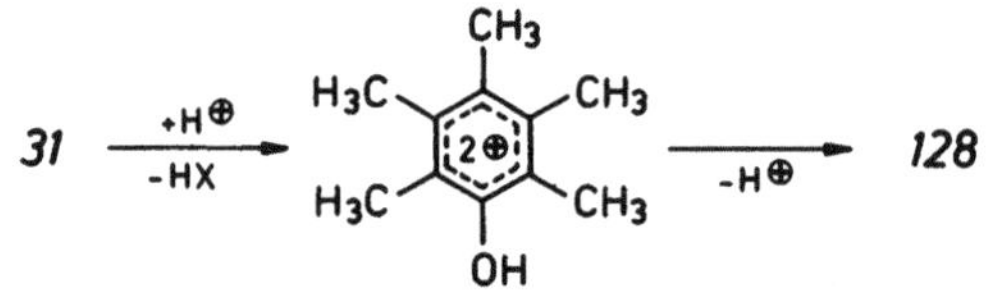

131

The (*131*)-type dications were postulated as intermediates in the substitution of the CH_3 group bonded with the benzene ring by an acetoxyl or acetamide group Y in the course of the electrochemical oxidation of methyl-substituted derivatives of benzene in acetic acid and acetonitrile [647,648]:

$$ArCH_3 \xrightarrow[-2e]{} ArCH_3^{2+} \xrightarrow[-H^+]{} ArCH_2^+ \xrightarrow{+Y^-} ArCH_2Y\,.$$

A more probable scheme for the electrochemical substitution implies an intermediate radical cation losing a proton with conversion into a benzyl radical oxidized on the anode to a benzyl cation [649] (cf. [648]):

$$ArCH_3 \xrightarrow[-e]{} ArCH_3^{+\cdot} \xrightarrow[-H^+]{} ArCH_2^{\cdot} \xrightarrow[-e]{} ArCH_2^+ \xrightarrow{+Y^-} ArCH_2Y\,.$$

For other possible mechanisms of these processes see [650].

Schemes (A) and (B) are close to those proposed for the electrophilic halogenation of alkylbenzene into the side chain to form benzylhalogenides [651–655] as well as for the reactions of polymethyl-substituted benzenes with sulphonating [656] and nitrating [657] agents yielding benzyl derivatives. Recently the schemes of the (B) type have been prefered (cf. [657]).

Besides, the conversion of 4-hydroxy-1-X-1,2,3,5,6-pentamethylbenzenium ions into 4-hydroxybenzyl cations could be explained by the quinobenzylic rearrangement

69 For the generation of dications from polycyclic aromatic hydrocarbons and their derivatives see Sect. IV.6. The interconversions between dications $[ArCH_2R]^{2+}$ and benzyl-type cations $[ArCHR]^+$ are discussed in [405]. The proton was only lost from the dication corresponding to 9-ethylanthracene ($R = CH_3$).

of the X-substituted cyclohexa-2,5-dienones. As shown on 1-halogeno-1-methyl-3,5-di-tert-butylcyclohexa-2,5-dienones [658–660] the rearrangement proceeds via an intermediate phenoxyl radical:

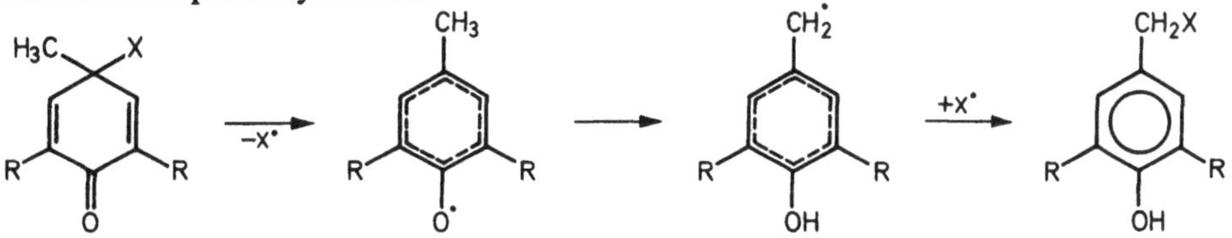

R = C(CH₃)₃

However, 4-hydroxy-1-nitro-1,2,3,5,6-pentamethylbenzenium ion is converted into a benzyl cation at a temperature at which the corresponding dienone undergoes no rearrangement into a benzyl derivative [223].

The 1-X-1,2,3,4,5,6-hexamethylbenzenium ions are not rearranged into a benzyl cation in strongly acid media, e.g. in HSO_3F. The ions generated by adding SO_3 and NO_2^+ to hexamethylbenzene, with the temperature rising, respectively, to —25 °C and 0 °C, are rearranged into ions with the oxygen function at the ring sp³-hybridized carbon atom [169–171]; the latter, similar to 1-methoxy- and 1-hydroxy-1,2,3,4,5,6-hexamethylbenzenium ions [243], are further isomerized by the 1,2-shift of the CH_3 group:

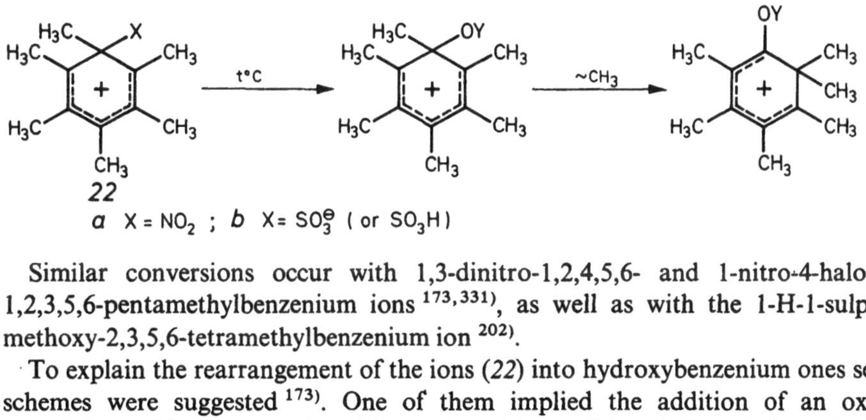

22

a X = NO_2 ; *b* X = $SO_3^{\ominus}$ (or SO_3H)

Similar conversions occur with 1,3-dinitro-1,2,4,5,6- and 1-nitro-4-halogeno-1,2,3,5,6-pentamethylbenzenium ions [173,331], as well as with the 1-H-1-sulpho-4-methoxy-2,3,5,6-tetramethylbenzenium ion [202].

To explain the rearrangement of the ions (*22*) into hydroxybenzenium ones several schemes were suggested [173]. One of them implied the addition of an oxygen-containing acid anion to the 2(6)- or 4-positions of ion (*22*) and the subsequent cleavage of NO_2^- from the formed cyclohexadiene. In the case of 1-nitro-^{18}O-1,2,3,4,5,6-hexamethylbenzenium ion, however, the oxygen function of the 2,3,4,5,6,6-hexamethylcyclohexa-2,4-dienone from the rearrangement is only formed at the expense of the oxygen atoms of the nitro group [331]. The rearrangement can therefore proceed either by homolytic scission of the C—NO_2 bond with the subsequent recombination of the hexamethylbenzene radical cation and NO_2:

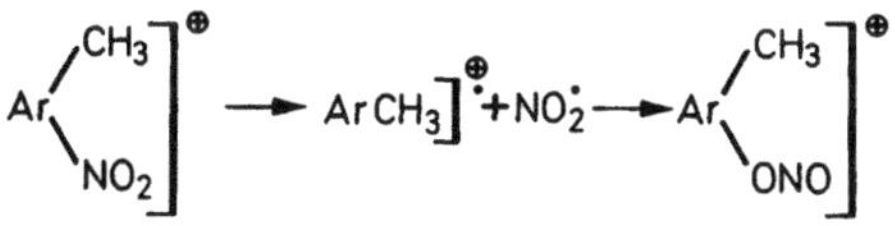

or via a cyclic intermediate (or the transition state) with a bond between the nitro-group oxygen and the carbon carrying a positive charge. The intramolecular interaction of this type has been observed for o-nitrobenzyl cations[661]:

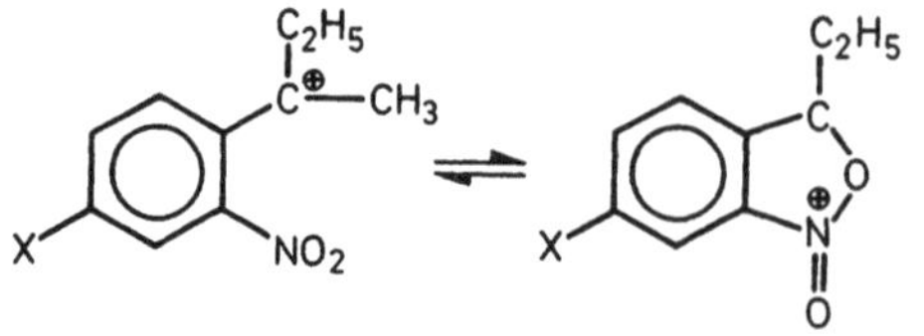

If a benzyl cations is formed from benzenium ions according to scheme (B) a decrease in the medium acidity should accelerate such conversions facilitating the formation of a conjugate base. Some observations are consistent with this suggestion. For example, the 1-nitro-1,2,3,4,5,6-hexamethylbenzenium ion is not converted into the benzyl cation in HSO_3F; when the acid solution is poured into diethylamine/methanol, it gives a high yield of methoxymethylpentamethylbenzene[169] (cf. [171, 243]). Similarly, the 1-nitro-1,2,3,4,6-pentamethylbenzenium ion (*24*) in acid solution with methyl alcohol/diethyl ether at −110 °C yields[173], as the main product, 1-methoxymethyl-2,3,4,5-tetramethylbenzene:

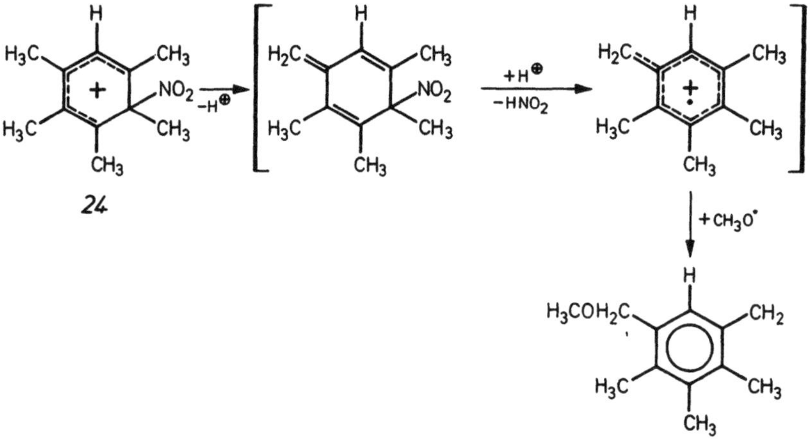

Triene (*132*), a conjugate base of the 1-hydroxy-1,2,3,4,5,6-hexamethylbenzenium ion, in HSO_3F at −70 °C is more than half-converted into the 2,3,4,5,6-pentamethylbenzyl cation[243]:

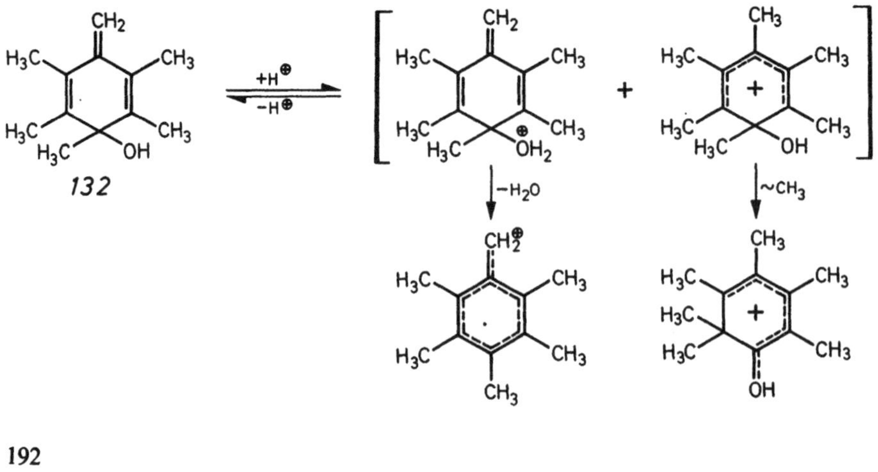

For the conversion of chloro-substituted polymethylbenzenium ions into benzyl chlorides see [176].

In the author's opinion the conversion mechanism of 1-X-1-CH_3-substituted benzenium ions into benzyl cations is closely connected with the formation of the 2,3,4,5,6-pentamethylbenzyl cation from hexamethylbenzene in concentrated sulphuric acid at elevated temperature [261,262] (see also the data for hexasubstituted benzenes with polymethylene cycles [422]). Since the benzenium ion, formed by adding a sulphonating agent to hexamethylbenzene in HSO_3F, is converted into the 2-hydroxy-hexamethylbenzenium ion [171], the mechanism with intermediate formation of the radical-cation (scheme C) should be preferred. The oxidation rate of hexamethylbenzene to form a pentamethylbenzyl cation sharply increases in conc. H_2SO_4/polyphosphoric acid [261], as well as in conc. H_2SO_4/diphenoquinone [262] whose diprotonated forms is an effective acceptor of hydrogen atoms providing for polycyclic arenium ions their complete conversion into radical-cations. The radical cations of polycyclic aromatic hydrocarbons act similar to the diprotonated diphenoquinone (see Sect. IV.6).

5 Arenium Ions as Electrophiles

Arenium ions possessing a ring fragment of the CHR type or ortho(para)-positioned OH-, NHR- and CHR-type substituents mainly react with rising nucleophilicity of the medium by loss of the proton, as marked earlier. In some cases the loss of the proton from the substituent competes with the addition of a nucleophile to the ion. For example, arenium ions formed by ipso-addition of NO_2^+ "capture" an acetate-anion (nitroacetoxylation [662–665]):

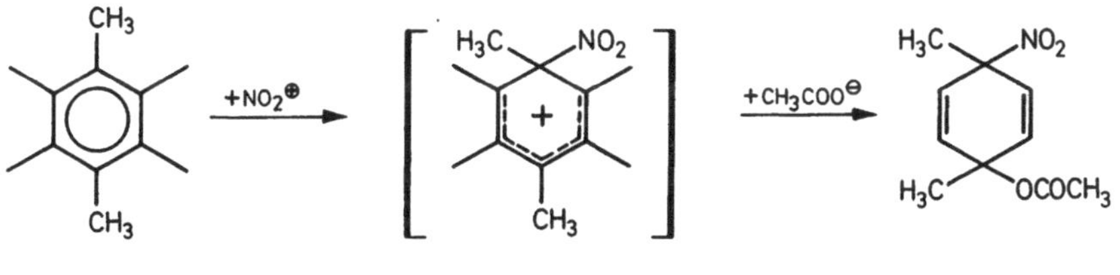

Ions having no substituents of the above type exhibit an increased feasibility of such processes. An example is the nitrofluorination of polyfluorinated aromatic compounds with HNO_3 in HF or with nitronium tetrafluoroborate [441,666,667]:

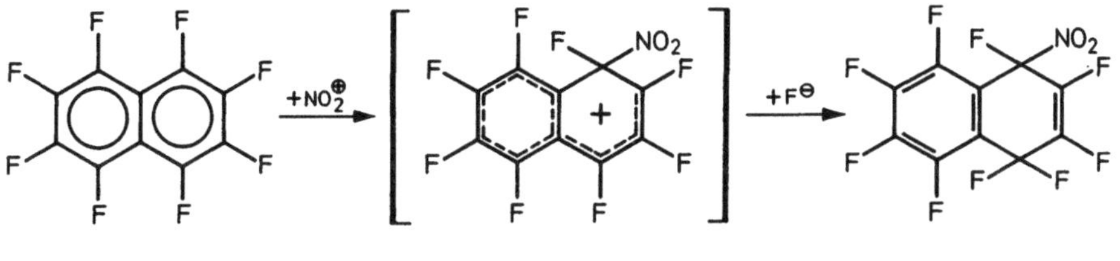

The addition of a nucleophile is also predominant when quenching acid solutions of stable polyfluorinated arenium ions. Careful pouring into a frozen mixture of HCl with CH_3OH effects the addition of Cl^- [178,179]. When acid solutions are poured on ice the primary addition products of OH^- are further converted to form cyclohexadienones [178,180,237–241,413]:

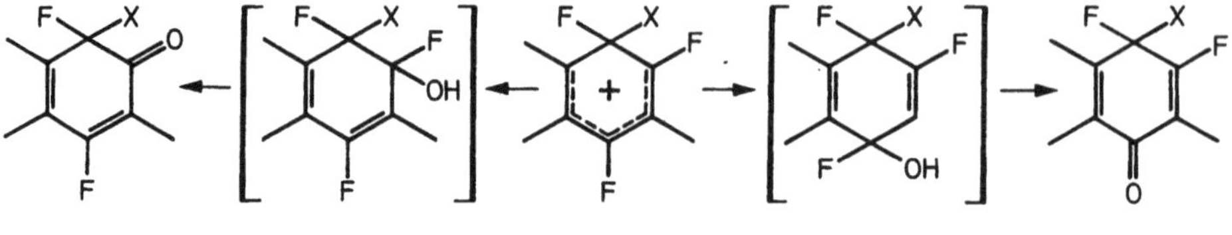

The addition and subsequent elimination of F^- from another carbon appear to account for the mentioned (see Sect. IV.1.B) isomerization of polyfluorinated arenium ions.

Arenium ions, just like the other carbenium ions, can serve as alkylating agents to the other aromatic compounds. This seems to be realized in the condensation of phenols in $HF—SbF_5$ [668] and of their tautomeric complexes (see Sect. III.1.B) [669–672] with benzene and its derivatives. Also polyfluorinated arenium ions react with incompletely substituted polyfluorinated compounds [412,413,673] e.g., in the consecutive conversions:

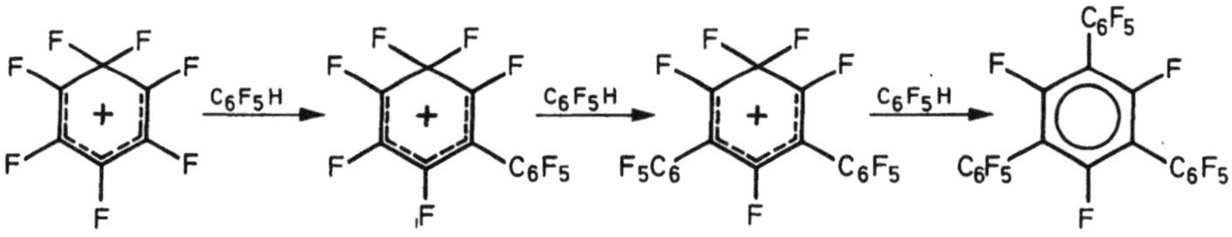

However, other explanations of the reactions observed are not precluded. Thus, for polyfluorinated arenium ions a preliminary conversion of the ions into radical cations of the respective aromatic structures is discussed. Indeed, the radical cation of octafluoronaphthalene reacts with pentafluorobenzene [674] to yield the same products as the perfluoro-α-naphthalenium ion does.

Benzene in the $HF—TaF_5$ system is hydrogenated by molecular hydrogen in the presence of isoalkanes at 50 °C to yield cyclohexane which is further partly converted into methylcyclopentane [675]. It is assumed that the first formed benzenium ion takes a hydride-ion from the H_2 molecule or isoalkane. For example, mesitylene has been hydrogenated in $HF—TaF_5$ with Pt, Pd or Ir without adding aliphatic hydrocarbons [676].

6 Interconversions of Arenium Ions and Radical Cations of Aromatic Compounds

The reactivity and the orientation effects in aromatic electrophilic substitutions are usually estimated from the relative facility of formation of σ-complexes (arenium ions) by comparing such calculation parameters as the index of free valence, the π-electron density, the localization energy of a pair of electrons etc. [11]. S. Nagakura and J. Tanaka [677–679] and, independently, R. Brown [680] have emphasized the role of charge transfer in the σ-complex formation (see also [681]): By analogy with the donor-acceptor interaction of particles in the formation of molecular complexes the interaction of the aromatic molecule ArH and the electrophile X^+ can be formally described in terms of Mulliken's concept [682,683] (see also [684,685]) by the combination of the wave functions of a "structure without bond formation" (ArH, X^+) and a "structure with a charge transfer" ($ArH^{+\cdot}X^{\cdot}$).

Figure 21 shown the potential energy changes for each of these "structures" versus the reaction coordinate represented by the internuclear distance C ... X. The curve for the "structure with a charge transfer" (CT) has a minimum for the binding between the aromatic molecule and the electrophile as they approach each other. The "no bond" function (NB) impulses an increase of repulsion with decreasing distance.

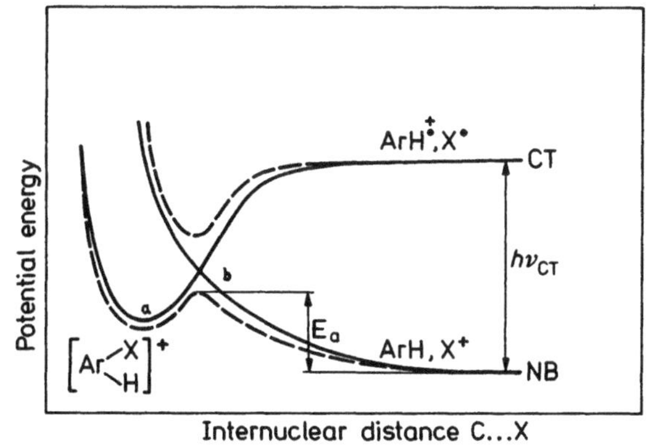

Fig. 21. Interaction between the "structure without bonding" (NB) and the "structure with charge transfer" (CT) [679,681]

If the highest occupied level of an aromatic molecule is located higher than the lower vacant level of the electrophile (a necessary condition for the σ-complexe formation [677–679]) the curves of CT and NB intersect. In this case, considering the noncrossing rule, the curves should be substituted by the dotted lines; the lower line characterizes the energetic profile of transition from noninteracting particles ArH and X^+ to the σ-complex (a) via a transition state (b). With decreasing intermolecular distance C ... X the ArH—X^+ interactions is first determined by the NB wave function and then by the CT function; the configuration of the transition state correlates with about an equal contribution of either "structure".

A more detailed analysis of the ArH—X^+ interaction leads to the conclusion that the activation energy of the electrophilic substitution must be linearly related with the difference in the energies of the two "structures" for an infinitely great intermolecular distance, i.e. with the charge-transfer energy ($h\nu_{CT}$) [679,681]; for a fixed electrophile it must be connected with the ionization potentials of aromatic molecules [681]. In addition, since the "structure with a charge transfer" essentially contributes to the transition state, one can except that the relative reactivities of different positions of the aromatic molecule ArH are determined by finding the unpaired electron (spin density) at the respective positions of the $ArH^{+\cdot}$ radical cation [677–679,681]. Indeed, there is a qualitative relation between the relative reactivity of individual positions of aromatic molecules and the hyperfine splitting constants a_i^H in the EPR spectra of corresponding radical cations (for π-systems the values of a_i^H are proportional to the spin density at respective positions) [681,686].

To describe the relative reactivity of aromatic compounds with electrophiles it is suggested [687] to use the equation:

$$\log \frac{k_{ArH_i}}{k_{ArH_0}} = A \cdot \Delta I + B \cdot \Delta L^+$$

where ΔI is the difference of ionization potentials of the substrates being compared and ΔL^+ is the difference in the stability of the corresponding σ-complexes.

The estimation of the relative facility of formation of the transition state in the interaction of the aromatic ArH molecule and an electrophile by parameters characterizing the $ArH^{+\cdot}$ radical cation and the traditional approach in which for the same purpose the parameters of the parent aromatic molecule are used are the two extreme approximate descriptions of electrophilic substitution. Either approach has its advantages and disadvantages. The due regard for the wave function of the "structure with a charge transfer" in describing the $ArH—X^+$ interaction reveals the interrelation between the electrophilic substitution and the oxidation of aromatic compounds. Thus, in an energy diagram like Fig. 21, when the maximum of lower curve and the minimum of the upper curve are close to each other along the axis of energies a thermally induced excitation with a transfer to the upper level may occur [681]. This can lead to the formation of an $ArH^{+\cdot}$ radical cation and an $X^\cdot$ radical. Indeed, radical cations are frequently formed under conditions close to those of electrophilic substitutions. Even an opinion has been expressed that the transfer of an electron from the substrate to the electrophile with formation of the radical pair $ArH^{+\cdot}$, $X^\cdot$ is a necessary intermediate stage of electrophilic substitutions (see, e.g., [688,689]). Therefore, it is reasonable to discuss experimental data on the interrelations between radical cations and arenium ions.

Polycyclic aromatic hydrocarbons (perylene, tetracene, anthracene, naphthalene), their alkyl- and aryl-substituted derivatives and compounds of the benzene series having strong electron-releasing substituents (p-dialkoxybenzenes, p-bisdialkylaminobenzenes etc.) are oxidized to yield radical cations upon action of Lewis-type acids ($SbCl_5$ [191,690–695], SbF_5 [696,697], $AlCl_3$ [698], $AlCl_3$ combined with RNO_2 [699–701], SO_3 [191] etc.), as well as in strong protonic oxidizing acids (conc. H_2SO_4 [191,690,702,703], HSO_3Cl [694], HNO_3 and H_2SO_4 [699]). In not oxidizing protonic acids the formation of radical cations is observed in the presence of oxygen (HF [191,416,704], $CF_3COOH—H_2O—BF_3$ [705]), or added oxidizers (nitrogen oxides [705]), as well as upon UV irradiation (boric acid [706,707], $CH_3COOH—H_2SO_4$ [708]). For example, perylene in HF is transformed, judging by the electronic absorption spectrum [416], into the perylenium ion. With oxygen the shape of the absorption spectrum [191,416] approaches that of the solution of perylene in conc. H_2SO_4; here the corresponding radical cation has been detected by ESR. The conversion of protonated polycyclic aromatic hydrocarbons into radical cations with oxygen is reversible since an increase of oxygen pressure is accompanied by a growing radical cation concentration; with its decrease the protonated hydrocabon is accumulated [191,705]. Formally the equilibrium can be written:

$$Ar^{+}\genfrac{}{}{0pt}{}{\diagup H}{\diagdown R} + A \rightleftarrows AR^{+\cdot} + A^{-\cdot} + H^{+},$$

where A is the electron acceptor (in this case O_2).

This formal scheme does not mean that the arenium ion reacts with oxygen although the increase in the medium acidity seems to affect favourably the formation of radical cations (perylene does not yield any detectable concentration of radical cation in $CH_3COOH—C_6H_6$ even at high oxygen pressure). In the reaction free hydrocarbons may participate, and the acidity effects the facilitated decay of the short-lived molecular complex $[ArH \cdot O_2]$ into two radicals when oxygen is protonated [705]:

$$ArR + O_2 \rightleftarrows [ArR \cdot O_2] \underset{-H^+}{\overset{+H^+}{\rightleftarrows}} ArR^{+\cdot} + HO_2^{\cdot}.$$

A powerful oxidizer relative to aromatic compounds are salts of oxygen radical cations, e.g., $O_2^{+}AsF_6^{-}$ [709].

Lewis-type acids can serve as electron acceptors, as already pointed out, as well. When polycyclic aromatic hydrocarbons interact with BF_3 in 1,2-dichlorethane the respective radical cations and particles which are supposedly bipolar arenium ions of type (*133*) are recognized in the electronic absorption spectra [191]. These particles are formed rapidly, the radical cations slowly. The accumulation of the latter is accelerated by illumination.

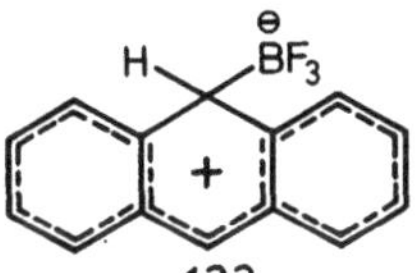

133

Similarly tetracene and anthracene interact with $AlCl_3$ [698]; for bipolar complexes with SbF_5 see [710].

Effective oxidizers of aromatic hydrocarbons in equilibrium with arenium ions are the diprotonated quinones particularly diphenoquinone and its 3,3′,5,5′-tetrabromo-derivative [262,711,712]. In this case the unsubstituted polycyclic arenium ions are converted into radical cations:

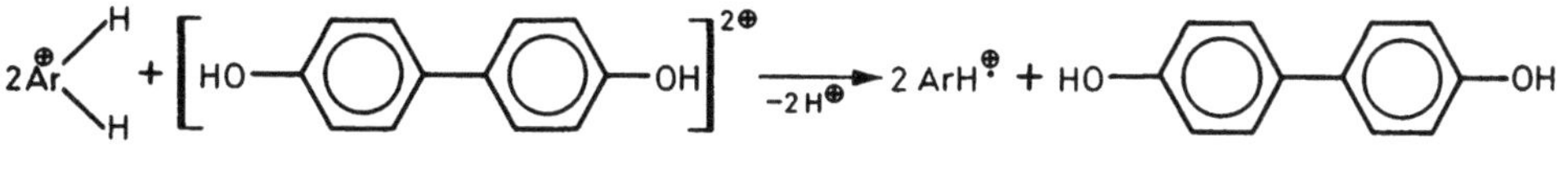

whereas, the hexamethylbenzenium ion is turned into the 2,3,4,5,6-pentamethylbenzyl cation. With rising acidity the rate of this process first rises and then falls [712]. For the study on the kinetics of such processes see [713].

Radical cations of polycyclic aromatic hydrocarbons are rather strong hydrogen acceptors. They take hydrogen from the hexamethylbenzenium ion converting it into the 2,3,4,5,6-pentamethylbenzyl cation [262,712] as well as from alkanes [262,712,714].

$$ArH^{+\cdot} \xrightarrow{RH} \overset{+}{Ar}<^{H}_{H}$$

Thus, in acid solutions of polycyclic hydrocarbons reversible interconversions of arenium ions and radical cations occur. Oxidizers in the system shifts the equilibrium to the side of radical cations. Reducing agents reverse the conversion. Other things being equal, the position of equilibrium between an arenium ion and a radical cation depends on the basicity of the ArR hydrocarbon and its oxidizing potential [191]. The data on degree of conversion of protonated polycyclic aromatics into radical cations show that both of the above factors are essential, so no simple relation exists between the basicity constant (K_B) of hydrocarbon and the position of the equilibrium. For instance, tetracene and perylene (log K_B is equal to 5.8 and 4.4, respectively) are practically completely converted into radical cations in CF_3COOH—H_2O—BF_3 and in HF-oxygen, as well as in H_2SO_4 without oxygen; but for 3,4-benzpyrene and anthracene (log K_B equals 6.5 and 3.8) a complete conversion is only observed in the first system. However, the formation rates of radical cations from arenium ions interacting with diprotonated quinones [262,711] showed the basicity of hydrocarbon to be the determining factor (the protonated anthracene and perylene react rapidly, tetracene and 3,4-benzpyrene slowly). For the reaction

$$\text{perylene} \cdot H^+ + \text{tetracene}^+ \rightleftarrows \text{perylene}^+ + \text{tetracene} \cdot H^+$$

the equilibrium is greatly shifted to the right [711].

Strong oxidizers turn the radical cations generated from polycyclic aromatic hydrocarbons into diamagnetic dications. This was first established from the electronic absorption spectra for tetracene with SO_3 in dimethylsulphate [191] and then for tetracene, 1,2-benzanthracene [423] and perylene [714] in HSO_3F—SbF_5 (without SbF_5 protonation yields only arenium ions). Addition of SbF_5 to acidic solutions of radical cations also results in the formation of dications [714]. The strongest oxidizers are antimony pentafluoride and its solution in SO_2ClF. For example, in SbF_5 and in SbF_5—SO_2ClF anthracene is converted into the corresponding dication (*134*) generated also from the 9,10-dibromo-9,10-dihydroanthracene:

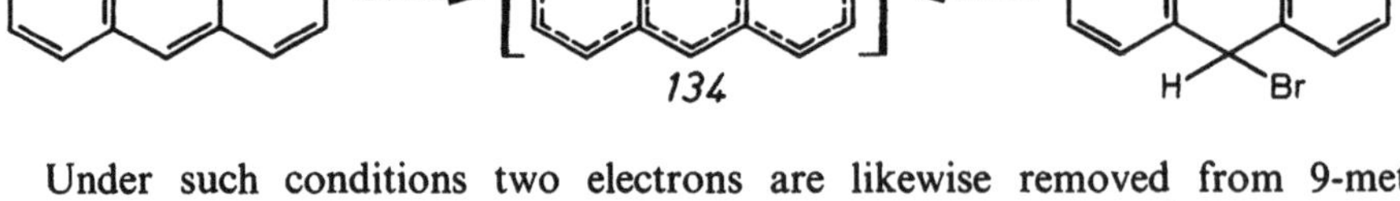

134

Under such conditions two electrons are likewise removed from 9-methyl-, 9,10-dimethyl-, 2,6-dimethyl-, 9-chloro- and 9-bromoanthracenes [714], 9-ethylanthracene [405], biphenylene [402,715] and methylated biphenylenes [715]. In SbF_5—SO_2ClF [403] dications are formed from the polycyclic aromatics for which the energy of the highest occupied molecular orbital (HOMO) calculated by the simple Hückel method is less than 0.5 β (pentacene, naphthacene, perylene, 1,2-benzpyrene, anthracene, 1,12-benzperylene, pyrene, 1,2-benzanthracene, 1,2; 5,6-dibenzanthracene). At higher values of the HOMO energy the oxidation stops at the stage of radical cation (chrysene, coronene, phenanthrene, naphthalene, benzene). Accumulating substituents in these hydrocarbons may facilitate further oxidation. Thus, hexa-, hepta- and octamethylnaphthalenes [403–405] in SbF_5—SO_2ClF are completely converted into dications. The action of SbF_5 and Cl_2 on hexachlorobenzene at low temperature yields the dication $C_6Cl_6^{2+}$ [716] whose ground state, as distinct from the above dications is triplet. No dications have been obtained from hexamethylbenzene with SbF_5 [710].

Convenient tools for diamagnetic dications of aromatic compounds are PMR and particularly NMR-^{13}C [402–405]. Line broadening in the NMR spectra informs on the electron exchange between dications and radical cations [403].

$$ArH^{2+} + ArH^{+\cdot} \rightleftarrows ArH^{+\cdot} + ArH^{2+}$$

The radical cations of the benzene series are essentially less investigated than those of polycyclic hydrocarbons. The radical cation of benzene has been generated by the photoionization of benzene in a frozen sulphuric acid matrix at −110 ÷ ÷ −190 °C [717]. No formation of this particle has been observed in a solution. On the contrary the UV irradiation of a solution of hexamethylbenzene in 98% H_2SO_4 at about 25 °C yields [718] the respective radical cation (*135*)[70] which was also generated by a short-time heating of hexamethylbenzene in 20% oleum [719]. A yellow solution of hexamethylbenzene in 98–100% H_2SO_4 with the protonated hydrocarbon after being kept for a few hours at about 25 °C turns red (λ_{max} = 520 nm) and gives a time-stable ESR spectrum. This spectrum has led to the conclusion [318,720] that it corresponds rather to radical (*136*) which might be formed by intermolecular transfer of a CH_3 group and the subsequent oxidation of the heptamethylbenzenium ion (removal of a hydrogen atom).

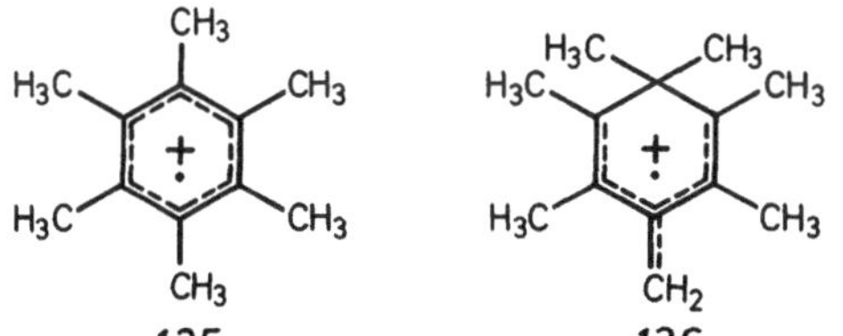

135 *136*

Hexamethylbenzene conversions have a complicated character in such media since heating it in conc. H_2SO_4 results, in the formation of the 2,3,4,5,6-pentamethylbenzyl cation [261,262] (see also [710]).

Radical cations of hexa-, penta- and tetramethylbenzenes, as well as 1,3,5-tri- and 1,4-di-tert-butylbenzenes have been detected by ESR in a flow system after interaction with cobalt triacetate in CF_3COOH [721].

The radical cations of polyfluorinated aromatic compounds proved to be quite stable. Hexafluorobenzene interacted with $O_2^+AsF_6^-$ in WF_5 to yield the salt of the hexafluorobenzene radical cation $C_6F_6^+AsF_6^-$ [709]. Octafluoronaphthalene, decafluorobiphenyl, hexa-, penta- and 1,2,4,5-tetrafluorobenzenes yield radical cations

70 A long-time irradiation results in the formation of the radical cation of duroquinone.

when dissolved in SbF_5 [696,697] and octafluoronaphthalene does so when interacting with SO_3 as well [696]. Radical cations are also formed upon heating or keeping the solutions of the salts of α-nonafluoronaphthalenium and heptafluorobenzenium ions in SbF_5 [237,238,722]. In the latter case the accumulation of radical cation is accompanied by the disproportionation products — hexafluorobenzene and perfluorocyclohexene [722]:

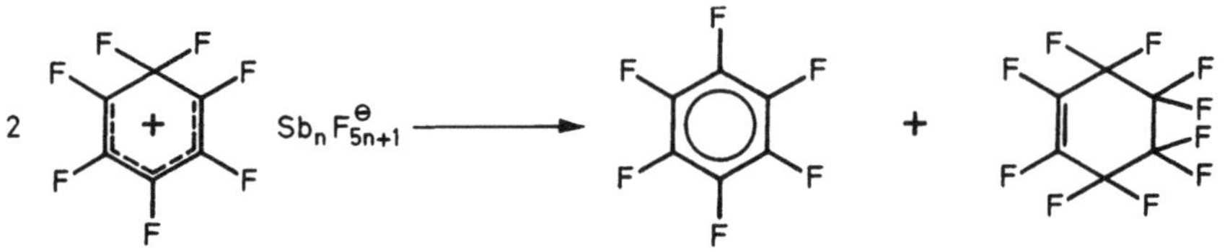

It remains unclear, however, if the radical cation is the primary conversion product of the heptafluorobenzenium ion responsible for the disproportionation reaction or whether it is formed from hexafluorobenzene appearing in the system due to this reaction.

p-Dialkoxybenzenes and their radical cations: The ESR spectrum of p-dimethoxybenzene (*137*) in conc. H_2SO_4, at 20 °C shows the radical cations (*138a*) and (*138b*) differing in the spatial orientation of the CH_3O group [723]. The concentration of interconverting radical cations at about 20 °C accounts for no more than 0.025% of the theoretically possible amount.

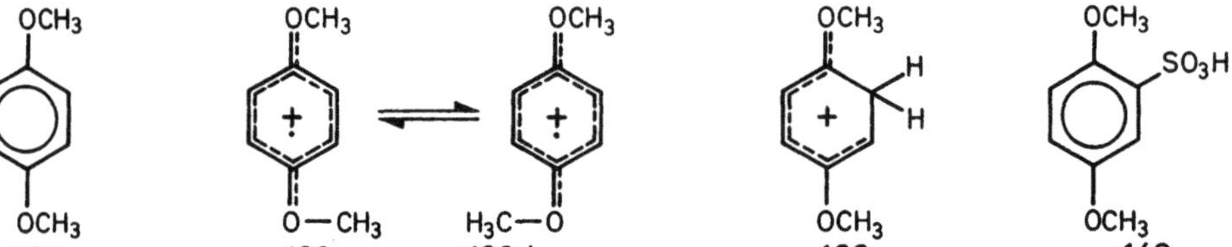

137 138a 138b 139 140

The electronic absorption spectrum, the PMR spectrum and the exchange of ring hydrogen for deuterium in passing from H_2SO_4 to D_2SO_4 indicate these particles to be in equilibrium with the parent p-dimethoxybenzene and the corresponding benzenium ion (*139*). Keeping the solution at room temperature for a long-time or heating it to 90 °C results in the sulphonic acid (*140*). However, p-dimethoxybenzene in conc. H_2SO_4 is sulphonated very fast (within a few minutes at about 25 °C); the low concentration of radical cations (*138a*) and (*138b*) is probably due to a shift of the equilibrium between p-dimethoxybenzene and its sulphonic acid to the side of the acid [724].

The role of radical cations in the alkylation of p-dialkoxybenzenes and dealkylation of alkyl-substituted p-dialkoxybenzenes is elucidated [724,725]. When p-dimethoxybenzene is dissolved in conc. H_2SO_4 with some isobutene or tert-butyl alcohol the observed ESR spectrum corresponds to the radical cation of di-tert-butyl-p-dimethoxybenzene (apparently 2,3-di-tert-butyl-isomer [726] rather than 2,5-isomer as originally assumed [724,725]). If, however, isobutene is passed through a solution of p-dimethoxybenzene in conc. H_2SO_4 in which radical cations (*138a*) and (*138b*) are present, no radical cation of di-tert-butyl-p-dimethoxybenzene is formed. On this basis it was concluded that radical cations, in contrast to neutral molecules, are not involved in the alkylation and hence cannot be intermediates of the reaction product. A similar conclusion was made an the dealkylation, but the reasons given have to be reconsidered after the publication of [726].

The ESR spectra [727] show radical cations of some tert-butylated phenols when they interact with CH_3COCl and $(CH_3)_3CCl$ in the presence of $AlCl_3$ as well as in the conditions of nitration and bromination. In a number of cases the phenoxyl radicals are formed, probably by the proton loss from the radical cation. Therefore the reactions of tert-butylated phenols with electrophiles have been assumed to follow the scheme of electron transfer:

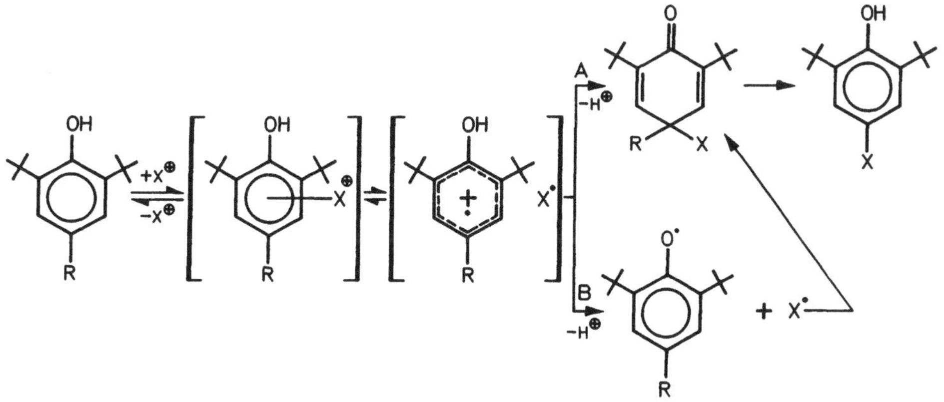

where A is the "process in the cage" and B is the reaction of radicals in solution.

Radical cations are formed by interaction of hexa-, penta- and tetramethylbenzenes with Cl_2 in CH_3COOH [728] and in the reaction of phenothiazine, phenoxazine and 5,10-dihydrophenazine derivatives with nitrosonium tetrafluoroborate [729].

Having determined the oxidation potential of $NO_2^{\cdot}$ in CH_3CN and compared it with that of aromatic compounds Ch. Perrin [730] concluded that for all compounds more reactive than toluene the electron transfer to NO_2^+ is exothermal; so the nitration follows the scheme of electron transfer forming a radical cation intermediate and $NO_2^{\cdot}$.

$$ArH + NO_2^+ \rightarrow [ArH^{+\cdot} + NO_2^{\cdot}] \rightarrow Ar^+\langle^{H}_{NO_2}$$

This scheme explains the loss of substrate selectivity in nitration with nitronium salts and the high position selectivity [731] which is determined by the spin density distribution in the radical cation (see above) and by the relative stability of the formed σ-complexes. In favour of this scheme naphthalene is electrochemically oxidized in CH_3CN with $NO_2^{\cdot}$ at a lower oxidation potential than that of $NO_2^{\cdot}$ but sufficient to form the radical cation of naphthalene [730]; 1- and 2-nitronaphthalenes are formed in the same ratio (9:1) as from nitration by HNO_3 and H_2SO_4 in CH_3CN. Later, however, the formation of nitronaphthalenes was shown [732] to be mainly, if not completely, due to the nitration of naphthalene by N_2O_4 catalyzed by the acid formed on the anode. This should also be considered when generating radical cations to study their role in electrophilic substitution through oxidation of hydrocarbons with salts of tetravalent cerium (cf. [733]), because the acid is formed during the reaction.

The generation and behaviour in acid media of the radical cations of aromatic compounds investigated by ESR and absorption spectroscopy in the UV and visible regions do not allow us to draw unequivocal conclusions for the role of similar particles in the formation and conversions of arenium ions and hence for electrophilic aromatic substitutions. At a certain ratio of energetic levels the interaction of ArH and X^+ may be accompanied by the formation of the radical pair [ArH^{+}, $X^{\cdot}$] which is in equilibrium with the free radical ArH^{+} and $X^{\cdot}$, the binding of the $X^{\cdot}$ radicals leading to accumulation in the system of the $ArH^{+\cdot}$ radical cation. It remains unclear, however, whether the formation and recombination of the radicals $ArH^{+\cdot}$ and $X^{\cdot}$ (route b, c) may be more favourable than the direct route (a) of σ-complex formation.

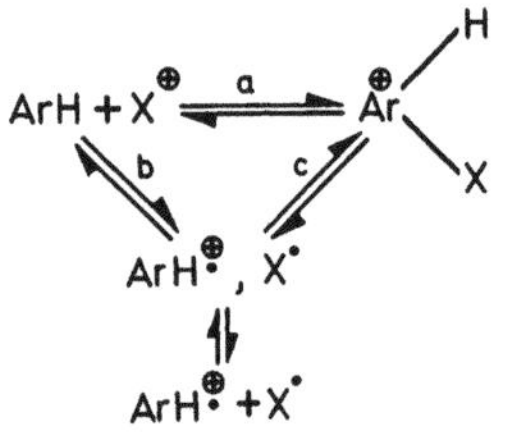

In keeping with the microscopic reversibility principle this problem is also concerned with the routes of decomposition of σ-complexes (routes a and c, b).

Attempts have been made to establish the intermediate formation of the radicals ArH^{+} and $X^{\cdot}$ by using the phenomenon of chemically induced dynamic nuclear polarization [734,735]. For example, the interaction of benzenediazonium tetrafluoroborate prepared from aniline-^{15}N and $H^{15}NO_2$ with sodium phenolate in methanol is accompanied by the polarization of the nuclei ^{15}N of diazonium salt and the formed azocompound [736]. This implies that the reaction following the scheme:

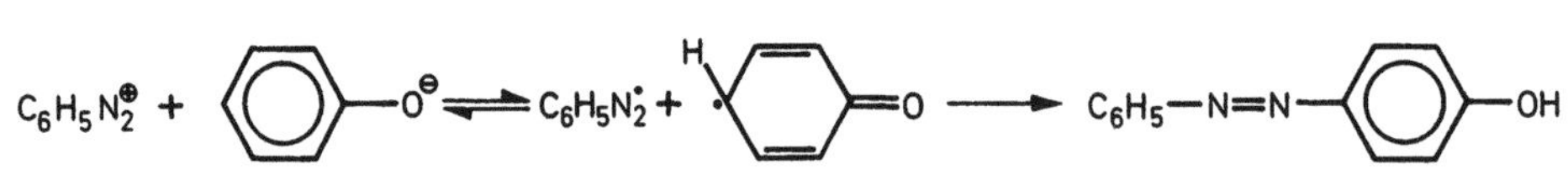

and the polarization of the ^{15}N nuclei of diazonium cation points to the reversibility of the electron transfer. However, in this scheme one might expect a simultaneous polarization of carbon and hydrogen in the phenol component which is not reported [736]. Consequently, the polarization results from the side radical reactions of the diazocomponent [737].

Attempts to reveal the nuclei polarization effect in the nitration of benzene by nitronium tetrafluoroborate failed [738].

In conclusion one should recall the triaminobenzenium ion conversions (Section IV.4) in which the radical cations seem to take part [197,266,267]. The acid-catalized transfer of the nitro group to the ring in the rearrangement of aromatic N-nitroamines [739,740], probably proceeds via the intermediate radical pair [$Ar(R)N^{+\cdot}$, $NO_2^{\cdot}$].

7 Formation and Conversion of Arenium Ions in Reactions of Aromatic Compounds with Electrophiles

The combination of the electrophilic substitution of hydrogen with the nucleophilic substitution of groups other than hydrogen supplies many methods for processing aromatic raw materials. Therefore, the interaction of aromatic compounds with electrophiles has been dealt with in many works. Nevertheless the mechanism of these reactions and the quantitative reactivity of aromatic compounds is not yet clear, as shown by recent studies.

The previous sections leave no doubts that aromatic compounds react with positively charged electrophiles to form σ-complexes-arenium ions. But are they the primary intermediates? It is not by accident that the problem of preliminary formation of radical cations has arisen. Its statement is an attempts to explain the orientational peculiarities of electrophilic aromatic substitution of hydrogen. The widespread view [9,11] that the orientation in the reactions of aromatic compounds with electrophiles is dictated by the relative stabilities of the σ-complexes explains but a part of the accumulated material. In the first place this refers to the meta- and para-orienting effects of electron-releasing substituents in benzene in terms of the $\varrho\sigma^+$-approach [741] and to that of the relative reactivity of various aromatic substrates by comparing it with the measured or calculated stability of the respective σ-complexes [11].

In recent years, however, large evidence indicates that the orientation effects in the reactions of aromatic compounds with electrophiles are not only determined by the thermodynamic characteristics of σ-complexes. This is evidenced, in particular, by kinetic control in the protonation of aromatic compounds, i.e. cases of the primary formation of less stable arenium ions which are then rearranged into thermodynamically more favourable ions [81,201,203,204,265] (see the preceding sections).

In many cases the reactions with electrophiles of monosubstituted benzenes containing electron-releasing substituents yield comparable amounts of ortho and para isomers whereas the corresponding σ-complexes essentially differ in their stability. Thus, in the hydrogen exchange for toluene the ratio f_0/f_p increasing with rising medium acidity [742] approaches unity. At the same time in the protonation of monosubstituted benzenes (R = CH_3, $CH_2Si(CH_3)_3$, OH, OCH_3, F) σ-complexes of para-structure are strongly preferable (see Sect. II and IV.1).

Of particular interest are ipso-additions [641,743,744] of an electrophile, since the respective σ-complexes are in most cases unfavourable from the thermodynamic point of view. Therefore, one must be careful to draw conclusions on the initial ratio of isomeric σ-complexes from the ratio of isomeric reaction products. Thus, the nitration of 4-isopropyltoluene (p-cymene) by nitronium tetrafluoroborate in sulpholane at 25 °C gives 2-nitro- and 3-nitro-4-isopropyltoluenes as well as p-nitrotoluene with the yields 85.2, 5.3 and 9.5%, respectively [745]. Accordingly the reactivity ratio of the 2- and 3-positions of p-cymene in the nitration is 16:1. However, in the nitration of p-cymene by acetylnitrate in acetic anhydride at 0 °C among the above reaction products (with 41.8 and 10% yields, respectively) there is a considerable amount (41%) of cis- and trans-isomers of 1-nitro-4-acetoxy-1-methyl-4-isopropylcyclohexa-2,5-diene [746]. This may be due to the "capture" of the acetate-anion by the σ-complex from the addition of the nitronium cation at the 1-position of p-cymene (for more details on nitroacetoxylation see [747,748]). With acidic agents the nitroacetoxy derivative is smoothly converted into 2-nitro-4-isopropyltoluene.

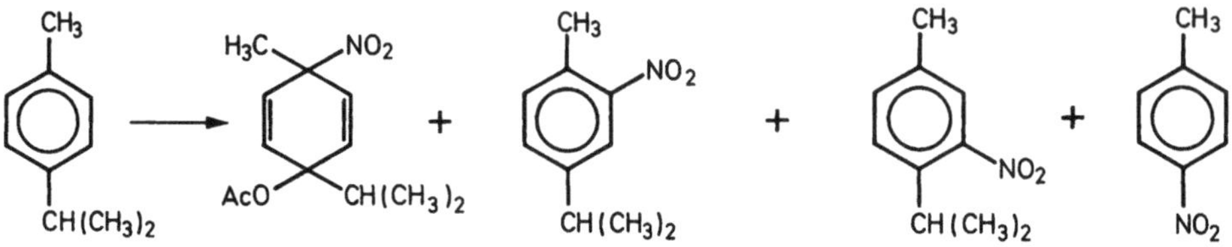

This allows one to assume that even by nitrating with nitronium tetrafluoroborate 2-nitro-4-isopropyltoluene is to a great extent formed by the initial addition of the nitronium cation at the 1-position and the subsequent shift of the nitro group to the 2-position. In this case discussing the relative "reactivity" of the 2- and 3-positions from the ratio of the formed 2- and 3-nitro derivatives is meaningless.

A still more striking example is the interaction of pentamethylbenzene with the nitronium cation [173]. The initial product is only ion (*24* X = CH_3) by additing the nitronium cation at the 2-position. Further 1,2-shifts of the NO_2 group to the unsubstituted carbon finally yields pentamethylnitrobenzene.

Similarly, in 2,3,5,6-tetramethylanisole the nitronium cation first attaches the 2-position and then is transferred to the unsubstituted 4-position with the formation, of 4-nitro-2,3,5,6-tetramethyanisole [202].

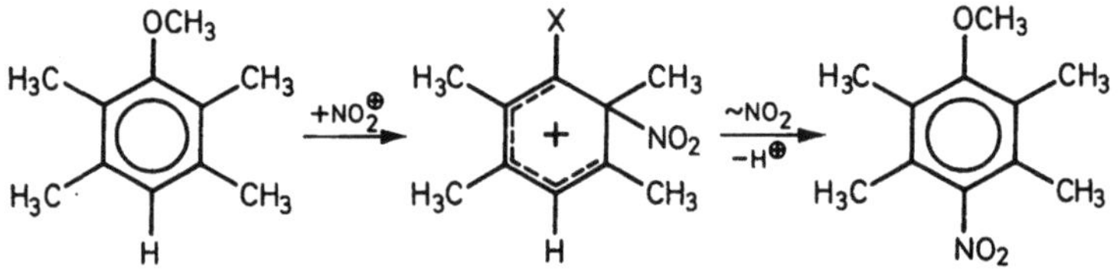

24, X = OCH_3

Recently [749,750] peculiarities of orientational effects observed when aromatic compounds interact with electrophiles were explained by assuming the π-complexes [751] formation being able to affect the ratio of the further formed σ-complexes. Such a conclusion was earlier made by G. Olah [532,731]; he showed that the reactions with active electrophiles fail to comply with the well-known selectivity relationship [741]. The latter implies that the kinetics of electrophilic substitution is determined by the thermodynamic characteristics of intermediate σ-complexes, but the concurrent nitration of toluene and benzene by nitronium salts in polar solvents shows an extremely low substrate selectivity ($k_{C_6H_5CH_2}:k_{C_6H_6} \approx 1.6$) with a high positional selectivity retained. In such cases the rate-determining stage seems to be the π-complex formation. Since by π-complex formation the π-electron system is not strongly perturbed, the substituents cannot display the electronic effect to the same extent as in the formation of σ-complexes; thereby the difference in the reactivity between substituted and unsubstituted aromatic compounds is decreased. At the same time the substituents may strongly affect the barriers on the routes leading to isomeric σ-complexes determining thereby the positional selectivity.

This interpretation [532,731] has been doubted [752] since the use of very active electrophiles makes it difficult to ensure the required completeness of mixing the reagents [753,754]. Moreover, when the reaction rate approaches the limiting value determined by the probability of substrate molecules encountering the electrophile the notion "substrate selectivity" itself loses its sense [755,756]. However, British scientists have demonstrated that the positional selectivity is retained in the nitration of methylbenzenes when the encounter rate has been attained [185,757]. They believe that at the first stage at which the rate is controlled by diffusion a certain intermediate, an encounter pair, is formed; this intermediate is more converted into σ-complexes than it is "split" into components (ArH and NO_2^+) separated by the molecules of the medium. This point of view leaves the intermediate preceding the formation of σ-complexes open to question.

In connection with the possible role of π-complexes attempt was made [749] to explain the orientation in electrophilic hydroxylation by ipso-attack. Table 57 lists the benzene derivatives hydroxylated by hydrogen peroxide in HSO_3F-SO_2ClF at −78 °C [758]. Part of the reaction products are formed with a shift of the CH_3 group. Since the parent compounds do not isomerize under reaction conditions the shift of the CH_3 group must be realized in hydrobenzenium ions formed as HO^+ is added to the substituted ring carbons.

Table 57. Isomer Distribution for the Hydroxylation of Substituted Benzenes ($H_2O_2-HSO_3F-SO_2ClF$, −78 °C) [758]

Substituted benzenes	Ratio of isomeric phenols, % (in parentheses position of substituents)		
C_6H_5F	24 (2)	3 (3)	73 (4)
$C_6H_5CH_3$	71 (2)	6 (3)	23 (4)
1,2-$C_6H_4(CH_3)_2$	59 (2,3)	29 (3,4)	12 (2,6)
1,3-$C_6H_4(CH_3)_2$	1 (2,3)	82 (2,4)	2 (2,5), 16 (2,6)
1,4-$C_6H_4(CH_3)_2$	64 (2,5)	36 (2,4)	
1,2,3-$C_6H_3(CH_3)_3$	3 (2,3,6)	91 (2,3,4)	6 (3,4,5)
1,2,4-$C_6H_3(CH_3)_3$	9 (2,4,6)	30 (2,3,6)	61 (2,3,5+3,4,6)
1,3,5-$C_6H_3(CH_3)_3$	100 (2,4,6)		

The composition of the hydroxylation products of substituted benzenes contradicts the assumption that primarily σ-complexes are formed, controlling the process by their relative stability, i.e. by the thermodynamic factors. E.g., in the hydroxylation of toluene the amount of the formed o-cresol is three times as large as that of p-isomer although $\sigma^+_{o\text{-}CH_3} = -0.25$ and $\sigma^+_{p\text{-}CH_3} = -0.31$.

Therefore it was suggested [749] that the hydroxylating electrophile is first added to the C^i-----C^j bond of the benzene ring forming a π-complex of the A type[71] or a

---

71 Below the structural denotation of the π-complex will be used conventionally.

protonated oxide B which are further converted into σ-complexes with the OH group at the C^i or C^j atom.

(A) (B)

The initial addition of the electrophile must be controlled by its electrostatic interaction with the atoms C^i and C^j carrying the partial charges q_i and q_j as well as by the occupied π-molecular orbitals of the substrate in the zone of the C^i-----C^j bond. Therefore the energy of electrophile bonding at the initial stage of interaction can be expressed by the following equation (cf. [759]):

$$E^{\pi}_{ij} = A_1(q_i + q_j) + B_1 \sum \frac{(c^k_i - c^k_j)^2}{E_k - E_e}$$

where c^k_i is the coefficient of the atomic orbital of the C^i atom in the k-th occupied molecular orbital of the substrate possessing the energy E_k; A_1 and B_1 depend on the nature of the electrophile whose vacant orbital has the energy E_e.

The concurrent conversion of the π-complex or the protonation oxide into the i- and j-σ-complexes is most likely controlled by their relative thermodynamic stabilities which can be described by the modified Hammett equation of the $\varrho\sigma^+$-type (Sect. IV.1). Assuming the main contribution to orbital interaction to be made by the highest occupied π-molecular orbital (HOMO) of the substrate (see [760]) and neglecting for a first approximation, the difference in the HOMO energies of various substrates one can describe the positional selectivity of electrophile addition to the i- and m-positions of the aromatic substrate by the expression:

$$\lg \frac{k_{i(i,j)}}{k_{m(m,n)}} = A[(q_i + q_j) - (q_m + q_n)] + B[(c_i + c_j)^2 - (c_m + c_n)^2] + D[\Sigma\sigma^+(i) - \Sigma\sigma^+(m)]$$

where $\Sigma\sigma^+(i)$ is the sum of the σ^+-constants of the substituents in the σ-complex formed upon addition of HO^+ to the C^i atom, and c_i, c_j, c_m and c_n are the coefficients relating to the HOMO.

In the methylbenzene hydroxylation the ions having the OH or CH_3 groups and a hydrogen atom at the ring sp^3-hybridized carbon are assumed to be converted via rapid 1,2-hydrogen shifts into the most stable protonated phenol while the ions carrying at the indicated carbon the OH and CH_3 groups are rearranged via a 1,2-shift of the CH_3 group.

An example is the scheme of the toluene hydroxylation:

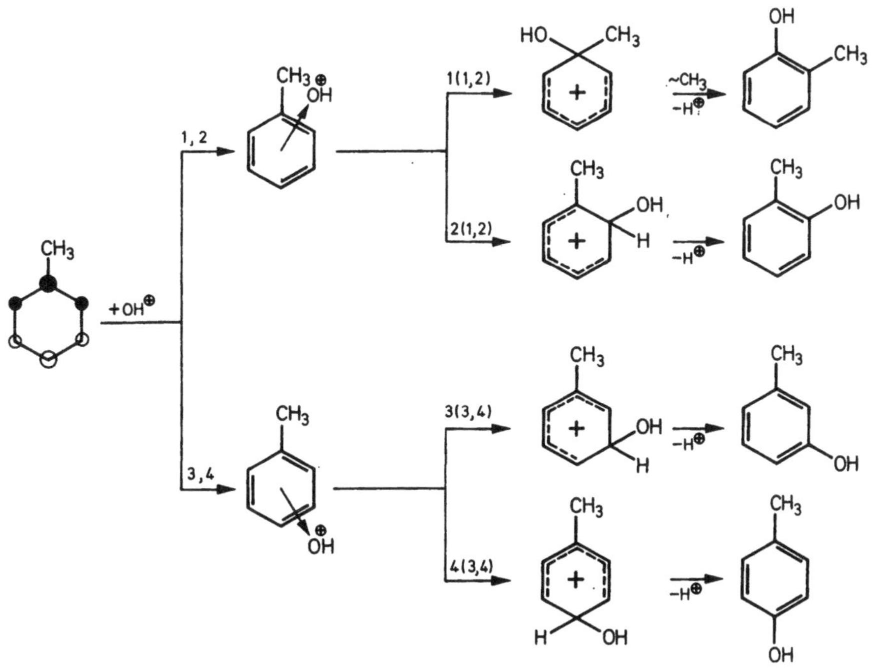

Since the HOMO of toluene has a node in the zone of the C^2—C^3(C^5—C^6) bond, the formation of complexes along the 2,3-route can be neglected. Consequently p- and m-cresols are formed through channels 4(3,4) and 3(3,4), respectively, o-cresol through 2(1,2) and 1(1,2).

Applying the values of q and c calculated by CNDO/2 and those of the σ^+-constants cited in Table 44 the authors of [749], from the composition of the hydroxylation products to toluene and fluorobenzene having a similar HOMO, found the values of $A = -3.8$, $B = +5.8$ and $D = -2.4$. The reverse calculation using these coefficient values gave the following relative probabilities for the toluene hydroxylation — ipso:ortho:meta:para = 9.0:62.1:6.1:22.8 ($\Sigma = 100$). Thus about 12% of the ortho-cresol obtained is, according to the calculation, formed through channel 1(1,2) with the CH_3-group shift. This conclusion has been experimentally confirmed by hydroxylation of toluene-1-^{13}C [749].

The high fraction of ortho-isomer in the toluene hydroxylation product as compared with para-isomer is due to orbital control ($c_1 + c_2 > c_3 + c_4$) at the initial binding of the hydroxylating agent. The reversed ratio of the formed ortho- and para-isomers in passing from toluene to fluorobenzene accounts for the difference of the thermodynamic and the charge factors for the reactions of these compounds.

The above values of the A, B and D coefficients can explain the principal regularities of orientation in the hydroxylation of other compounds. Thus, in p-xylene the orbitally controlled addition of HO^+ has to be realized only by one route with the subsequent formation of 2,4- and 2,5-xylenols in a ratio which is determined by the thermodynamic term of the equation mentioned on the page 206.

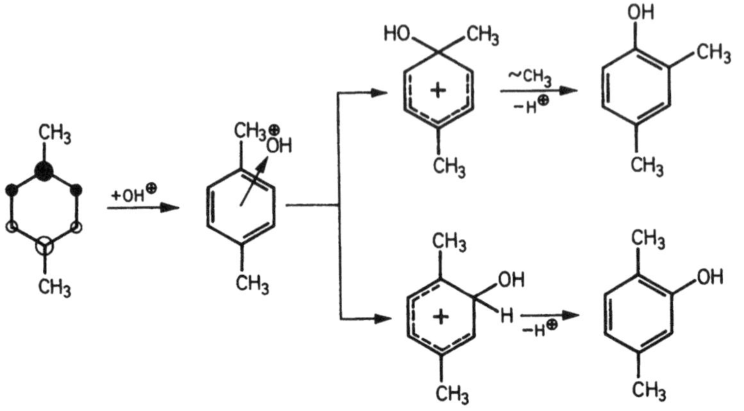

This ratio is calculated 35:65 which agrees well with experimental data (36:64).

For pentamethylbenzene having a different symmetry of the HOMO the hydroxylation is calculated to proceed practically only with the participation of the $C^1—C^2$ ($C^4—C^5$) bonds to form σ-complexes corresponding with the addition of HO^+ at the 1(5)- and 2(4)-positions in a ratio of 27:73. The expected routes of further conversions of these σ-complexes are shown in the following scheme:

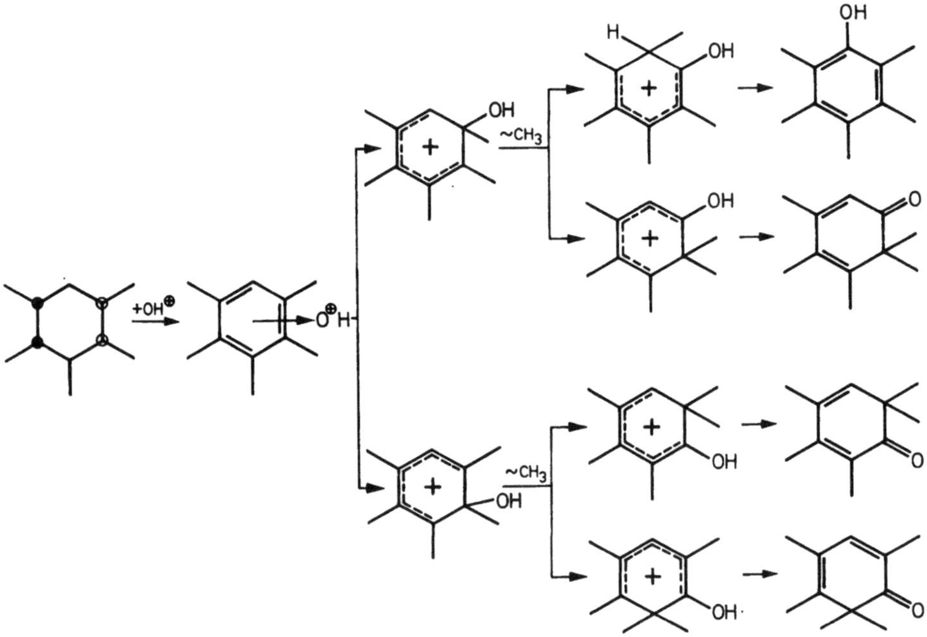

Hydroxylation of pentamethylbenzene gave 12% of pentamethylphenol, 21% of 3,4,5,6,6-, 30% of 2,3,4,6,6- and 34% of 2,4,5,6,6-pentamethylcyclohexa-2,4-dienones which agrees fairly well (33:67) with the predicted directions of the reaction.

An attempt was made [750] to consider from the same positions the reactions of benzene derivatives with other electrophiles.

The reactions of monosubstituted benzenes should above all be analysed because so far their comparison has been limited to the meta- and para-orientations while the ortho-substitution was assumed to be affected by sterical factors. However, this seems to be exaggerated since the amount of the formed ortho-isomer often exceeds that of the para-isomer by more than twice. It is essential, therefore, to describe the whole complex of data on orientation in monosubstituted benzenes.

The highest π-molecular orbital of monosubstituted benzenes with substituents of the ortho- and para-orienting type usually has the symmetry b_1, and with those of the meta-orienting type the symmetry a_2.

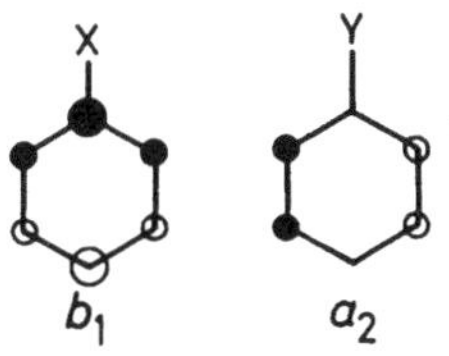

In the first case the π-addition of the electrophile must mainly involve the C^1—C^2 (C^1—C^6) and C^3—C^4 (C^4—C^5) bonds, the first route leading to the formation of 1- and 2(6)-σ-complexes (ipso- and ortho-addition). The ratio of meta- and para-products of substitution is only determined by the thermodynamic term, and that of ortho- and para-products is also affected by orbital and charge factors. Predominant for the a_2 orbital is, on the contrary, the addition to the C^2—C^3 (C^5—C^6) bond with the subsequent formation of 2(6)- and 3(5)-σ-complexes in a ratio determined by their relative thermodynamic stability. The low probability of electrophile addition to the C^3—C^4 (C^4—C^5) bond precludes the formation of significant amounts of para-isomer.

The ratio of ortho-, and para-isomers formed in reactions of monosubstituted benzenes with electrophiles is presented in Table 58. The values of q and c required for this analysis were calculated by the CNDO/2 method. For the +I- and +M-type substituents missing in Table 44 the σ^+_{ortho} values are calculated from the relationship $\sigma^+_{ortho} = 0.07 + 0.95\,\sigma^+_{para}$ [515]; for the —M-type substituents, as a first approximation, it is taken that $\sigma^+_{ortho} \approx \sigma^+_{para}$.

Table 58 shows that on the whole the approach suggested reflects rather well the main regularities of orientation in the considered reactions.

For reactions in which the formation of π-complexes does not essentially affect the ratio of the further formed isomeric σ-complexes the equation is transformed, by neglecting the first two terms (charge and orbital control), into a conventional Hammett-type equation. This may occur either in the formation of very labile π-complexes (readily formed and as readily decomposed) or if the formed σ-complex is readily rearranged into isomeric σ-complexes via the 1,2-shift of the added electrophile. The latter is only observed for X groups with high migration ability since at a sufficiently high activation barrier for the 1,2-shift of the added electrophile the concurrently more effective route is the migration or loss of the proton. The groups of high migration ability involve the bromine atom (see Sect. IV.2.B).

Assumably the electrophilic bromination, at least if the formation of the σ-bond C—Br precedes the proton loss (i.e. the process is not synchronous), can be described by a conventional Hammett-type equation with the σ^+-constants of substituents. Indeed, for the bromination of, say, benzene derivatives in NaBr—$HClO_4$ [764] the logarithms of the rate constants correlate well with the sums of the σ^+-constants of substituents including the ortho-substitution.

Table 58. Observed and Calculated (in parentheses) Isomer Distribution in Electrophilic Substitution of Monosubstituted Benzenes [749] (Data used for determining the parameters A, B and D are given by italics)

Substituents	ortho	meta	para
	Hydroxylation, HOF + HF, 25 °C [761] A = −2.8, B = +7.1, D = −2.9 [749]		
CH_3	*77.5* (77.5)	*3.8* (3.8)	*18.7* (18.7)
F	*34.4* (31.0)	0.0 (9.2)	*65.6* (59.8)
Cl	61.2 (58.9)	2.5 (5.2)	36.3 (35.9)
OCH_3	69.5 (61.0)	2.8 (0.1)	27.7 (38.9)
NO_2	37.7 (31.0)	63.3 (69)	0.0 (0.0)
	Chlorination, Cl_2 + $HClO_4$ + $AgClO_4$, 25 °C [762] A = −8.6, B = +8.9, D = −5.0 [749]		
CH_3	*74.7* (75.9)	*2.2* (1.4)	*23.1* (22.7)
Cl	*36.4* (36.1)	*1.3* (2.2)	*62.3* (61.7)
OCH_3	34.9 (29.6)	0.0 (0.0)	65.1 (73.4)
NO_2	17.6 (19.9)	80.9 (79.1)	1.5 (0.0)
CN	23.2 (24.0)	73.9 (80.0)	2.9 (0.0)
CHO	30.7 (22.0)	63.5 (78.0)	5.8 (0.0)
CF_3	15.7 (16.6)	80.2 (83.4)	4.1 (0.0)
	Nitration, CH_3ONO_2 + BF_3, 25 °C [763] A = −3.8, B = +5.8, D = −6.5 [749]		
CH_3	63.9 (60.1)	3.4 (1.1)	32.7 (38.8)
F	11.1 (8.7)	0.3 (0.2)	88.6 (91.1)
Cl	31.4 (36.7)	1.0 (0.8)	67.6 (62.5)
OCH_3	27.9 (23.3)	1.2 (0.0)	70.9 (76.7)

It is difficult to decide whether the above approach is generally applicable to the reactions of aromatic compounds with electrophiles. It seems desirable to compare it further with the intermediate formation of radical cations of aromatic substrates (see Sect. IV.6) whose conversion into σ-complex is controlled by the spin density distribution and by the relative stability of σ-complexes (cf. [730]). In the latter case the charge term of the equation vanishes and in the orbital term the squares of the sums of atomic orbital coefficients $(c_i + c_j)^2$ and $(c_m + c_n)^2$ should be replaced by c_i^2 and c_m^2. Qualitatively these two approaches must yield close results, but according to the tentative conclusion the extent of the quantitative agreement of calculated and experimental data on the distribution of isomers is higher for the scheme with the intermediate formation of π-complexes.

8 Photochemical Conversions of Arenium Ions and the Generation of their Valence Isomers

The Woodward-Hoffmann concept for concerted electrocyclic reactions guided Childs and Winstein to conclude that benzenium ions should rearrange photochemically to bicyclo[3.1.0]hexenyl cations [253,277,767,771]. When they irradiated solutions of penta-, hexa-, and heptamethylbenzenium ions in HSO_3F at −78 °C with UV light of the long-wave absorption maximum of benzenium ion they observed the PMR spectra of the expected methyl-substituted bicyclo[3.1.0]hexenyl cations (*142*) (*143*). The cyclopropane ring in (*142*) and (*143*) was corroborated by the measurement of the $J_{^{13}C-^{1}H}$ constant for the $C(CH_3)$—H fragment [766].

Similar photochemical conversions were later carried out for other benzenium ions.

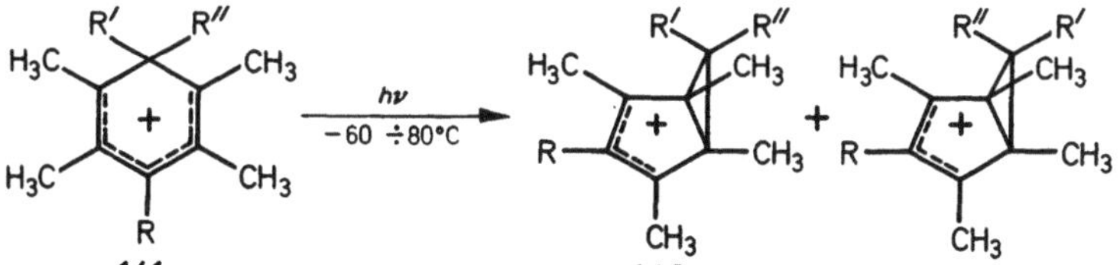

141 *142* *143*

a) $R=CH_3$, $R'=R''=H$ [277,767]; b) $R=R'=CH_3$, $R''=H$ [253];
c) $R=R'=R''=CH_3$ [253,771]; d) $R=R'=CH_3$, $R''=C_2H_5$ [254];
e) $R=R'=CH_3$, $R''=CH_2Cl$ [254]; f) $R=R'=R''=H$ [255,767].

The photochemical conversion of the protonated isodurene into a respective bicyclo[3.1.0]hexenyl cation ended in failure [767]. An attempt to generate this ion indirectly has shown it to be unstable (since it contains a CH_3 group only at one end of the enylic fragment) and to rapidly undergo reverse rearrangement.

Since protonated durene (*141f*) yields the 1,2,4,5-tetramethylbicyclo[3.1.0]hexenyl cation (*142f*) and not the 1,2,3,5-isomer [255,767] the photochemical conversion must be a disrotatory electrocyclic process rather than a reaction of the $[\sigma 2a + \pi 2a]$-cycloaddition.

The above scheme shows two isomeric bicyclo[3.1.0]hexenyl cations (*142*) and (*143*) if $R' \neq R''$. In photochemical conversion products of ions (*141d*) and (*141e*) the predominant isomer has the bulkier substituent located in the endo position, i.e. on the same side of the cyclopropane ring as the allylic part of the cyclopentenyl fragment. The signals of isomeric bicyclo[3.1.0]hexenyl cations were assigned on the basis of assumption that protons of exo-R-groups undergo a stronger deshielding influence of the electric field than those of endo-R-groups (see also [254,277,768])[72].

UV irradiation of protonated hexamethylbenzene (*141b*) yields only one of the two possible isomeric bicyclo[3.1.0]hexenyl cations [253] namely the 6-exo-H-1,2,3,4,5,6-hexamethylbicyclo[3.1.0]hexenyl cation (*142b*) [256,768].

---

72 Contrary to this, the signal of the exo-H-atom of the non-substituted bicyclo[3.1.0]hexenyl cation turned out to be in a stronger field than that of the endo-H-atom [257].

The preferable formation of bicyclo[3.1.0]hexenyl cations with the less bulky substituent located in the exo position is explained by smaller steric hindrance to its passage between the two CH_3-groups as the ring sp^3-hybridized carbon leaves the plane of the rest of the ring carbons [254,769,770].

Bicyclo[3.1.0]hexenyl cations can also be generated by protonation of 4-methylene-bicyclo[3.1.0]hexenes-2 at low temperatures

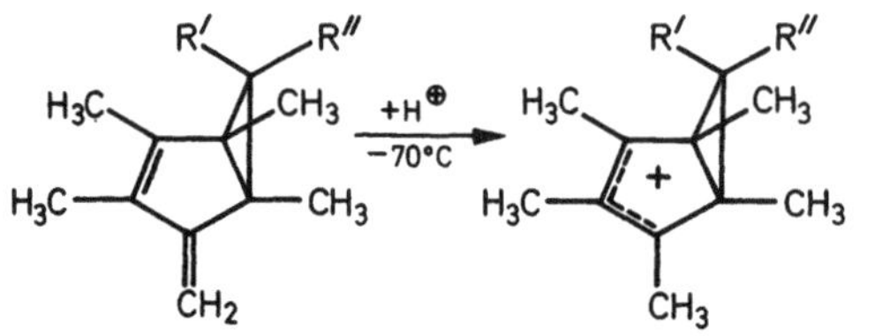

R′ = R″ = H [277] ; R′ = H ; R″ = CH_3 [256,768,771]

or by the detachment of X from 4-X-bicyclo[3.1.0]hexenes-2:

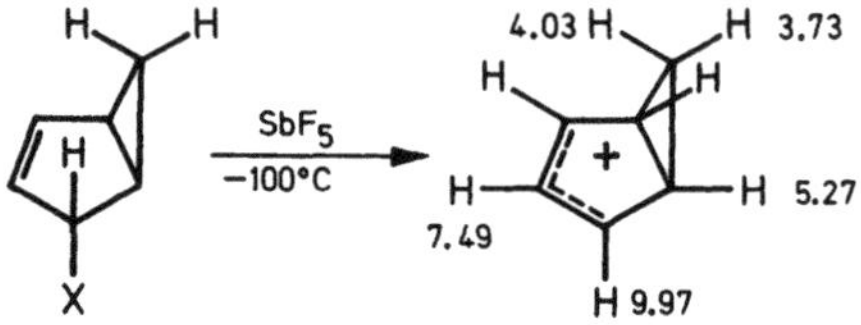

X = Cl [257] , OCH_3 [395]

Representatives of benzbicyclo[3.1.0]hexenyl cations [395] were generated in a similar way.

The 6-exo-H-1,2,3,4,5,6-hexamethylbicyclo[3.1.0]hexenyl cation (*142b*) results from the interaction of 1-(α-chlorethyl)-1,2,3,4,5-pentamethylcyclopentadiene with $AlCl_3$ at −70 °C [256,768]:

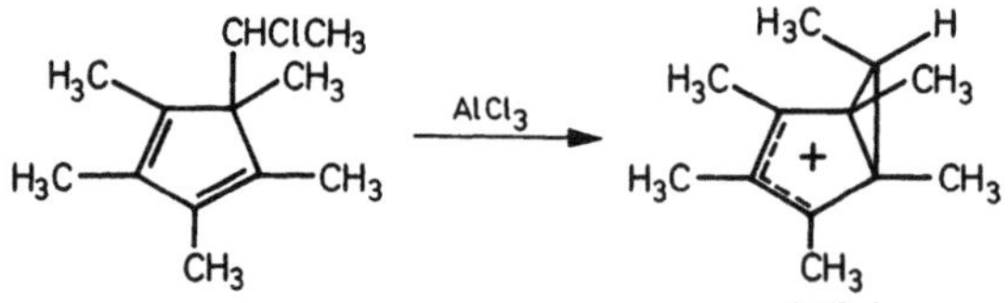

142 b

The reasons for the preferable formation of one isomer are discussed in [268,769].

The PMR spectra of methyl-substituted bicyclo[3.1.0]hexenyl cations are listed in [253–256,277,767,768,771][73]. The ranges of the chemical shifts observed (ppm) are:

6-endo-CH_3	1.1–1.3	6-endo-H	2.5–3.2
6-exo-CH_3	1.4–1.9	6-exo-H	3.4–3.9
1(5)-CH_3	1.7–1.9	1(5)-H	
2(4)-CH_3	2.5–2.9	2(4)-H	...
3-CH_3	1.8–2.1	3-H	6.9–1.1

73 For the non-substituted ion see [257].

The shift of hydrogens at C_6 is stronger down-field in comparison with the position typical of cyclopropane hydrocarbons (by 2–3 ppm). It is largely due to the conjugation of cyclopropane ring σ-bonds with the allylic part of the ion [277,767,768]. This is also concluded by comparing the chemical shifts of the ring carbons of 6-endo-H-1,2,3,4,5,6-hexamethylbicyclo[3.1.0]hexenyl cation (*143b*) and 6-endo-H-4-methylene-1,2,3,5,6-pentamethylbicyclo[3.1.0]hexene-2 [772] (which also reports data on 6-exo isomers), cf. [395].

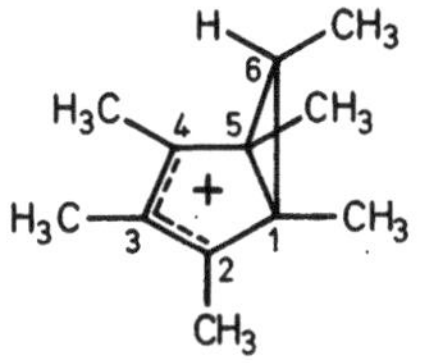

143 b

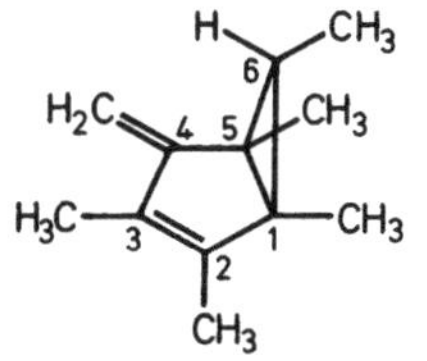

C_1, C_5	55.6	30.2
C_2, C_4	224.4	148.1
C_3	144.6	128.1
C_6	118.9	38.2

The NMR-^{13}C spectra of the non-substituted bicyclo[3.1.0]hexenyl cation led to the same conclusion [282,395].

From the study of selectively deuterium labeled hexamethylbicyclo[3.1.0]hex-3-en-2-one in conc. H_2SO_4 Swatton and Hart concluded that bicyclo[3.1.0]hexenyl cations can rearrange by migration of the $C{<}^{R'}_{R''}$ fragment relative to the five-membered cycle [545]. For 6-R'-6-R''-1,2,3,4,5-pentamethylbicyclo[3.1.0]hexenyl cations this rearrangement has been observed through the merging of CH_3 group signals with rising temperature [253,254,256,771,773,774].

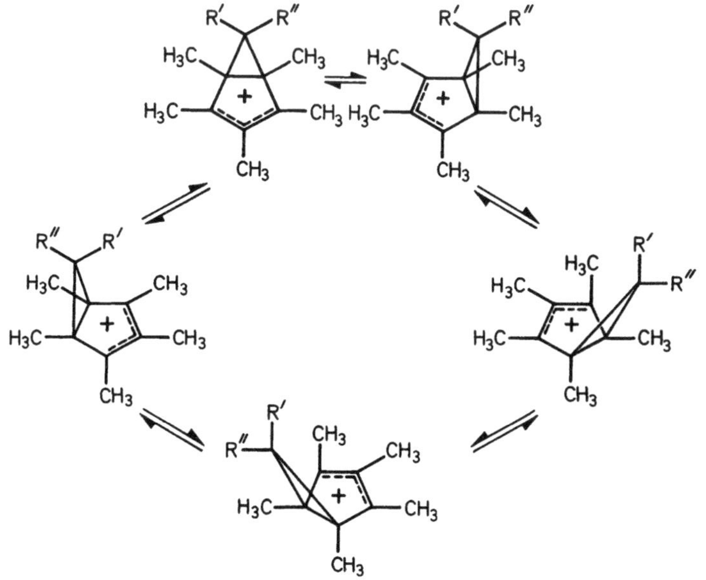

The kinetic characteristics of this rearrangement depend on the nature of R′ and R″ [253–255,771,774].

For the non-substituted bicyclo[3.1.0]hexenyl cation the degenerate rearrangement proceeds far more slowly and it could only be observed by labeled atoms [257]. The quantum chemical analysis of this process is made in [777].

Another rearrangement of bicyclo[3.1.0]hexenyl cations is the C_1—C_5 bond rupture to the benzenium ions [254–256,277,767,771,774]:

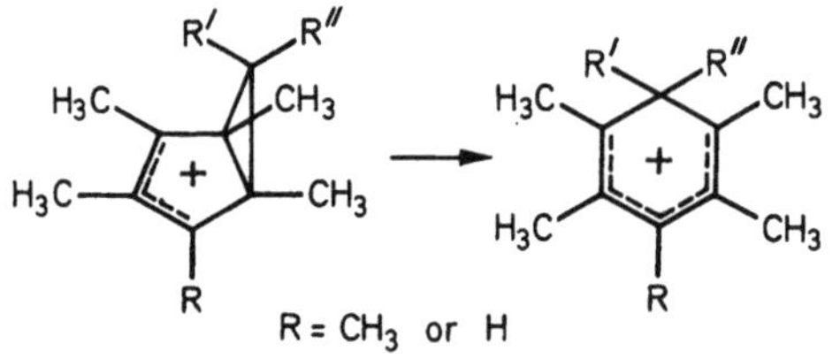

The rate of this rearrangement is determined by the volume of R″ passing between the two CH_3 groups. For R″ = H the free activation energy is 17–17.5 kcal/mole and the reaction at a marked rate ($k = 10^{-3} - 10^{-4}\ s^{-1}$) proceeds at —30 ÷ —40 °C. But if R″ = CH_3, the value of ΔG^{*} increases by 3–4 kcal/mole and the same rate is reached at about 0 °C or higher.

The rather high energetic barrier preventing the isomerization of bicyclo[3.1.0]-hexenyl cations into benzenium ions should not surprise since, according to Woodward-Hoffmann, the disrotatory electrocyclic reactions are "forbidden" for the ground state of the cations in question. Arguments in favour of this conversion type for rearrangement of bicyclo[3.1.0]hexenyl cations into benzenium ions are given in [767].

The benzbicyclo[3.1.0]hexenyl cation converts into the 1-H-naphthalenium ion at —50 °C in HSO_3F—SO_2ClF [395].

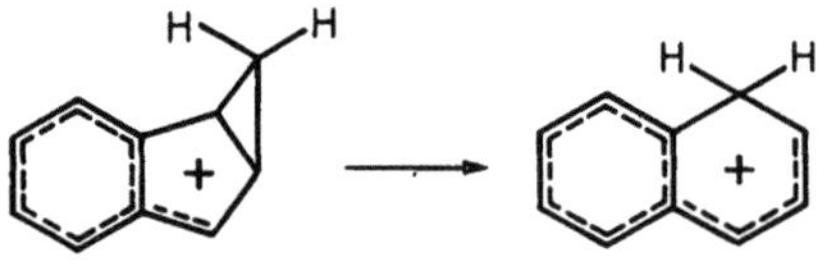

The photochemical conversions of methylbenzenium ions and the properties of the resulting bicyclo[3.1.0]hexenyl cations explain the results obtained from the photoisomerization of hydroxybenzenium ions. For example, when 1,1-dimethyl-2-hydroxy- and 1,1-dimethyl-4-hydroxybenzenium ions (*144*) and (*114*) (λ_{max} 363 and 298 nm) are irradiated by UV light with $\lambda > 325$ nm and $\lambda > 275$ nm resp. in HSO_3F at below —60 °C, the C-prótonated 3,4- and 2,4-dimethylphenols are observed [325]. Since both the initial and the resulting cations do not rearrange without irradiation, ions (*144*) and (*114*) were assumed to convert photochemically into the 1- and 3-hydroxy-6,6-dimethylbicyclo[3.1.0]hexenyl cations (*145*) and (*146*) which are isomerized, through the $C(CH_3)_2$ fragment transfer, into the more stable 2-hydroxy isomer (*147*) in which the hydroxy group occupies the most electron-deficient position. The subsequent opening of the cyclopropane ring through the C_1—C_5 bond rupture (disrotatory reaction permitted by the conservation of orbital symmetry for the first excited state) forms the 1,2-dimethyl-3-hydroxybenzenium ion (*148*) which further isomerizes into the more stable ions (*149*) and (*41*).

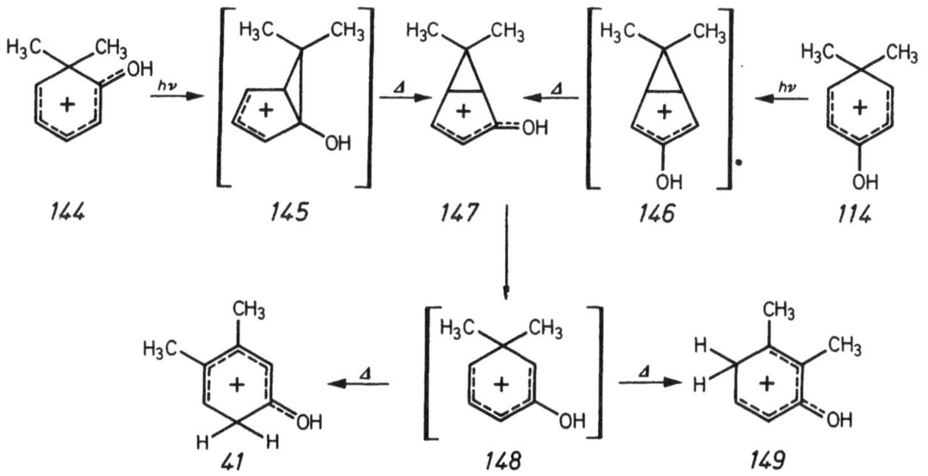

The intermediate cation (*147*) was recorded by PMR.

Similarly, 1-methyl-1-dichloromethyl-2-hydroxy- and 4-hydroxybenzenium ions are photochemically isomerized into 2-methyl-5-hydroxybenzaldehyde [427, 776].

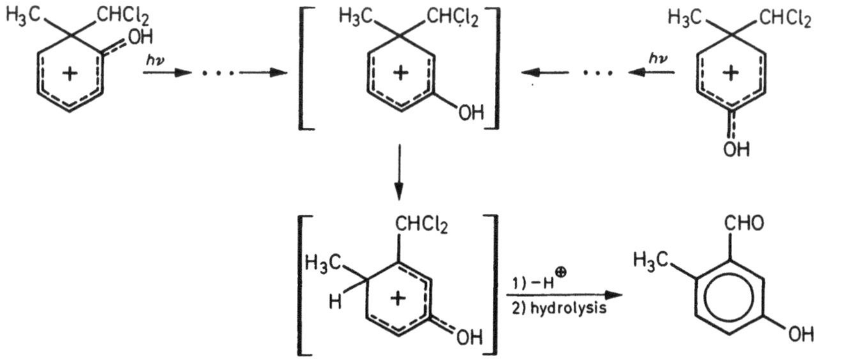

One of the conversions products of dimethylhydroxybenzenium ions — the protonated 3,4-dimethylphenol (*41*) — is further transformed when irradiated by UV light with $\lambda > 275$ nm to the 2,6-dimethyl-4-hydroxybenzenium ion (*150*) [325]. This conversion must be due to the photoisomerization of ion (*42*) present in a small amount (see [319]) in the equilibrium with ion (*41*).

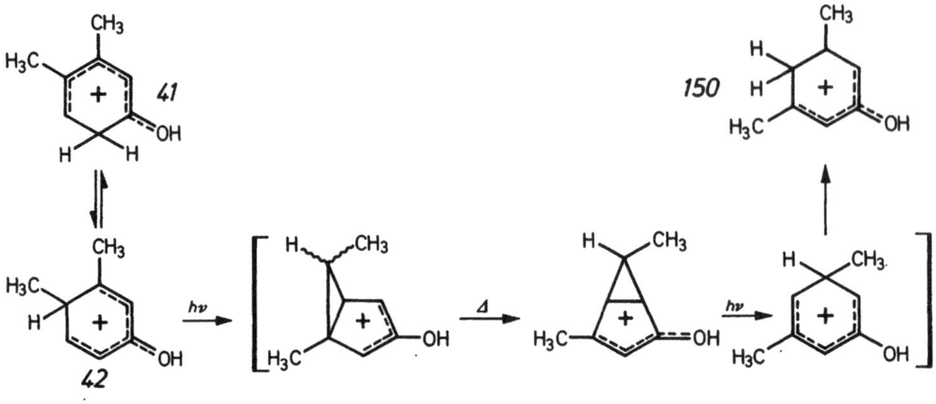

The para-protonated methylated phenols, when irradiated in their absorption band ($\lambda_{max} \leqq 330$ nm), were converted into the respective 2-hydroxybicyclo[3.1.0]hexenyl cations, for example:

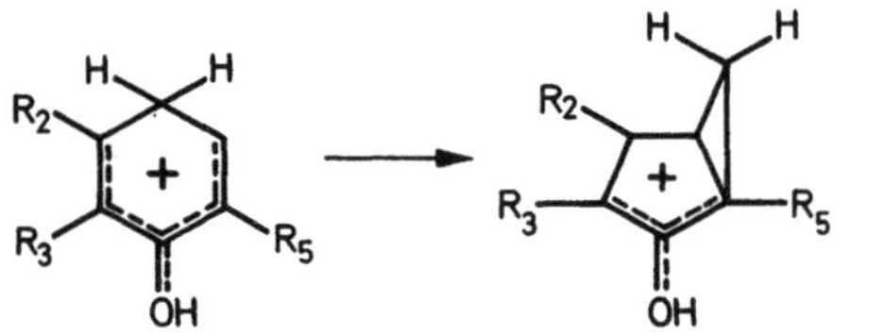

a $R_2 = R_3 = CH_3$, $R_5 = H$ *b* $R_3 = R_5 = CH_3$, $R_2 = H$

The resulting cations, in addition to the thermal isomerization due to the cyclopropane ring shift, can rearrange, when irradiated in their absorption band ($\geqq 330$ nm), into hydroxybenzenium ions. The combination of these processes by irradiating with a wide band of UV frequencies accounts for the isomerization of methylated phenols. This has been most carefully traced on the rearrangement of 2,4,5-trimethylphenol into 2,3,5-isomer:

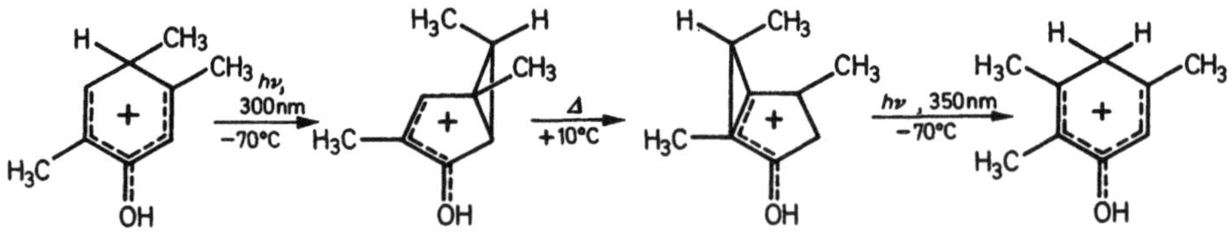

In a similar manner 2,4,6-trimethylphenol is converted into 2,3,6-isomer; additionally the meta-protonated form is likely to be photochemically converted.

Irradiation of 2- and 4-hydroxyhexamethylbenzenium ions (*97*) and (*98*) in HSO_3F at a low temperature provides a photostationary equilibrium with similar amounts of ions (*151*) and (*98*) [325].

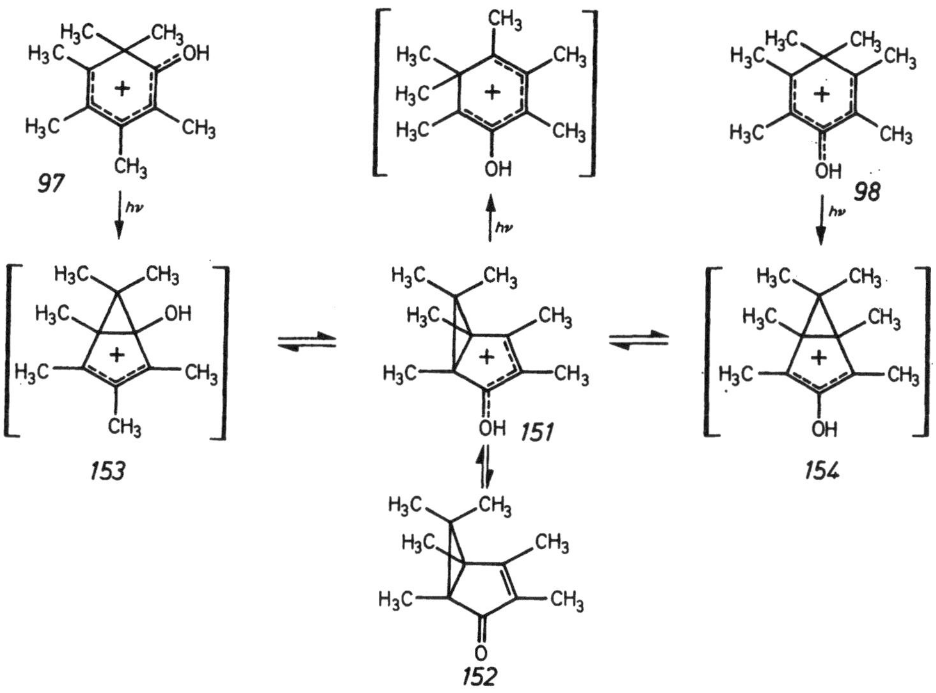

That ion (*151*) is thermally rearranged into ions (*153*) and (*154*) has been proved with ketone (*152*) selectively labeled by deuterium [545,771,775].

The opening of the cyclopropane ring in ion (*151*) "forbidden" for the ground state by the orbital symmetry conservation proceeds at a marked rate only at above 0 °C [545,771]. The main product was here the ion (*97*) and not (*98*).

Bicyclo[3.1.0]hexenyl cations are one (B) of the possible valence isomers of benzenium ions — which can be formed, at least theoretically, by redistribution of σ- and π-bonds between the 6 ring carbon atoms of benzenium ions. For example, for protonated hexamethylbenzene such isomers should contain five C—CH_3 fragments and one C(H)CH_3 fragment. The number of possible combinations of these fragments is rather large; only some of them are represented:

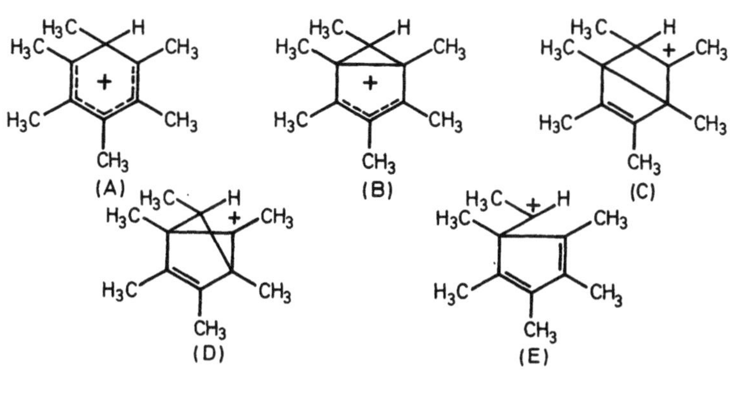

The best studied groups of valence isomers of benzenium ions are those of type D — bicyclo[2.1.1]hexenyl cations. This type derives from the interaction of the Dewar hexamethylbenzene with electrophilic agents, in particular with acids. Thus, with liquid HCl in CH_2Cl_2 at −100 °C a proton is added to form the 5-endo-H-1,2,3,4,5,6-hexamethylbicyclo[2.1.1]hexenyl cation (*155*) which at a higher temperature is partially converted into the 5-exo-H-isomer (*156*) [778,779].

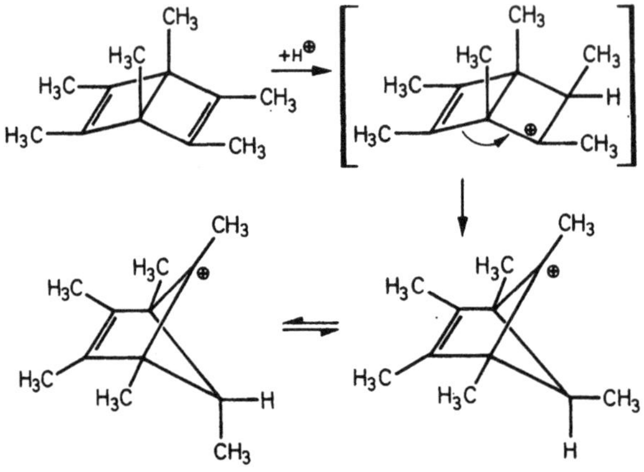

156 *155*

The solution of the Dewar hexamethylbenzene in strong acids ($HF—BF_3$ [780,781], HSO_3F [164]) at —80 °C results in an equilibrated mixture of isomeric ions (*155*) and (*156*) in the ratio ~1:3 (cf. the data for $HAlCl_4$ [782]).

The same mixture of isomeric ions is formed from hexamethylprismane (*157*) in $HSO_3F—SbF_5$ at —78 °C [258], as well as from 1,2,4,5,6-pentamethyl-3-methylene-tricyclo[2.2.0.0^{2,6}]hexane (*158*) in HSO_3F at —80 °C.

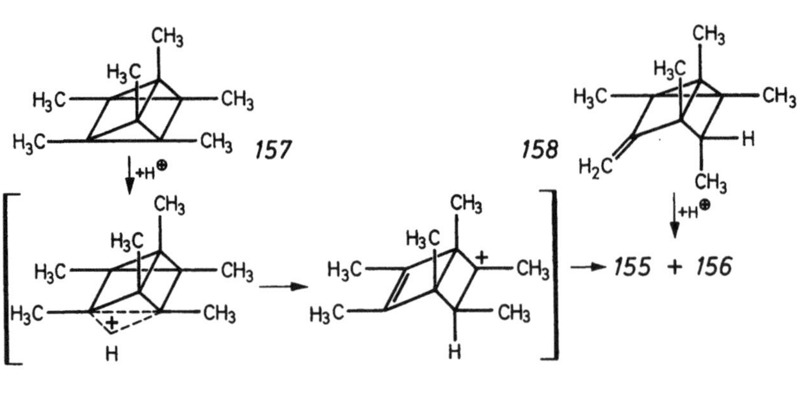

As the temperature rises above —60 °C the PMR signals of these two ions are observed to widen and then to merge reflecting their rapid interconversion following the scheme [781]:

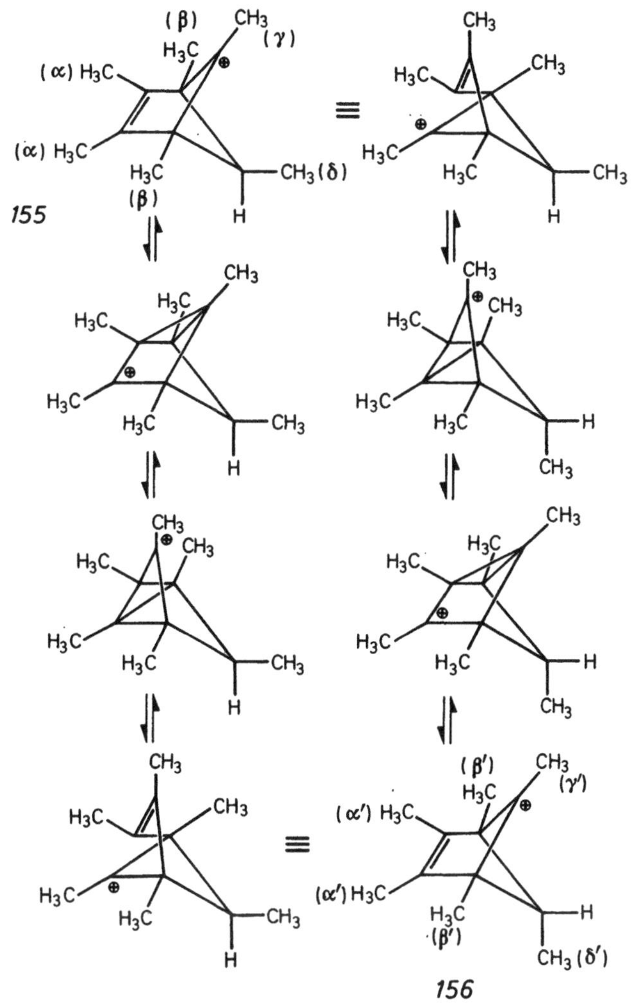

At above 0 °C the ions (*155*) and (*156*) are irreversible rearranged into the 1-H-1,2,3,4,5,6-hexamethylbenzenium ion [780, 781, 783] according to the scheme [778, 785]:

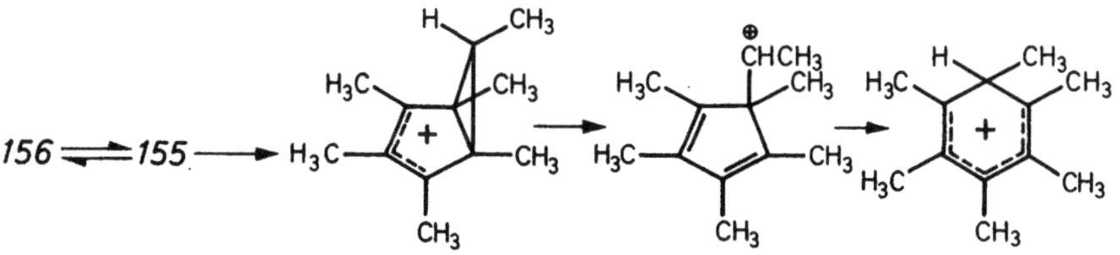

with intermediate formation of two other valence isomers.

Electrophilic addition of halogens to Dewar hexamethylbenzene at —80 ÷ —100 °C leads to the formation of 5-endo-halogen-1,2,3,4,5,6-hexamethylbicyclo[2.1.1]hexenyl cations (*159*).

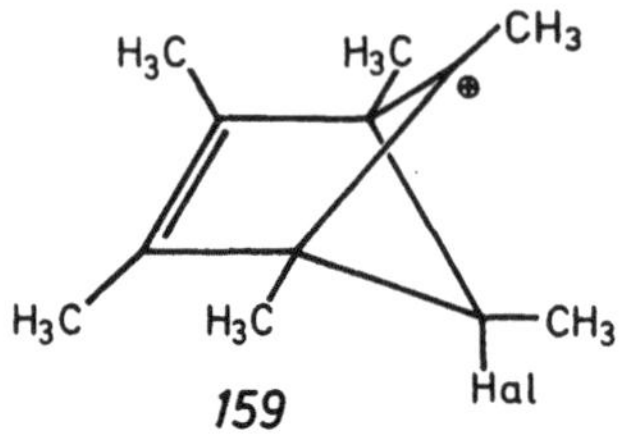

159

Hal = Cl [172, 259, 782, 784, 785]
Br [172, 259, 779, 785]

If the 5-exo-halogen isomer is present in equilibrium with 5-endo-isomer, its amount is insufficient to be observed by PMR.

The conversions of ions (*159*) with rising temperature depends on the nature of the anion [172, 259, 779]. For halogenaluminates at above 25 °C ions (*159*) isomerize into halogenhexamethylbenzenium ions [172, 259].

The PMR and NMR-^{13}C spectra [772, 786] of methylated bicyclo[2.1.1]hexenyl cations as well as the ESCA method [782] testify that in these ions the positive charge is largely transferred to the C_2 and C_3 atoms. The ions have the structure of rapidly rearranging cations (*160*), or of a nonclassical structure (*161*) with a three-centre two-electron orbital.

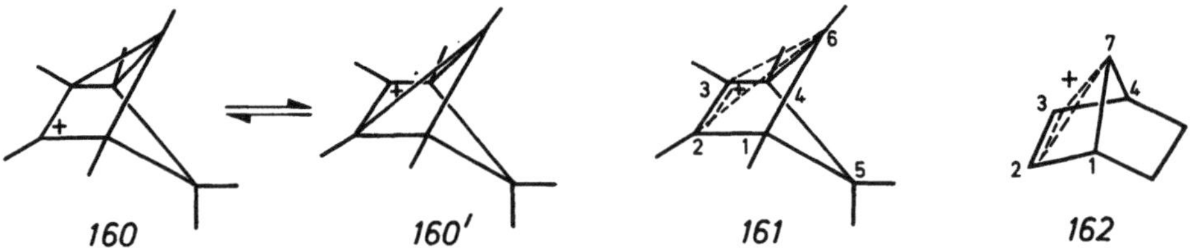

160 *160′* *161* *162*

The chemical shifts of ring carbons in 5-endo- and 5-exo-H-pentamethylbicyclo-[2.1.1]hexenyl cations are similar (65.9 and 66.8 ppm for $C_{1(4)}$, 121.0 and 118.1 ppm for $C_{2(3)}$ and 41.8 and 37.3 ppm for C_6 [772] to chemical shifts of the 7-norbornenyl cation (54.8 ppm for $C_{1(4)}$, 124.1 ppm for $C_{2(3)}$ and 30.6 ppm for C_7 [446] for wich the nonclassical structure (*162*) is assumed [1]. This is why one prefers structure (*161*) for bicyclo[2.1.1]hexenyl cations.

The proton chemical shifts of the nonsubstituted bicyclo[2.1.1]hexenyl cation (*163*) generated by dissolving 5-exo-acetoxybicyclo[2.1.1]hex-2-ene (*164*) in HSO_3F at —60 °C [787, 788] are comparable with the shifts of the 2-norbornyl cation (*165*) [789].

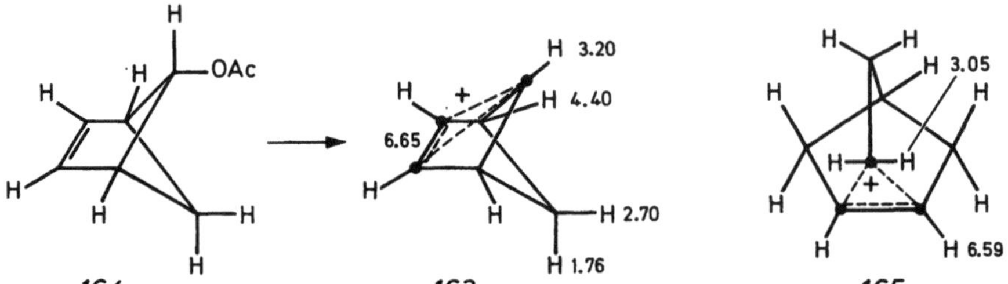

164 *163* *165*

The positive charge distribution in bicyclo[2.1.1]hexenyl cations determines considerably the directions of interaction with nucleophiles [785,786,790].

5-endo-halogen- and 5-exo-hydroxy-1,2,3,4,5,6-hexamethylbicyclo[2.1.1]hexenyl cations in strong acids convert into the stable cation $(CCH_3)_6^{2+}$ [791,792] which seems to have a pyramidal structure (*166*).

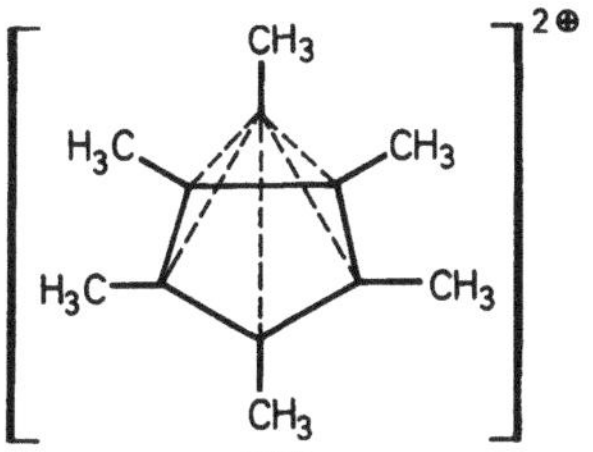

166

For the quantum chemical analysis of its electronic structure see [793].

The accumulated data on the reactivity of valence isomers of arenium ions permit one to estimate with sufficient reliability the possibility or impossibility of their participation in conversions of arenium ions.

V Additions

In the span of time that has elapsed since the manuscript was written in Russian a number of new publications have appeared dealing with the structure and reactivity of arenium ions. The results reported therein render it possible to make the following additions to the respective sections of the book.

Chapter II

Evidence has been published on the values of the acidity function H_0 for the HSO_3F—SbF_5 system with different component ratios [794].

In using ^{13}C NMR spectra to estimate the degree of benzene protonation in various acid systems it was possible to estimate the acid strength of these systems as well [795,796]. By this method the HBr—$AlBr_3$ system has been shown [795] to be a stronger acid than HF—TaF_5 and to be about as strong as HF—SbF_5. At 0 °C in HBr—$AlBr_3$ benzene is protonated completely if the molar ratio $AlBr_3$:HBr $\geqq 2$.

In [797] the measurements of electrical conductivity, of UV and PMR spectra have served as basis for the conclusion that benzene, toluene, xylenes and mesitylene are completely protonated in CF_3SO_3H. The latter is a very strong acid ($H_0 = -14.6$ [798], see also Table 3), but it is weaker than the acid system HF—TaF_5 ($H_0 = -18.9$ for the 167:1 ratio) in which benzene is only half protonated [795]. Since even in the case of mesitylene the PMR spectrum fails to show any signals of the ring protons due to the fast exchange with CF_3SO_3H, the authors' conclusion is questionable. The protonation of anthracene alone has been unequivocally recorded.

The MINDO/3 method has been applied to calculate the proton affinity (basicity) of all anisole positions [799,800], and the change of bond lengths in passing from naphthalene to a 1-H-naphthalenium ion [801].

A detailed analysis of experimental data of the basicity of nitrogen-bearing compounds in the gas phase [802] has confirmed that aniline in the gas phase is protonated at the nitrogen atom. A similar conclusion has been drawn for a number of substituted anilines (p-CH_3, m-CH_3, m-F, m-Cl, m-CF_3, m-CN). But in case the substituents are the groups m-OH, m-OCH_3, m-NH_2, the predominant process is the C-protonation.

New evidence has been obtained on the character of interaction between $Hg(OCOCF_3)_2$ and hexamethylbenzene [803]. The interaction of these components in CH_2Cl_2 results in the formation of an adduct of a 1:1 composition which was isolated in crystalline form. An X-ray diffraction analysis has shown the mercury atom to be located over one of the C—C bonds of the ring, the longer Hg—C distance (2.57 Å) pointing to comparatively weak π-binding (the σ-bond length of Hg—C is usually 2.1–2.2 Å). The PMR and ^{13}C NMR spectra of the crystalline adduct recorded with magic-angle spinning proved to be very similar to those observed for solutions in CH_2Cl_2 and CF_3COOH. An answer is thereby given to the question of the structure of the complex formed from hexamethylbenzene and $Hg(OCOCF_3)_2$ in solutions.

A circumstantial review has been published [804] dealing with the methods of generation and the properties of polyfluorinated arenium ions.

Besides the earlier demonstrated possibilities for the polyfluorinated aromatic compounds to add the electrophiles Cl^+ and CH_3^+, the addition of F^+ and Br^+ has also been effected [805–807]. Thus the interaction of hexafluorobenzene with the salt $XeF^+ \cdot SbF_6^-$ [805] and with the complex $XeF_2 \cdot 2\,SbF_5$ [806] resulted in the generation of a heptafluorobenzenium ion, and perfluorobiphenyl reacted with $XeF_2 \cdot 2\,SbF_5$ to form a perfluoro-3-phenylbenzenium ion. Pentafluorotoluene and tetrafluoro-m-xylene interact with Br_2 in the presence of SbF_5 in SO_2ClF to yield ions corresponding, seemingly, to the addition of Br^+ to a fluorine-substituted carbon atom which is located in a meta-position with respect to the CH_3 groups [807].

The reaction of pentafluorotoluene, tetrafluoro-m- and -p-xylenes with CH_3F and SbF_5 in SO_2ClF yielded mixtures of benzenium ions corresponding to the attachment of CH_3^+ in the positions substituted by fluorine or a CH_3 group [808].

Chapter III

An X-ray analysis has been made of the tetrachloroaluminate of the 1-phenyl-1,2,3,4,5,6-hexamethylbenzenium ion [809].

Section 1

The PMR spectral data have been reported for 1-X-1,2-dimethyl-3,4,5,6-tetrafluoro- and -tetrachlorobenzenium ions (X=H [810], Cl [811]); 1-benzyl-1,2,3,4,5,6-hexamethylbenzenium ions substituted in the benzyl group [812]; 1-X-1,3,4,5,6-pentamethyl-2-hydroxy- and 1-X-1,2,3,5,6-pentamethyl-4-hydroxybenzenium ions (X=Cl [813], C_6H_4Y [814]); 2- and 4-β-chloroethyl-1-H-naphthalenium ions [815]; isomeric tri- and tetramethylnaphthalenium ions [816]; mono-, di-, tri- and tetramethyl-10-H-anthracenium ions [817]; α-chlorosubstituted 9-alkyl-10-H-anthracenium ions [818]; 1,2,2,3,4-pentamethylbiphenylenium ion [819].

Octamethylbiphenylene in strong acids (HSO_3F, CF_3SO_3H, $HCl—AlCl_3—CH_2Cl_2$) are protonated, as concluded from the PMR and ^{13}C NMR spectra, into 2-position [820]; in CF_3COOH the NMR and EPR spectra have revealed the formation of radical-cation of this hydrocarbon [821].

Interesting results have been obtained by studying the protonation of the tetrafluoroborate of 1-dimethylamino-8-trimethylammoniumnaphthalenium ion and of its analogues [822,823]. The exceptional protonation of this cation at the nitrogen atom of the dimethylamino-group initially observed in CF_3SO_3H in the PMR and ^{13}C NMR spectra has proved to be a kinetically controlled process; the equilibrium is attained at 75% of 2-C-, 20% of 4-C- and 5% of N-protonated forms. In weaker acids (e.g., in CF_3COOH) considerable amounts of C-protonated forms are immediately generated.

Section 2

^{13}C NMR spectra have been recorded for 1-X-1,2-dimethyl-3,4,5,6-tetrafluoro- and -tetrachlorobenzenium ions (X=H [810], Cl [811]); 1-(X-phenyl)-1,3,4,5,6-pentamethyl-2-hydroxy- and 1-(X-phenyl)-1,2,3,5,6-pentamethyl-4-hydroxybenzenium ions [814]; isomeric tri- and tetramethvl-1-H-naphthalenium ions [816]; 2- and 4-β-chloroethyl-1-H-naphthalenium ions [815]; α-naphtonium ions of the (167) type [815]; 2- and 3-substituted 10-H-9-methylanthracenium ions [824]; 9-R-9,10-di-(p-X-phenyl)phenanthrenium ions (R=CH_3, C_6H_4Y) [825]; 1,2,2,3,4-pentamethyl- and 2-H-octamethylbiphenylenium ions [819,820].

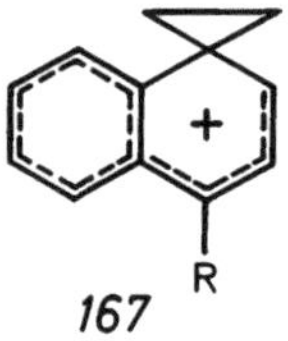

167

R = H, CH_3, C_6H_5, OH, OCH_3

The INDO method has been used to analyse various contributions for fluorine-substituted ions to the values $J_{^{13}C-^{19}F}$ [826].

Section 3

^{19}F NMR spectral data have been published for 1-X-1,2-dimethyl-3,4,5,6-tetrafluorobenzenium ion (X=H [810], Cl [811]); ions corresponding to the attachment of Br^+ to the perfluorinated (at the ring) toluene, m-xylene and mesitylene [807]; perfluorinated 3-phenylbenzenium ion [806]; isomeric dimethylpentafluoro- and trimethyltetrafluorobenzenium ions [808].

Section 4

The electronic absorption spectra (EAS) of the protonated forms of benzene and its methylated derivatives in CF_3SO_3H has been reported [797]. Since the completeness of protonation in this acid raises doubts and the spectra change with time. this data should be used with care. Characteristics of the EAS have been published for 1-(X-phenyl)-1,3,4,5,6-pentamethyl-2-hydroxy- and 1-(X-phenyl)-1,2,3,5,6-pentamethyl-4-hydroxybenzenium ions [814].

Section 5

An IR spectrum has been recorded for the tetrabromoaluminate (with an admixture of heptabromodialuminate) of a 10-H-anthracenium ion in the range of 90–3100 cm^{-1} [827]. Based on the results of deuteration at 9- and 10-positions and of comparison with the spectra of anthracene, 9- and 10-substituted anthracenes and of an acridinium cation, the authors have performed the assignment of the observed absorption bands of the anthracenium ion and of the anions $AlBr_4^-$ and $Al_2Br_7^-$. The stretching vibrations of the ring group CH_2 correlate with the bands at 2938 (ν_a, weak), 2850 (ν_s, the most intensive) and 2832 cm^{-1} (δ, shoulder). The computational analysis [828] of the C—H stretching vibrations of the CH_2 group in arenium ions has shown that the increase in the force constant by passing from the benzenium ion to the 1-H-naphthalenium and the 10-H-anthracenium ions can be accounted for by a decreasing interaction of the 1s orbital of the CH_2 group hydrogen with the $2p_z$ orbitals of the adjacent carbon atoms. This relates to the earlier suggestion [21] about the decrease of hyperconjugation of the CH_2 group and the electron-deficient π-system in the above series of the arenium ion.

Chapter IV

Section 1

Note has been made of the nonconformity of the kinetic and thermodynamic control in the protonation of 1,4,5-tri- and 1,4,5,8-tetramethylnaphthalenes [816].

With ^{13}C NMR spectroscopy m-xylene and m-diethylbenzene were found [829] to be protonated in $HF—SbF_5$ not only into 4(6)-, but also into 2-position. The fraction of 2-H-1,3-dimethylbenzenium ion in protonated m-xylene amounts to 2.5% at −60 °C and to 6.7% at +95 °C. Note that the calculation with the aid of ϱ and σ^+ values mentioned in Section 1.A the concentration of 2-H-isomer to be 6 and 13.5%, respectively, for these two temperatures. To ensure the agreement between calculated and experimental data the value of $\sigma^+_{ortho\text{-}CH_3}$ should be increased from −0.25 to −0.22.

3-Ethyltoluene is equally protonated into 4- and 6-positions [796] which points to the similarity of the electronic effects of CH_3 and CH_2CH_3.

Hexahydropyrene is shown to be protonated in HSO_3F at the β-position of the naphthalene ring; diprotonation is achieved in the system $HSO_3F—SbF_5$ [816].

Section 2

Additional data have been obtained on the kinetics of the degenerate rearrangements of arenium ions by the 1,2-shift of various groups. Measurements have been made of the characteristics of the hydrogen migration between 1- and 2-positions of 1-H-1,2-dimethyl-3,4,5,6-tetrafluoro- and -tetrachlorobenzenium ions [810], and between 2- and 3-positions of a 2-H-octamethylbiphenylenium ion [820] (cf. the data for the 2-H-biphenylenium ion [314] and its 1,2,3,4-tetramethyl derivative [556]), as well as the migration of the CH_3 group between 2- and 3-positions of the 1,2,2,3,4-pentamethylbiphenylenium ion [819].

According to the data obtained for 1-chloro-1,2,3,4,5,6-hexamethyl- and 1,3,4,5,6-pentachloro-1,2-dimethylbenzenium ions, as well as for 9-chloro-9,10-dimethyl- and 9-chloro-3,6,9,10-tetramethylphenanthrenium ions, the dependence of the free activation energy of the 1,2-shift of chlorine on the chemical shift of the carbon atom to which the migrant travels is described by the following equation [811]:

$$\Delta G^{\ddagger}_{Cl}(-110\ °C) = 70.1 - 0.29\ \delta_C \qquad (r = -0.970,\ s = 1.2)$$

On the strength of these and the earlier available data it can be concluded that the sensitivity of the free activation energy of the 1,2-shift of migrants to the electron deficiency changes in the position to which the migrant shifts decreases in the series:

$$Cl > C_6H_5 > NO_2 > CH_3 > H$$ [811]

The dependence of the migrating ability of various groups (H, CH_3, NO_2, Cl) on the electronic structure of ions has been analyzed by CNDO/2 and MINDO/3 in the framework of the perturbation theory. The analysis has shown [830] the initial step of rearrangement is characterized by the moring of the migrant toward the carbenium centre, the migrant-skeleton bond remaining basically unchanged. What does change essentially is the contribution to this bond made by the interaction with the vacant orbital of the carbenium centre.

The logarithms of the 1,2-shift rate coefficients of the substituted benzyl groups in 1-(X-benzyl)-1,2,3,4,5,6-hexamethylbenzenium ions (X=H, p-CH_3, p-CF_3, p-Cl, m-F) are connected by a linear dependence with the σ_X^+ constants [812]:

$$\log k^x\,(-110\,^\circ C) = -2.53 - 5.95\,\sigma_x^+ \qquad (r = -0.997,\ s = 0.18)$$
$$\log k^x\,(25\,^\circ C) = 4.76 - 3.38\,\sigma_x^+ \qquad (r = -0.995,\ s = 0.14)$$

In studying the degenerate 1,2-shifts of ethyl and p-chlorobenzyl groups (R) in 1-R-1,2-dimethylacenaphthenium ions the configuration of the migrating tetragonal carbon atom was found to remain unchanged in the process of rearrangement [831].

The tracer method has been used to illustrate the possibility of rearranging 1-phenyl-1,2,3,4,5,6-hexamethylbenzenium ion in the solid phase (crystalline salt with the $AlCl_4^-$ anion) by the 1,2-shift of the C_6H_5 group; in comparison with the solution the rearrangement rate at 22 °C is 10^4 times smaller [832].

The PMR method has been used to study the kinetics of rearrangement of 1-(X-phenyl)-1,3,4,5,6-pentamethyl-2-hydroxybenzenium ions (X = p-CH_3, p-OCH_3, p-Cl, p-CF_3, m-CF_3) into 1-(X-phenyl)-1,2,3,5,6-pentamethyl-4-hydroxybenzenium ions [814]. The logarithms of the XC_6H_4 group shift rate coefficients are connected by a linear dependence with the σ_x^+ constants:

$$\log k^{XC_6H_4}\,(-50\,^\circ C) = -10.18 - 3.37\,\sigma_x^+ \qquad (r = -0.990,\ s = 0.18)$$

Comparison with the rearrangements of 1-(X-phenyl)-1,2,3,4,5,6-hexamethylbenzenium ions and 9-(X-phenyl)-9,10-dimethylphenanthrenium ions shows the relative migrating ability of the substituted phenyl groups is little sensitive to electron deficiency changes in the carbonium centre, although the absolute values of the kinetic characteristics for the 1,2-shift of the XC_6H_4 groups change drastically as the migrant-carrying rest varies.

For the kinetic characteristics of the rearrangement of the 1-chloro-1,3,4,5,6-pentamethyl-2-hydroxybenzenium ion into the 1-chloro-1,2,3,5,6-pentamethyl-4-hydroxybenzenium ion see [813].

Section 3

The case of 9-methyl-, 9-ethyl- and 9-propylanthracenes has confirmed [833] that in the system $CF_3COOD—CDCl_3$, besides the ring hydrogen atoms, only the α-positions of the alkyl groups are involved in the isotope exchange.

Section 8

The photolysis of the salts of a heptamethylbenzenium ion ($\lambda > 350$ nm) in aqueous acids (24% H_2SO_4, 50% HBF_4, 70% $HClO_4$) results in a good yield of 1,2,3,4,5-pentamethylcyclopentadiene [834], which is conveniantly prepared this way. Apparently, the heptamethylbicyclo[3.1.0]hexenyl cation resulting from the photoreaction interacts with water to yield 1-(α-hydroxyisopropyl)-1,2,3,4,5-pentamethylcyclopentadiene which in the acid medium fragmentates to acetone and pentamethylcyclopentadiene.

The photolysis of 4-methylene-1,1,2,3,5,6-hexamethylcyclohexa-2,5-diene adsorbed on a polymer film of polyfluorinated aliphatic sulphoacids (Nafion 125 and 501), strong enough ($H_0 < -6.5$) to convert the diene into a heptamethylbenzenium ion, yields 1-isopropenyl-1,2,3,4,5-pentamethylcyclopentadiene as the main reaction product [835].

The durenol disolved in CF_3SO_3H on irradiation ($\lambda = 300$ nm, 35 °C) and subsequent neutralization, yields 42% of 1,3,4,5-tetramethylbicyclo[3.1.0]hex-3-en-2-on, while phenol ($\lambda = 254$ nm) gives 24% of bicyclo[3.1.0]hex-3-en-2-on [836].

VI References

1. Olah, G. A.: J. Am. Chem. Soc. *94*, 808 (1972)
2. Stolyarov, B. V., in: Sovremennye problemy organicheskoj khimii (Eds. Dyakonov, I. A., Ioffe, B. V.), Leningrad, Izd. LGU 1969, p. 5
3. Doering, E., Saunders, M., Boyton, H. G., Earhart, H. W., Wodley, E. F., Edwards, W. R., Laber, G.: Tetrahedron *4*, 178 (1958)
4. Brouwer, D. M., Mackor, E. L., MacLean, C., in: Carbonium Ions. Vol. II (Ed. Olah, G. A., Schleyer, P. von R.) Wiley-Interscience 1970, p. 837
5. Vorozhtsov, N. N.: Osnovy sinteza promezhytochnykh produktov i krasitelej, Moscow, Goskhimizdat 1955
6. Vorozhtsov, N. N. Jr., in: The Chemistry of Synthetic Dyes (Ed. K. Venkataraman. Vol. 3., N.Y., Academic Press Inc. 1970
7. Shatenshtejn, A. I.: Izotopnyj obmen i zameshchenie vodoroda v organicheskikh soedineniyakh, Moscow, Izdatel'stvo Akademii Nauk SSSR 1960
8. Berliner, E., in: Sovremennye problemy fizicheskoj organicheskoj khimii, Moscow, Mir 1967, p. 444
9. Norman, R. O. C., Taylor, R.: Electrophilic Substitution in Benzenoid Compounds. Elsevier, 1965
10. Olah, G. A., in: Organic Reaction Mechanisms, Chem. Soc. Special Publication N. 19. London 1965, p. 21
11. Streijtvizer, E.: Teoriya molekulyarnykh orbit dlya khimikov-organikov, Moscow, Mir 1965
12. Comprehensive Chemical Kinetics. Vol. 13. Reaction of Aromatic Compounds (Ed. C. H. Bamford, C. F. H. Tipper) Elsevier, 1972
13. Koptyug, V. A.: Izomerizatsiya aromaticheskikh soedinenij, Novosibirsk, Izdatel'stvo Sibirsk. Otd. Akademii Nauk SSSR 1963
14. Shine, H. J.: Aromatic Rearrangements, Elsevier, 1967
15. Olah, G. A., Mo, Y. K., in: Carbonium Ions, Vol. V, (Ed. Olah, G. A., Schleyer, P. von R.) Wiley-Interscience 1976, p. 2135
16. Perkampus, G. G., in: Novye problemy fizicheskoj organicheskoj khimii, Moscow, Mir 1969, p. 257
17. Koptyug, V. A.: Izv. Akad. Nauk SSSR, Otd. Khim. Nauk *1974*, 1081
18. Koptyug, V. A., in: Sovremennye problemy khimii karbonievykh ionov (Ed. Koptyug, V. A.), Novosibirsk, Nauka 1975, p. 5
19. Shtejngarts, V. D.: Izv. Sibirsk. Otd. Akad. Nauk SSSR *7* (3), 53 (1980)
20. Brown, H. C., Pearsall, H. W.: J. Am. Chem. Soc. *74*, 191 (1952)
21. Koptyug, V. A., Korobejnicheva, I. K., Andreeva, T. P., Bushmelev, V. A.: Zh. Obshch. Khim. *38*, 1979 (1968)
22. Koptyug, V. A., Rezvukhin, A. I., Zaev, E. E., Molin, Yu. N.: Izv. Akad. Nauk SSSR, Otd. Khim. Nauk *1963*, 1700; Zh. Obshch. Khim. *34*, 3999 (1964)
23. Lieser, K. H., Pfluger, C. E.: Chem. Ber. *93*, 181 (1960)
24. Brown, H. C., Wallace, W. J.: J. Am. Chem. Soc. *75*, 6268 (1953)
25. Baddeley, G., Voos, D., in: Cationic Polymerisation and Related Complexes (Ed. Plesh, P. H.), Cambridge, 1953
26. Olah, G. A., Kuhn, S. J.: Nature *178*, 693 (1956); J. Am. Chem. Soc. *80*, 6535 (1958)
27. Koptyug, V. A., Rezvukhin, A. I., Shubin, V. G., Korchagina, D. V.: Zh. Obshch. Khim. *35*, 864 (1965)

28. Olah, G. A.: J. Am. Chem. Soc. *87*, 1103 (1965)
29. Gustavson, G.: C. r. *140*, 940 (1905); J. pr. Chem. *72*, 57 (1905)
30. Volkov, B. V.: Khim. Prom. *1*, 20 (1957)
31. Volkov, B. V.: Zh. Prikl. Khim. *34*, 456 (1961)
32. Lieser, K. H., Pfluger, C. E.: Chem. Ber. *93*, 176 (1960)
33. Eley, D. D., King, P. J.: J. Chem. Soc. *1952*, 2517, 4972
34. MacLean, C., Mackor, E. L.: Disc. Faraday Soc. *34*, 165 (1962)
35. Okami, Y., Nambu, N., Okuda, Sh., Hamanaka, S., Ogawa, M.: Tetrahedron Lett. *1972*, 5259
36. Okami, Y., Otani, N., Katoh, D., Hamanaka, S., Ogawa, M.: Bull. Chem. Soc. Japan *46*, 1860 (1973)
37. Nambu, N., Yamamoto, K., Hamanaka, S., Ogawa, M.: Nippon Kagaku Kaishi *1979*, 925
38. Nambu, N., Hamanaka, S., Ogawa, M.: Bull. Chem. Soc. Japan *51*, 1978 (1978)
39. US Pat. 2.481.843 — Ch. A. *46*, 3563 (1952); US Pat. 2.838.583 — RZhKhim *1960*, 1964
40. McCaulay, D. A., Highley, W. S., Lien, A. P.: J. Am. Chem. Soc. *78*, 3009 (1956)
41. McCaulay, D. A., Shoemaker, B. H., Lien, A. P.: Ind. Eng. Chem. *42*, 2103 (1950)
42. US Pat. 2.728.803 — Ch. A. *50*, 6034 (1956)
43. US Pat. 2.726.275 — Ch. A. *50*, 6034 (1956)
44. US Pat. 2.739.992 — Ch. A. *50*, 15065 (1956)
45. Brit. Pat. 690.687 — Ch. A. *47*, 9605 (1953)
46. Brit. Pat. 734.038 — Ch. A. *50*, 7854 (1956)
47. US Pat. 2.727.078 — Ch. A. *50*, 6035 (1956)
48. Luther, H., Pockels, G.: Z. Electrochem. *59*, 159 (1958)
49. Korshak, V. V., Lebedev, N. N., Fedoseev, S. D.: Zh. Obshch. Khim. *17*, 575 (1947)
50. Koptyug, V. A., Shubin, V. G., Baeva, I. K., Korchagina, D. V., Komarorov, A. M., Rezvukhin, A. I.: Izv. Akad. Nauk SSSR, Otd. Khim. Nauk *1964*, 948: Zh Obshch. Khim. *35*, 1111 (1965)
51. Norris, J. F., Ingraham, J. N.: J. Am. Chem. Soc. *62*, 1298 (1940)
52. Baddeley, G., Holt, G., Voos, D.: J. Chem. Soc. *1952*, 100
53. Korobejnicheva, I. K., Bochkarev, V. S., Koptyug, V. A.: Zh. Obshch. Khim. *40*, 854 (1970)
54. Norris, J. F., Wood, J. E.: J. Am. Chem. Soc. *62*, 1428 (1940)
55. Brown, H. C., Brady, J.: ibid. *74*, 3570 (1952)
56. Nelson, K. Le-Rua, Braun, G. K., in: Khimiya uglevodorov nefti (Ed. Bruks, B. T.), Vol. 3, Moscow, Gostoptekhizdat *1959*, 391
57. McCaulay, D. A., Lien, A. P.: J. Am. Chem. Soc. *73*, 2013 (1951)
58. Gold, V., Tye, F. L.: J. Chem. Soc. *1952*, 2172
59. MacLean, C., van der Waals, J. H., Mackor, E. L.: Mol. Phys. *1*, 247 (1958)
60. MacLean, C., Mackor, E. L.: Mol. Phys. *4*, 241 (1961)
61. MacLean, C., Mackor, E. L.: J. Chem. Phys. *34*, 2208 (1961)
62. Brouwer, D. M., Doorn, J. A.: Rec. trav. chim *89*, 88 (1970)
63. Olah, G. A., Porter, R. D.: J. Am. Chem. Soc. *93*, 6877 (1971)
64. Repinskaya, I. B., Rezvukhin, A. I., Koptyug, V. A.: Zh. Org. Khim. *8*, 1765 (1972)
65. Olah, G. A., Schlosberg, R. H., Porter, R. D., Mo, Y. K., Kelly, D. P., Mateescu, G. D.: J. Am. Chem. Soc. *94*, 2034 (1972)
66. Repinskaya, I. B., Rezvukhin, A. I., Koptyug, V. A.: Zh. Org. Khim. *8*, 1647 (1972)
67. Olah, G. A., Kiovsky, T. E.: J. Am. Chem. Soc. *89*, 5692 (1967)
68. Birchall, T., Bourns, A. N., Gillespie, R. J., Smith, P. J.: Can. J. Chem. *42*, 1433 (1964)
69. Golounin, A. V., Koptyug, V. A.: Zh. Org. Khim. *8*, 2555 (1972)
70. Brouwer, D. M., Mackor, E. L., MacLean, C. : Rec. trav. chim. *85*, 109 (1966)
71. Brouwer, D. M., Mackor, E. L., MacLean, C.: ibid. *85*, 114 (1966)
72. Farcaşiu, D., Melhior, M. T., Craine, L.: Angew. Chem. *89*, 323 (1977)
73. Arnett, E. M., in: Sovremennye problemy fizicheskoj organicheskoj khimii, Moscow, Mir 1967, p. 195
74. Birchall, T., Gillespie, R. J.: Canad. J. Chem. *43*, 1045 (1965)
75. Eian, G. L., Kingsbury, C. A.: J. Org. Chem. *32*, 1864 (1967)
76. Van der Linde, R.: Tetrahedron Lett. *1968*, 525
77. Olah, G. A., Kiovsky, T. E.: J. Am. Chem. Soc. *90*, 6461 (1968)

78. Beistel, D. W., Atkinson, E. K.: J. Mol. Spectr. *29*, 244 (1969)
79. Mamatyuk, V. I., Detsina, A. N., Koptyug, V. A.: Zh. Org. Khim. *11*, 2382 (1974)
80. Mamatyuk, V. I., Krysin, A. P., Bodoev, N. V., Koptyug, V. A.: Izv. Akad. Nauk SSSR, Otd. Khim. Nauk *1974*, 2392
81. Bodoev, N. V., Mamatyuk, V. I., Krysin, A. P., Koptyug, V. A.: Zh. Org. Khim. *14*, 1929 (1978)
82. Rodionov, V. I., Shakirov, M. M., Koptyug, V. A.: ibid. *16*, 1002 (1980)
83. Mackor, E. L., Hofstra, A., van der Waals, J. H.: Trans. Faraday Soc. *54*, 66, 186 (1958)
84. Chong, Sh. L., Frankling, J. L.: J. Am. Chem. Soc. *94*, 6630 (1972)
85. Hehre, W. J., McIver, R. T., Pople, J. A., Schleyer, P. von R.: ibid. *96*, 7162 (1974)
86. Wolf, J. F., Dalvin, J. L., DeFrees, D. J., Taft, R. W., Hehre, W. J.: ibid. *98*, 5097 (1976)
87. Yamdagni, R., Kebarle, P.: ibid. *98*, 1320 (1976)
88. Lau, Y. K., Kebarle, P.: ibid. *98*, 7452 (1976)
89. Pollack, S. K., Dalvin, J. L., Summerhays, K. D.: Taft, R. W., Hehre, W. J., ibid. *99*, 4583 (1977)
90. Summerhays, K. D., Pollack, S. K., Taft, R. W., Hehre, W. J.: ibid. *99*, 4585 (1977)
91. DeFrees, D. J., McIver, R. T., Hehre, W. J.: ibid. *99*, 3853 (1977)
92. Colpa, J. P., MacLean, C., Mackor, E. L.: Tetrahedron *19*, Suppl. 2, 65 (1963)
93. Ehrenson, S.: J. Am. Chem. Soc. *84*, 2681 (1962)
94. Ehrenson, S.: ibid. *83*, 4493 (1961)
95. Flurry, R. L., Lycos, P. G.: ibid. *85*, 1033 (1963)
96. Clark, D. T., Fairweather, D. J.: Tetrahedron *25*, 4083 (1969)
97. Jakubetz, W., Schuster, P.: Angew. Chem. *83*, 499 (1971)
98. Helbstrand, E.: Acta Chem. Scand. *24*, 3687 (1970); *26*, 2024 (1972)
99. Abronin, I. A., Burshtejn, K. Ya., Gagarin, S. G., Zhidomirov, G. M.: Zh. Strukt, Khim. *18*, 250 (1977)
100. Heidrich, D., Grimmer, M., Sommer, B.: Tetrahedron *32*, 202 (1976)
101. Motell, E. L., Fink, W. H., Dallas, J. L.: ibid. *29*, 350 (1973)
102. Streitwieser, A., Mowery, P. C., Jesaitis, R. G., Lewis, A.: J. Am. Chem. Soc. *92*, 6529 (1970)
103. Howe, G. R.: J. Chem. Soc. (B) *1971*, 984
104. Isaacs, N. S., Cvitas, D.: Tetrahedron *27*, 4139 (1971)
105. Tuyen, M., Neshev, N., Zidarov, D., Dimitrov, Ch.: Comptes Rendus de l'Academic Bulgare des Sciences *31*, N2, 193 (1978)
106. Bursey, M. M., Greenberg, R. S., Pedersen, L. G.: Chem. Phys. Lett. *36*, 470 (1975)
107. Sordo, T., Bertran, J.: J. Chem. Soc. Perkin II *1979*, 1486
108. Gleghorn, J. T., McConkey, F. W.: ibid. *1976*, 1078
109. Hehre, W. J., Pople, J. A.: J. Am. Chem. Soc. *94*, 6901 (1972)
110. Palmer, M. H., Findlay, R. H., Moyes, W., Gaskell, A. J.: J. Chem. Soc. Perkin II *1975*, 841
111. Fratev, F., Janoschek R., Preuss, H.: Communications of the Department of Chemistry, Bulgarian Academy of Sciences, *7*, 205 (1974)
112. Delvin, J. L., Wolf, J. F., Taft, R. W., Hehre, W. J.: J. Am. Chem. Soc. *98*, 1990 (1976)
113. Greenberg, R. S., Bursey, M. M., Pedersen, L. G.: ibid. *98*, 4061 (1976)
114. Hehre, W. J., Hiberty, Ph. C.: ibid. *96*, 7163 (1974)
115. Ermler, W. C., Mulliken, R. S., Clementi, E.: ibid. *98*, 388 (1976)
116. Ermler, W. C., Mulliken, R. S.: ibid. *100*, 1647 (1978)
117. Janoschek, R.: Z. Naturforsch. *32A*, 119 (1977)
118. Binning, R. C., Sando, K. M.: J. Am. Chem. Soc. *102*, 2948 (1980)
119. McKelvey, J. M., Alexandratos, S., Streitwieser, A.: ibid. *98*, 244 (1976)
120. Deno, N. C., Groves, P. T., Saines, G.: ibid. *81*, 5790 (1959)
121. Jorgensen, M. J., Harter, D. R.: ibid. *85*, 878 (1963)
122. Kresge, A. J., Barry, G. W., Charles, K. R., Chiang, Y.: ibid. *84*, 4343 (1962)
123. Köhler, H., Scheibe, G.: Z. anorg. allg. Chem. *285*, 221 (1956)
124. Yamaoka, T., Hosoya, H., Nagakura, S.: Tetrahedron *24*, 6203 (1968)
125. Commeyras, A., Olah, G.: J. Am. Chem. Soc. *91*, 2929 (1969)
126. Gillespie, R. J., Peel, T. E., Robinson, E. A.: ibid. *93*, 5083 (1971)
127. Gillespie, R. J.: Accounts Chem. Res. *1*, 202 (1968)
128. Gillespie, R. J., Peel, T. E.: J. Am. Chem. Soc. *95*, 5173 (1973)
129. Grondin, J., Sagnes, R., Commeyras, A.: Bull. Soc. Chim. France *1976*, 1779

130. Pal'm, V. A.: Dokl. Akad. Nauk SSSR *108*, 270 (1956)
131. Hyman, H., Kilpatrick, M., Katz, J. J.: J. Am. Chem. Soc. *79*, 3668 (1957)
132. Dallinga, G., Gaaf, J., Mackor, E. L.: Rec. trav. chim. *89*, 1068 (1970)
133. Kramer, G. M.: J. Org. Chem. *40*, 302 (1975)
134. Hyman, H. H., Garber, R. A.: J. Am. Chem. Soc. *81*, 1847 (1959)
135. Spitzer, U. A., Toone, T. W., Stewart, R. S.: Canad. J. Chem. *54*, 440 (1976)
136. Olah, G. A., White, A. M.: Chem. Rev. *70*, 561 (1970)
137. Sommer, J., Rimmelin, P., Drakenberg, T.: J. Am. Chem. Soc. *98*, 2671 (1976)
138. Brouwer, D. M., van Doorn, J. A.: Rec. trav. chim. *91*, 895 (1972)
139. Olah, G. A., Schlosberg, R. H., Kelly, D. P., Mateescu, G. D.: J. Am. Chem. Soc. *92*, 2546 (1970)
140. Gillespie, R. J., Moss, K. C.: J. Chem. Soc. (A) *1966*, 1170
141. Dean, P. A. W., Gillespie, R. J.: J. Am. Chem. Soc. *92*, 2362 (1970)
142. Dean, P. A. W., Gillespie, R. J.: ibid. *91*, 7260, 7264 (1969)
143. Brouwer, D. M.: Rec. trav. chim. *88*, 9 (1969)
144. Harris, M. G., Milne, J. B.: Canad. J. Chem. *49*, 2937 (1971)
145. Brookhart, M., Anet, F. A. L., Winstein, S.: J. Am. Chem. Soc. *88*, 5657 (1966)
146. Olah, G. A., Kiovsky, T. E.: ibid. *90*, 2583 (1968)
147. Ramsey, B. R.: ibid. *88*, 5358 (1966)
148. Svanholm, U., Parker, V. D.: J. Chem. Soc. Perkin II *1972*, 962
149. Hefelfinger, D. T., Cram, D. J.: J. Am. Chem. Soc. *93*, 4755 (1971)
150. Hart, H., Oku, A.: J. Org. Chem. *37*, 4269 (1972)
151. US Pat. 3.371.113 — Ch. A. *69*, 51810 (1968)
152. Shubin, V. G., Tabatskaya, A. A., Derendyaev, B. G., Koptyug, V. A.: Izv. Akad. Nauk SSSR, Otd. Khim. Nauk *1968*, 2417
153. Shubin, V. G., Tabatskaya, A. A., Derendyaev, B. G., Korchagina, D. V., Koptyug, V. A.: Zh. Org. Khim. *6*, 2072 (1970)
154. Koptyug, V. A., Shubin, V. G., Rezvukhin, A. I.: Izv. Akad. Nauk SSSR, Otd. Khim. Nauk *1965*, 201
155. Tabatskaya, A. A.: Thesis, Novosibirsk 1972
156. Korchagina, D. V., Derendyaev, B. G., Shubin, V. G.: Zh. Org. Khim. *10*, 1457 (1974)
157. Olah, G. A., Spear, R. J., Messina, G., Westerman, P. W.: J. Am. Chem. Soc. *97*, 4051 (1975)
158. Olah, G. A., Kuhn, S. J.: Nature *178*, 1344 (1956); ibid. *80*, 6541 (1958)
159. Garmonov, V. I., Sokolov, V. I., Yukel'son, I. I.: Dokl. Akad. Nauk SSSR, Ser. Khim. *171*, 867 (1966)
160. Nakane, R., Natsubori, A., Kurihara, O.: J. Am. Chem. Soc. *87*, 3597 (1965)
161. Nakane, R., Oyama, T., Natsubori, A.: J. Org. Chem. *33*, 275 (1968)
162. Nakane, R., Natsubori, A.: J. Am. Chem. Soc. *88*, 3011 (1966)
163. Oyama, T., Nakane, R.: J. Org. Chem. *34*, 949 (1969)
164. Olah, G., Noszko, L., Pavlath, A.: Nature *179*, 146 (1957)
165. Olah, G., Pavlath, A., Olah, J.: J. Am. Chem. Soc. *80*, 654 (1958)
166. Olah, G. A., Lin, H. C., Mo, Y. K.: ibid. *94*, 3667 (1972)
167. Kreienbühl, P., Zollinger, H.: Tetrahedron Lett. *1965*, 1739
168. Hunziker, E., Penton, J. R., Zollinger, H.: Helv. chim. acta *54*, 2043 (1971)
169. Detsina, A. N., Koptyug, V. A.: Zh. Org. Khim. *8*, 2215 (1972)
170. Detsina, A. N., Mamatyuk, V. I.: Izv. Akad. Nauk SSSR, Otd. Khim. Nauk *1973*, 2337
171. Detsina, A. N., Koptyug, V. A.: Zh. Org. Khim. *8*, 2158 (1972)
172. Isaev, I. S., Buraev, V. I., Novikov, G. P., Sosnovskaya, L. M., Koptyug, V. A.: ibid. *9*, 2430 (1973)
173. Detsina, A. N., Mamatyuk, V. I., Koptyug, V. A.: Izv. Akad. Nauk SSSR, Otd. Khim. Nauk *1973*, 2163
174. Olah, G. A., Lin, H. C., Forsyth, D. A.: J. Am. Chem. Soc. *96*, 6908 (1974)
175. Ostashevskaya, L. A., Detsina, A. N., Mamatyuk, V. I., Isaev, I. S., Koptyug, V. A.: Zh. Org. Khim. *10*, 2374 (1974)
176. Ostashevskaya, L. A., Shakirov, M. M., Isaev, I. S., Koptyug, V. A.: ibid. *13*, 2362 (1977)
177. Loktev, V. F., Korchagina, D. V., Shubin, V. G.: Izv. Sibirsk. Otd. Nauk SSSR, Otd. Khim. Nauk *12*(5), 135 (1978)
178. Shtejngarts, V. D., Dobronravov, P. N.: Zh. Org. Khim. *12*, 2005 (1976)

179. Dobronrarov, P. N., Shtejngarts, V. D.: ibid. *13*, 461 (1977)
180. Dobronravov, P. N., Shtejngarts, V. D.: ibid. *13*, 1679 (1977)
181. Dobronravov, P. N., Shtejngarts, V. D., in: Khimiya karbokationov, Novosibirsk *1979*, 126
182. Myhre, P. C.: Acta Chem. Scand. *14*, 219 (1960)
183. Baciocchi, E., Illuminati, G., Sleiter, G., Stegel, F.: J. Am. Chem. Soc. *89*, 125 (1967)
184. Morton, J.: Acta Chem. Scand. *23*, 3321, 3329 (1969)
185. Hoggett, J. G., Moodie, R. B., Penton, J. R., Schoffield, K.: Nitration and aromatic reactivity. Cambridge University Press. 1971
186. Cerfontain, H., Telder, A.: Rec. trav. chim. *86*, 371 (1967)
187. Sokolov, V. I., Bashilov, V. V., Reutov, O. A.: Dokl. Akad. Nauk SSSR *197*, 101 (1971)
188. Olah, G. A., Yu, S. H., Parker, D. G.: J. Org. Chem. *41*, 1983 (1976)
189. Dean, Ph. A. W., Ibbott, D. G., Stothers, J. B.: J. Chem. Soc. Chem. Commun. *1973*, 626
190. Perkampus, H. H., Baumgarten, E.: Z. phys. Chem. (BRD) *40*, 144 (1964)
191. Aalbersberg, W. I., Hoijtink, G. J., Mackor, E. L., Weijland, W. P.: J. Chem. Soc. *1959*, 3049, 3055
192. Perkampus, H. H., Kranz, T.: Z. phys. Chem. (BRD) *34*, 213 (1962)
193. Morita, M., Hirosawa, K., Sato, T.: Bull. Chem. Soc. Japan *50*, 1256 (1977)
194. Perkampus, H. H., Baumgarten, E.: Angew. Chem. *76*, 965 (1964); Angew. Chem. Int. Ed. *3*, 776 (1964)
195. Romm, I. P., Gur'yanova, E. N., Kocheshkov, K. A.: Dokl. Akad. Nauk SSSR *189*, 810 (1969)
196. Niess, R., Nagel, K.: Tetrahedron Lett. *1968*, 4265
197. Menzel, P., Effenberger, F.: Angew. Chem. *84*, 954 (1972)
198. Menzel, P., Effenberger, F.: ibid. *87*, 71 (1975)
199. Effenberger, F., Menzel, P., Seufert, W.: Chem. Ber. *112*, 1660 (1979)
200. Effenberger, F., Mack, K. E., Nagel, K., Niess, R.: ibid. *110*, 165 (1977)
201. Koptyug, V. A., Kamshij, L. P., Mamatyuk, V. I.: Zh. Org. Khim. *11*, 1233 (1975)
202. Kamshij, L. P., Mamatyuk, V. I., Koptyug, V. A.: ibid. *13*, 810 (1977)
203. Koptyug, V. A., Kamshij, L. P., Mamatyuk, V. I.: Izv. Akad. Nauk SSSR, Otd. Khim. Nauk *1974*, 1440
204. Koptyug, V. A., Kamshij, L. P., Mamatyuk, V. I.: Zh. Org. Khim. *11*, 128 (1975)
205. Kamshij, L. P., Koptyug, V. A.: Izv. Akad. Nauk SSSR, Otd. Khim. Nauk *1974*, 236
206. Kamshij, L. P., Mamatyuk, V. I., Koptyug, V. A.: Zh. Org. Khim. *12*, 1781 (1976)
207. Kuura, Kh. I., Khaldna, Yu. L.: Reakts. sposobn. org. soed. *3*, (2), 162 (1966)
208. Koptyug, V. A., Mozulenko, L. M.: Zh. Org. Khim. *6*, 102 (1970)
209. Hünig, S., Schilling, P.: Chem. Ber. *108*, 3355 (1975)
210. Morozov, S. V., Shakirov, M. M., Shubin, V. G., Koptyug, V. A.: Zh. Org. Khim. *15*, 770 (1979)
211. Shubin, V. G., Korchagina, D. V., Rezvukhin, A. I., Koptyug, V. A.: Dokl. Akad. Nauk SSSR *179*, 119 (1968)
212. Shubin, V. G., Korchagina, D. V., Derendyaev, B. G., Mamatyuk, V. I., Koptyug, V. A.: Zh. Org. Khim. *6*, 2066 (1970)
213. Shubin, V. G., Korchagina, D. V., Borodkin, G. I., Derendjaev, B. G., Koptyug, V. A.: Chem. Commun. *1970*, 696
214. Korchagina, D. V., Derendyaev, B. G., Shubin, V. G.: Izv. Akad. Nauk SSSR, Otd. Khim. Nauk *1971*, 441
215. Korchagina, D. V., Derendyaev, B. G., Shubin, V. G., Koptyug, V. A.: Zh. Org. Khim. *12*, 384 (1976)
216. Loktev, V. F., Korchagina, D. V., Shubin, V. G.: Izv. Sibirsk. Otd. Akad. Nauk SSSR, Otd. Khim. Nauk *2*, (1), 83 (1977)
217. Loktev, V. F., Korchagina, D. V., Shubin, V. G.: ibid. *2* (1), 118 (1977)
218. Shubin, V. G., Chzhu, V. P., Rezvukhin, A. I., Tabatskaya, A. A., Koptyug, V. A.: Izv. Akad. Nauk SSSR, Otd. Khim. Nauk *1967*, 2365
219. Shubin, V. G., Chzhu, V. P., Korobejnicheva, I. K., Rezvukhin, A. I., Koptyug, V. A.: ibid. *1970*, 1742
220. Detsina, A. N., Koptyug, V. A.: Zh. Org. Khim. *7*, 2575 (1971)
221. Koptyug, V. A., Mozulenko, L. M.: ibid. *7*, 1419 (1971)
222. Mozulenko, L. M., Koptyug, V. A.: ibid. *8*, 2531 (1972)

223. Mozulenko, L. M., Rezvukhin, A. I., Koptyug, V. A.: ibid. *8*, 2535 (1972)
224. Vitullo, V. P.: J. Org. Chem. *34*, 224 (1969)
225. Vitullo, V. P.: ibid. *35*, 3976 (1970)
226. Cook, K. L., Waring, A. J.: Tetrahedron Lett. *1971*, 1675, 3359
227. Cook, K. L., Waring, A. J.: J. Chem. Soc. Perkin II *1973*, 84
228. Vitullo, V. P.: Chem. Commun. *1970*, 688
229. Waring, A. J.: J. Chem. Soc. Perkin II *1979*, 1029
230. Pilkington, J. W., Waring, A. J.: Tetrahedron Lett. *1973*, 4345
231. Vitullo, V. P., Logue, E. A.: J. Org. Chem. *37*, 3339 (1972)
232. Hughes, M. J., Waring, A. J.: J. Chem. Soc. Perkin II *1974*, 1043
233. Kostina, N. G., Shtejngarts, V. D.: Zh. Org. Khim. *10*, 1705 (1974)
234. Akhmetova, N. E., Shtejngarts, V. D.: ibid. *11*, 1226 (1975)
235. Akhmetova, N. E., Shtejngarts, V. D.: ibid. *13*, 1269 (1977)
236. Olah, G. A., Pavlath, A. E., Olah, J. A.: J. Am. Chem. Soc. *80*, 6540 (1958)
237. Shteingarts, V. D., Pozdnyakovich, Yu. V., Yakobson, G. G.: Chem. Commun. *1969*, 1264
238. Shtejngarts, V. D., Pozdnyakovich, Yu. V.: Zh. Org. Khim. *7*, 734 (1971)
239. Chujkova, T. V., Shtark, A. A., Shtejngarts, V. D.: ibid. *10*, 132, 1712 (1974)
240. Oksinenko, B. G., Shtejngarts, V. D.: ibid. *10*, 1190 (1974)
241. Oksinenko, B. G., Mamatyuk, V. I., Shtejngarts, V. D.: ibid. *12*, 1322 (1976)
242. Borodkin, G. I., Shakirov, M. M., Shubin, V. G.: ibid. *13*, 2152 (1977)
243. Mozulenko, L. M., Koptyug, V. A.: ibid. *8*, 2152 (1972)
244. Sazonova, L. I., Shakirov, M. M., Shubin, V. G.: ibid. *13*, 2456 (1977)
245. Holmes, J., Pettit, R.: J. Org. Chem. *28*, 1695 (1963)
246. Buck, H. M.: Rec. trav. chim. *89*, 794 (1970)
247. Koptyug, V. A., Bushmelev, V. A., Gerasimova, T. N.: Zh. Obshch. Khim. *37*, 140 (1967)
248. Volz, H., Zimmermann, G., Schelberger, B.: Tetrahedron Lett. *1970*, 2429
249. Shubin, V. G., Korchagina, D. V., Koptyug, V. A.: Izv. Akad. Nauk USSR, Otd. Khim. Nauk *1969*, 1200
250. Korchagina, D. V., Deredyaev, B. G., Shubin, V. G., Koptyug, V. A.: Zh. Org. Khim. *7*, 2582 (1971)
251. Loktev, V. F., Korchagina, D. V., Shubin, V. G., Koptyug, V. A.: ibid. *13*, 219 (1977)
252. Bushmelev, V. A., Shakirov, M. M., Derendyaev, B. G., Koptyug, V. A.: ibid. *15*, 1934 (1979)
253. Childs, R. F., Winstein, S.: J. Am. Chem. Soc. *90*, 7146 (1968)
254. Isaev, I. S., Mamatyuk, V. I., Kuzubova, L. I., Gordymova, T. A., Koptyug, V. A.: Zh. Org. Khim. *6*, 2482 (1970)
255. Childs, R. F., Parrington, B.: Chem. Commun. *1970*, 1540
256. Koptyug, V. A., Kuzubova, L. I., Isaev, I. S., Mamatyuk, V. I.: ibid. *1969*, 389
257. Vogel, P., Saunders, M., Hasty, N. M., Berson, J. A.: J. Am. Chem. Soc. *93*, 1551 (1971)
258. Paquette, L. A., Krow, G. R., Bollinger, J. M., Olah, G. A.: ibid. *90*, 7147 (1968)
259. Isaev, I. S., Buraev, V. I., Novikov, G. P., Sosnovskaya, L. M., Koptyug, V. A.: Zh. Org. Khim. *10*, 567 (1974)
260. Hine, K. E., Childs, R. F.: J. Am. Chem. Soc. *93*, 2323 (1971)
261. Buck, H. M., van der Sluys- van der Vlught, M. J., Dekkers, H. P. J. M., Brongersma, H. H., Oosterhoff, L. J.: Tetrahedron Lett. *1964*, 2987
262. Brongersma, H. H., Buck, H. M., Dekkers, H. P. J. M., Oosterhoff, L. J.: Catalysis *16*, 149 (1968)
263. van Pelt, P., Buck, H. M.: Rec. trav. chim. *93*, 206 (1974)
264. van Pelt, P., Buck, H. M.: ibid. *92*, 1057 (1973)
265. Ostashevskaya, L. A., Isaev, I. S., Koptyug, V. A.: Zh. Org. Khim. *12*, 1279 (1976)
266. Effenberger, F.: Angew. Chem. *84*, 37 (1972)
267. Effenberger, F., Stohrer, W. D., Steinbach, A.: ibid. *81*, 261 (1969)
268. Baenziger, N. C., Nelson, A. D.: J. Am. Chem. Soc. *90*, 6602 (1968)
269. Lyerla, J. R., Yannoni, C. S., Bruck, D., Fyfe, C. A.: ibid. *101*, 4770 (1975)
270. Brouwer, D. M., Mackor, E. L., MacLean, C.: Rec. trav. chim. *84*, 1564 (1965)
271. Birchall, T., Gillespie, R. J.: Canad. J. Chem. *42*, 502 (1964)
272. Rezvukhin, A. I.: Thesis, Novosibirsk 1968

273. Koptyug, V. A., Rezvukhin, A. I., Isaev, I. S., Shlejder, I. A.: Teor. Eksp. Khim. *6*, 509 (1970)
274. Strohmeyer, M., Witte, C.: Liebigs Ann. Chem. *729*, 21 (1969)
275. Brouwer, D. M., MacLean, C., Mackor, E. L.: Disc. Farad. Soc. *39*, 121 (1965)
276. Koptyug, V. A., Rezvukhin, A. I., Krysin, A. P., Isaev, I. S.: Zh. Strukt. Khim. *8*, 622 (1967)
277. Childs, R. F., Sakai, M., Winstein, S.: J. Am. Chem. Soc. *90*, 7144 (1968)
278. Isaev, I. S., Buraev, V. I., Rezvukhin, A. I., Koptyug, V. A.: Zh. Org. Khim. *8*, 1453 (1972)
279. Koptyug, V. A., Shubin, V. G., Rezvukhin, A. I., Korchagina, D. V., Tret'yakov, V. P., Rudakov, E. S.: Dokl. Akad. Nauk SSSR *171*, 1109 (1966)
280. Repinskaya, I. B., Rezvukhin, A. I., Koptyug, V. A.: Zh. Org. Khim. *7*, 2143 (1971)
281. Olah, G. A., Mo, Y. K.: J. Org. Chem. *38*, 3221 (1973)
282. Olah, G. A., Starol, J. S., Asencio, G., Liang, Gao, Forsyth, D. A., Mateescu, G. D.: J. Am. Chem. Soc. *100*, 6299 (1978)
283. Olah, G. A., Mo, Y. K.: J. Org. Chem. *38*, 3212 (1973)
284. Emsli, Dzh., Finej, Dzh., Satklif, L.: Spektroskopiya yadernogo magnitnogo rezonansa vysokogo razresheniya, Vol. 1, Moscow, Mir 1968
285. Muller, N., Pickett, L. W., Mulliken, R. S.: J. Am. Chem. Soc. *76*, 4770 (1954)
286. Rezvukhin, A. I., Isaev, I. S., Koptyug, V. A.: Izv. Akad. Nauk SSSR, Otd. Khim. Nauk *1969*, 279
287. Johnson, C. E., Bovey, F. A.: J. Chem. Phys. *29*, 1012 (1958)
288. Mamatyuk, V. I.: Thesis, Novosibirsk 1978
289. Haddon, B. C.: Tetrahedron *28*, 3613 (1972)
290. Buckingham, A. D.: Canad. J. Chem. *38*, 300 (1960)
291. Musher, J. I.: J. Chem. Phys. *37*, 34 (1962)
292. Goldstein, J. H., Reddy, G. S.: ibid. *36*, 2664 (1962)
293. Reddy, G. S., Goldstein, J. H.: ibid. *38*, 2736 (1963)
294. Pickett, L. W., Muller, N., Mulliken, R. S.: ibid. *21*, 1400 (1953)
295. Muller, N., Mulliken, R. S.: J. Am. Chem. Soc. *80*, 3489 (1958)
296. Buraev, V. I., Isaev, I. S., Koptyug, V. A.: Zh. Org. Khim. *15*, 782 (1979)
297. Schug, J. C., Deck, J. C.: J. Chem. Phys. *37*, 2618 (1962)
298. Aminova, R. M., Gubajdullina, R. Z.: Zh. Strukt. Khim. *10*, 253 (1969)
299. Mamatyuk, V. I., Detsina, A. N., Koptyug, V. A.: Zh. Org. Khim. *12*, 739 (1976)
300. Olah, G. A., Mateescu, G., Mo, Y. K.: J. Am. Chem. Soc. *95*, 1865 (1973)
301. Lammertsma, K., Cerfontain, H.: ibid. *101*, 3618 (1979)
302. Rodionov, V. I., Shakirov, M. M., Isaev, I. S., Koptyug, V. A.: Zh. Org. Khim. *16*, 1515 (1980)
303. Rezvukhin, A. I., Korchagina, D. V., Shubin, V. G.: Izv. Akad. Nauk SSSR, Otd. Khim. Nauk *1976*, 1253
304. Buchanan, A. C., Dworkin, A. S., Smith, G. P.: J. Am. Chem. Soc. *102*, 5262 (1980)
305. MacMillan, J., Walker, E. R.: Chem. Commun. *1969*, 1031
306. Hart, H., Bau-Chien Jiang, J., Gupta, R. K.: Tetrahedron Lett. *1975*, 4639
307. Borodkin, G. I., Shakirov, M. M., Shubin, V. G., Koptyug, V. A.: Zh. Org. Khim. *14*, 321 (1978)
308. Borodkin, G. I., Shakirov, M. M., Shubin, V. G., Koptyug, V. A.: ibid. *14*, 989 (1978)
309. Shubin, V. G., Korchagina, D. V., Borodkin, G. I., Derendyaev, B. G., Koptyug, V. A.: ibid. *9*, 1031 (1973)
310. Borodkin, G. I., Shakirov, M. M., Shubin, V. G., Koptyug, V. A.: ibid. *12*, 1303 (1976)
311. Derendyaev, B. G.: OMR *4*, 27 (1972)
312. Borodkin, G. I., Shakirov, M. M., Shubin, V. G.: Zh. Org. Khim. *14*, 374 (1978)
313. Borodkin, G. I., Korchagina, D. V., Derendjaev, B. G., Shubin, V. G.: Tetrahedron Lett. *1973*, 539
314. Bodoev, N. V., Mamatyuk, V. I., Krysin, A. P., Koptyug, V. A.: Izv. Akad. Nauk SSSR, Otd. Khim. Nauk *1978*, 1199
315. Olah, G. A., Liang, G., Westerman, Ph.: J. Am. Chem. Soc. *95*, 3698 (1973)
316. Bushmelev, V. A., Shakirov, M. M., Koptyug, V. A.: Zh. Org. Khim. *12*, 2480 (1976); *13*, 2161 (1977)

317. Cavalieri, E., Calvin, M.: J. Org. Chem. *41*, 2676 (1976)
318. Hulme, R., Symons, M. C. R.: J. Chem. Soc. A. *1965*, 446
319. Hartshorn, M. P., Richards, K. E., Vaughan, J., Wright, G. J.: J. Chem. Soc. B *1971*, 1624
320. Blackstock, S. M., Richards, K. E., Wright, G. J.: Canad. J. Chem. *52*, 3313 (1974)
321. Eckert-Maksič, M.: J. Org. Chem. *45*, 3355 (1980)
322. Furin, G. G., Yakobson, G. G.: Izv. Sibirsk. Otd. Akad. Nauk SSSR, Otd. Khim. Nauk *1*, 109 (1977)
323. TerBorg, A. P., Gersmann, H. R., Bickel, A. F.: Rec. trav. chim. *85*, 899 (1966)
324. Childs, R. F., Parrington, D. D.: Canad. J. Chem. *52*, 3303 (1974)
325. Parrington, D., Childs, R. F.: Chem. Commun. *1970*, 1581
326. Friedrich, E. C.: J. Org. Chem. *33*, 413 (1968)
327. Larsen, J. W., Eckert-Maksič, M.: Croat. Chem. Acta *45*, 503 (1973)
328. Larsen, J. W., Eckert-Maksič, M.: Tetrahedron Lett. *1972*, 1477
329. Shubin, V. G.: Thesis, Novosibirsk 1967
330. Berezina, R. N., Korchagina, D. V., Shubin, V. G.: Zh. Org. Khim. *16*, 371 (1980)
331. Detsina, A. N., Sidorova, N. V., Panova, E. B., Malykhin, E. V., Shakirov, M. M.: ibid. *15*, 1887 (1979)
332. Childs, R. F.: Chem. Commun. *1969*, 946
333. Olah, G. A., Mo, Y. K.: J. Org. Chem. *38*, 353 (1973)
334. Koptyug, V. A., Andreeva, T. P., Mamatyuk, V. I.: Izv. Akad. Nauk SSSR, Otd. Khim. Nauk *1968*, 2844; Zh. Org. Khim. *6*, 1848 (1970)
335. Koptyug, V. A., Golounin, A. V.: Zh. Org. Khim. *8*, 607 (1972)
336. Kamshij, L. P., Mamatyuk, V. I., Koptyug, V. A.: ibid. *11*, 344 (1975)
337. Andreeva, T. P., Tregub, G. P., Mamatyuk, V. I., Koptyug, V. A.: ibid. *8*, 1271 (1972)
338. Kamshij, L. P., Mamatyuk, V. I., Koptyug, V. A.: ibid. *10*, 2194 (1974)
339. Larsen, J. W., Eckert-Maksič, M.: J. Am. Chem. Soc. *96*, 4311 (1974)
340. Olah, G. A., Mo, Y. K.: J. Org. Chem. *38*, 2212 (1973)
341. Ingol'd, K.: Teoreticheskie osnovy organicheskoj khimii, Moscow, Mir 1973, 682
342. Bell, R. P.: The Proton in Chemistry, Methuen and Co. Ltd. 1973, ch. X
343. Kresge, A. J., Chen, H. J., Hakka, L. E., Kouba, J. E.: J. Am. Chem. Soc. *93*, 6174 (1971)
344. Kresge, A. J., Hakka, L. E.: ibid. *88*, 3868 (1966)
345. Arnett, E. M., Wu, C. Y.: ibid. *82*, 5660 (1960)
346. Kresge, A. J., Chiang, Y., Hakka, L. E.: ibid. *93*, 6167 (1971)
347. Alder, R. W., Taylor, F. J.: J. Chem. Soc. B *1970*, 845
348. Mamatyuk, V. I., Rezvukhin, A. I., Golounin, A. V., Koptyug, V. A.: Zh. Org. Khim. *9*, 2359 (1973)
349. Cook, K. L., Hughes, M. J., Waring, A. J.: J. Chem. Soc. Perkin II *1972*, 1506
350. de la Mare, P. B. D.: Tetrahedron Lett. *5*, 107 (1959)
351. Nesmeyanov, A. N., Lutsenko, I. F.: Dokl. Akad. Nauk SSSR *59*, 707 (1948)
352. Miller, B.: J. Am. Chem. Soc. *92*, 6246, 6252 (1970)
353. Miller, B.: Accounts Chem. Res. *8*, 245 (1975)
354. Olah, G. A., Mo, Y. K.: J. Am. Chem. Soc. *94*, 5341 (1972)
355. Mozulenko, L. M.: Thesis, Novosibirsk 1972
356. Schubert, W. M., Quacchia, R. H.: J. Am. Chem. Soc. *84*, 3778 (1962)
357. Schubert, W. M., Quacchia, R. H.: ibid. *85*, 1278 (1963)
358. Olah, G. A., Kobayashi, Sh., Mo, Y. K.: J. Org. Chem. *38*, 4056 (1973)
359. Effenberger, F., Niess, R.: Angew. Chem. *79*, 1100 (1967)
360. Olah, G. A., Dunne, K., Kelly, D. P., Mo, Y. K.: J. Am. Chem. Soc. *94*, 7438 (1972)
361. Zhdanov, Yu. A., Minkin, V. I.: Korrelyatsionnyj analiz v organicheskoj khimii, Izd. Rostovskogo Universiteta 1966
362. Brouwer, D. M.: Rec. trav. chim. *87*, 335 (1968)
363. Brouwer, D. M.: ibid. *87*, 342 (1968)
364. Olah, G. A., Yamada, Y., Spear, R. J.: J. Am. Chem. Soc. *97*, 680 (1975)
365. Erykalov, Yu. G., Belokurova, A. P., Isaev, I. S., Rezvukhin, A. I., Koptyug, V. A.: Zh. Org. Khim. *7*, 1195 (1971)
366. Spear, R. J., Forsyth, D. A., Olah, G. A.: J. Am. Chem. Soc. *98*, 2493 (1976)

367. Olah, G. A., Mo, Y. K.: ibid. *94*, 9421 (1972)
368. Mooney, E. F., Winson, P. H., in: Annual Review of NMR Spectroscopy. Ed. E. F. Mooney. Vol. I, Academic Press. 1968
369. Ref. 284: *2*, 82, 88
370. Hanstein, W., Berwin, H. J., Traylor, T. G.: J. Am. Chem. Soc. *92*, 829 (1970)
371. Detsina, A. N., Mamatyuk, V. I., Derendyaev, B. G., Koptyug, V. A.: Zh. Org. Khim. *12*, 610 (1976)
372. Mamatyuk, V. I., Derendyaev, B. G., Detsina, A. N., Koptyug, V. A.: ibid. *10*, 2487 (1974)
373. Muller, N., Pritchard, D. E.: J. Chem. Phys. *31*, 768, 1471 (1959)
374. Shoolery, J. N.: ibid. *31*, 1427 (1959)
375. Juan, C., Gutowsky, H. S.: ibid. *37*, 2198 (1962)
376. Stothers, J. B.: Carbon-13 NMR Spectroscopy. N.Y., Academic Press. 1972
377. Grant, D., Litchman, W.: J. Am. Chem. Soc. *87*, 3994 (1965)
378. Dreeskamp, H., Sackmann, E.: Z. phys. Chem. (BRD) *34*, 273 (1962)
379. Ranft, J.: Ann. Phys. *10*, 399 (1963)
380. Hammaker, R. M.: J. Chem. Phys. *43*, 1843 (1965)
381. Dichfield, R., Jensen, M., Murrel, J.: J. Chem. Soc. A. *1967*, 1674
382. Yue, C.: Canad. J. Chem. *46*, 2675 (1968)
383. Lunazzi, L., Taddei, F.: Spectrochim. acta *25A*, 533 (1969)
384. Leshina, T. V., Molin, Yu. N., Mamaev, V. P.: Reakts. sposobn. org. soed. *3* (1), 52 (1966)
385. Haake, P., Miller, W. B., Tyssee, D. A.: J. Am. Chem. Soc. *86*, 3577 (1964)
386. Spiesecke, H., Schneider, W. G.: Tetrahedron Lett. *1961*, 468
387. Olah, G. A., Baker, E. B., Evans, J. C., Tolgyesi, W. S., McIntyre, J. C., Bastien, I. J.: J. Am. Chem. Soc. *86*, 1360 (1964)
388. Olah, G. A., Comisarow, M. B.: ibid. *88*, 1818 (1966)
389. Koptyug, V. A., Isaev, I. S., Rezvukhin, A. I.: Tetrahedron Lett. *1967*, 823
390. Koptyug, V. A., Lippmaa, E. T., Rezvukhin, A. I., Pekhk, T. I.: Izv. Akad. Nauk SSSR, Otd. Khim. Nauk *1969*, 285
391. Olah, G. A., Spear, R. J., Forsyth, D. A.: J. Am. Chem. Soc. *98*, 6284 (1976)
392. Mamatyuk, V. I., Koptyug, V. A.: Zh. Org. Khim. *13*, 818 (1977)
393. Mamatyuk, V. I., Rezvukhin, A. I., Detsina, A. N., Buraev, V. I., Isaev, I. S., Koptyug, V. A.: ibid. *9*, 2429 (1973)
394. Koptyug, V., Rezvukhin, A., Lippmaa, E., Pehk, T.: Tetrahedron Lett. *1968*, 4009
395. Olah, G. A., Liang, G., Jindall, S. P.: J. Org. Chem. *40*, 3259 (1975)
396. Rezvukhin, A. I., Mamatyuk, V. I., Koptyug, V. A.: Zh. Org. Khim. *8*, 2443 (1972)
397. Mamatyuk, V. I., Pozdnyakovich, Yu. V., Oksinenko, B. G., Buraev, V. I., Malykhin, E. V., Shtejngarts, V. D.: Izv. Akad. Nauk SSSR, Otd. Khim. Nauk *1975*, 1626
398. Olah, G. A., Mo, Y. K.: J. Org. Chem. *38*, 3221 (1973)
399. Olah, G. A., Mateescu, G. D.: J. Am. Chem. Soc. *92*, 1430 (1970)
400. LaLancete, E. A.: ibid. *87*, 1491 (1968)
401. Abronin, I. A., Gagarin, S. G., Zhidomirov, G. M.: Izv. Akad. Nauk SSSR, Otd. Khim. Nauk *1974*, 2627
402. Bodoev, N. V., Mamatyuk, V. I., Krysin, A. P., Koptyug, V. A.: ibid. *1976*, 2644
403. Bodoev, N. V., Krysin, A. P., Mamatyuk, V. I., Koptyug, V. A.: ibid. *1976*, 1899
404. Forsyth, D. A., Olah, G. A.: J. Am. Chem. Soc. *98*, 4096 (1976)
405. Bodoev, N. V., Krysin, A. P., Koptyug, V. A.: Zh. Org. Khim. *16*, 1002 (1980)
406. Manner, J. A., Cook, J. A., Ramsey, B. G.: J. Org. Chem. *39*, 1199 (1974)
407. Mamatyuk, V. I., Kamshij, L. P., Bodoev, N. V., Ostashevskaya, L. A., Koptyug, V. A.: Zh. Org. Khim. *12*, 469 (1976)
408. Grossel, M. C., Perkins, M. J.: J. Chem. Soc. Perkin II *1976*, 1334
409. Borodkin, G. I., Shakirov, M. M., Shubin, V. G., Koptyug, V. A.: Zh. Org. Khim. *10*, 2622 (1974)
410. Borodkin, G. I., Shakirov, M. M., Shubin, V. G., Koptyug, V. A.: ibid. *12*, 1297 (1976)
411. Shtark, A. A., Shtejngarts, V. D.: ibid. *12*, 1499 (1976)
412. Pozdnyakovich, Yu. V., Shtejngarts, V. D.: ibid. *13*, 1911 (1977)
413. Pozdnyakovich, Yu. V., Chujkova, T. V., Shtejngarts, V. D.: ibid. *11*, 1689 (1975)

414. Dobronravov, P. N., Chujkova, T. V., Pozdnyakovich, Yu. V., Shtejngarts, V. D.: ibid. *16*, 796 (1980)
415. Mackor, E. L., MacLean, C.: Pure Appl. Chem. *8*, 393 (1964)
416. Dallinga, G., Mackor, E. L., Verrijn Stuart, A. A.: Mol. Phys. *1*, 123 (1968)
417. Reid, C.: J. Am. Chem. Soc. *76*, 3264 (1954)
418. Kitova, A. N., Varshavskij, Ya. M.: Dokl. Akad. Nauk SSSR *135*, 1395 (1960)
419. Freiser, B. S., Beauchamp, J. L.: J. Am. Chem. Soc. *98*, 3136 (1976)
420. Kilpatrick, M., Hyman, W. H.: ibid. *80*, 77 (1958)
421. Serebryanskaya, A. I.: Thesis, Moscow 1964
422. Ginzburg, A. G., Setkina, V. N., Kursanov, D. N.: Dokl. Akad. Nauk USSR *177*, 105 (1967)
423. van der Luft, W. T. A. M., Buck, H. M., Oosterhoff, L. J.: Tetrahedron *24*, 4941 (1968)
424. Salakhutdinov, N. F., Korobejnicheva, I. K., Koptyug, V. A.: Zh. Org. Khim. *15*, 1461 (1979)
425. Vitullo, V. P., Grossman, N.: Tetrahedron Lett. *1970*, 1559
426. Budzikiewicz, H.: ibid. *N. 7*, 12 (1960)
427. Filipescu, N., Pawlik, J. W.: J. Am. Chem. Soc. *92*, 6062 (1970)
428. Childs, R. F., Parrington, B. D., Zeya, M.: J. Org. Chem. *44*, 4912 (1979)
429. Koutecky, J., Paldus, J.: Coll. Czech. Chem. Commun. *28*, 1483 (1963)
430. Yamaoka, T., Hosoya, H., Nagakura, S.: Tetrahedron *26*, 4215 (1970)
431. Yamaoka, T.: Bull. Chem. Soc. Japan *43*, 3086 (1970)
432. Mataga, N.: ibid. *36*, 1109 (1963)
433. Mackor, E. L., Dallinga, G., Kruizinga, J. H., Hofstra, A.: Rec. trav. chim. *75*, 836 (1956)
434. Verrijn Stuart, A. A., Mackor, E. L.: J. Chem. Phys. *27*, 826 (1957)
435. Shriner, R. L., Geipel, L.: J. Am. Chem. Soc. *79*, 227 (1957)
436. Dallinga, G., Verrijn Stuart, A. A., Smit, P. J., Mackor, E. L.: Z. Electrochem. *61*, 1019 (1957)
437. Velthorst, N. H., Hoijtink, G. J.: J. Am. Chem. Soc. *87*, 4529 (1965)
438. Ola, G. A., Pittman, Ch., in: Novye problemy fizicheskoj organicheskoj khimii, Moscow, Mir 1969, 338
439. Olah, G. A., Backer, E. B., Evans, T. C., Tolgesi, W. S., McIntyre, J. S., Bastien, I. J.: J. Am. Chem. Soc. *86*, 1360 (1964)
440. Voronkova, E. M., Grechyshnikov, B. N., Distler, G. I., Petrov, I. P.: Opticheskie materialy dlya infrakrasnoj tekhniki, Moscow, Nauka 1965
441. Korobejnicheva, I. K.: Thesis, Novosibirsk 1972
442. Perkampus, H. H., Baumgarten, E.: Ber. Bunsenges. Physik. Chem. *67*, 576 (1963)
443. Perkampus, H. H., Baumgarten, E.: ibid. *68*, 70 (1964)
444. Bystrov, D. S.: Dokl. Akad. Nauk SSSR *154*, 407 (1964)
445. Olah, G. A., Commeyras, A., Lui, G. Y.: J. Am. Chem. Soc. *90*, 3882 (1968)
446. Olah, G. A., White, A. M., DeMember, J. B., Commeyras, A., Lui, G. Y.: ibid. *92*, 4627 (1970)
447. Olah, G. A., DeMember, J. R., Commeyras, A., Bribes, J. L.: ibid. *93*, 459 (1971)
448. Evans, J. C.: In: Carbonium Ions. (Ed. Olah, G. A. and Schleyer, P. von R.) Vol. 1, Wiley-Interscience *1968*, 223
449. Sharp, D. W. A., Sheppard, N.: J. Chem. Soc. *1957*, 674
450. Kresge, A. J., Lichtin, N. N., Rao, K. N., Weston, R. E.: J. Am. Chem. Soc. *87*, 437 (1965)
451. Weston, R. E., Tsukamoto, A., Lichtin, N. N.: Spectrochim. Acta *22*, 433 (1966)
452. Sharp, D. W. A.: J. Chem. Soc. *1957*, 4804
453. Gomes de Mesquita, A. H., MacGillavry, C. H., Erkis, K.: Acta Cryst. *18*, 437 (1965)
454. Deno, N. C., Richey, H. G., Lui, J. S., Hodge, J. D., Houser, J. J., Wisotsky, M. J.: J. Am. Chem. Soc. *84*, 2016 (1962)
455. West, R., Kwitowski, P. T.: ibid. *88*, 5280 (1966)
456. Katz, T. J., Gold, E. H.: ibid. *86*, 1600 (1964)
457. Deno, N. C., Richey, H. G., Hodge, J. D., Wistorsky, M. J.: ibid. *84*, 1498 (1962)
458. Breslow, R., Höver, H., Chang, H. W.: ibid. *84*, 3168 (1962)
459. Chatt, J., Guy, R. G.: Chem. Ind. *1963*, 212
460. West, R., Sado, A., Tobey, S. W.: J. Am. Chem. Soc. *88*, 2488 (1966)

461. Tobey, S. W., West, R.: ibid. *86*, 1459 (1964)
462. Scherer, J. R., Overend, J.: Spectrochim Acta *17*, 719 (1961)
463. Fateley, W. G., Lippincott, E. R.: J. Am. Chem. Soc. *77*, 249 (1955)
464. Fateley, W. G., Curnutte, B., Lippincott, E. R.: J. Chem. Phys. *26*, 1471 (1957)
465. Doering, W. E., Knox, L. H.: J. Am. Chem. Soc. *76*, 3203 (1954)
466. Aksnes, G., Songstad, J.: Acta Chem. Scand. *18*, 655 (1964)
467. Sendvina, L. B., Shejchenko, V. I., Shejnker, Yu. N., Dombrovskij, A. V., Shevchuk, M. I., Barazkov, L. I., Bergel'son, L. D.: Zh. Obshch. Khim. *37*, 499 (1967)
468. Sharp, D. W. A.: Chem. Ind. *1958*, 1235
469. Kemmit, R. D. W., Nuttall, R. H., Sharp, D. W. A.: J. Chem. Soc. *1960*, 46
470. Klages, F., Gordon, J. E., Jung, H. A.: Chem. Ber. *98*, 3748 (1965)
471. Basov, V. P., Lobanov, Yu. P., Shut'ko, A. P.: Zhurnal prikladnoj spektroskopii *25*, 1114 (1976)
472. Bellami, L.: Infrakrasnye spektry slozhnykh molekul, Moscow, IL 1963
473. Sycheva, I. M., Korobejnicheva, I. K.: Izv. Sibirsk. Otd. Akad. Nauk SSSR *9* (4), 112 (1979)
474. Korobejnicheva, I. K., Mitasov, M. M., Koptyug, V. A.: Zh. Org. Khim. *10*, 62 (1974)
475. Zajtsev, B. E., Kreshkov, Yu. D., Vol'in, M. E., Shejnker, Yu. N.: Dokl.-Akad. Nauk SSSR *139*, 1107 (1961)
476. Harmon, K. M., Coburn, T. T.: J. Am. Chem. Soc. *87*, 2499 (1965); J. Phys. Chem. *72*, 2950 (1968)
477. Harmon, K. M., Cummins, S. M., Coburn, T. T.: J. Phys. Chem. *73*, 2939 (1969)
478. Cook, D.: Spectrochim. Acta *22*, 415, 419 (1966); Canad. J. Chem. *42*, 2721 (1964)
479. Nakamoto, K.: Infrakrasnye spektry neorganicheskikh i koordinatsionnykh soedinenij, Moscow, Mir 1966
480. Susz, B. P., Wuhrmann, J. J.: Helv. chim. acta *40*, 971 (1957)
481. Mairesse, G., Barbier, P., Wignacourt, J. P.: Canad. J. Chem. *56*, 764 (1978)
482. Bradley, R. H., Brier, P. N., Jones, D. E. H.: J. Chem. Soc. A. *1971*, 1397
483. Jones, D. E. H., Wood, J. L.: Spectrochim. Acta *23A*, 2695 (1967)
484. Clark, R. J. H., Crociani, B., Wasserman, A.: J. Chem. Soc., A. *1970*, 2458
485. Carlson, G. L.: Spectrochim. Acta *19*, 129 (1963)
486. Kazitsina, L. A., Pasynkevich, S. V., Reutov, O. A.: Dokl. Akad. Nauk SSSR *141*, 624 (1961)
487. Gerding, H., Houtgraaf, H.: Rec. trav. chim. *72*, 21 (1953)
488. Balasubrahmanyam, K., Nanis, L.: J. Chem. Phys. *42*, 676 (1965)
489. Torsi, G., Mamontov, G., Begun, G. M.: Inorg. Nucl. Chem. Lett. *6*, 553 (1970)
490. Cyvin, S. J., Klaeboe, P., Rytter, E., Qye, H. A.: J. Chem. Phys. *52*, 2776 (1970)
491. Qye, H. A., Rytter, E., Klaeboe, P., Cyvin, S. J.: Acta Chem. Scand. *25*, 559 (1971)
492. Doorenbos, H. E., Evanc, J. C., Kagel, R. O.: J. Phys. Chem. *74*, 3385 (1970)
493. Gerding, H., Stufkens, D. J.: Rev. Chim. Min. *6*, 795 (1969)
494. Begun, G. M., Boston, C. R., Torsi, G., Mamontov, G.: Inorg. Chem. *10*, 866 (1971)
495. Brown, D. H., Stewarf, D. T.: Spectrochim. Acta *26A*, 1344 (1970)
496. Kinsella, E., Coward, J.: ibid. *24A*, 2139 (1968)
497. Kendall, Y., Crittendena, E. D., Miller, H. K.: J. Am. Chem. Soc. *45*, 963 (1923)
498. Filimonov, V. N., Terenin, A. N.: Dokl. Akad. Nauk SSSR *109*, 799 (1956)
499. Miller, A., Krebs, B.: J. Mol. Spectr. *24*, 180 (1967)
500. Woodward, L. A., Nord, A. A.: J. Chem. Soc. *1956*, 3721
501. Adams, D. M., Chatt, J., Davidson, J. M., Gerratt, J.: ibid. *1963*, 2189
502. Woodward, L. A., Nord, A. A.: ibid. *1955*, 2655
503. Schmulbach, C. D., Ahmed, I. Y.: J. Chem. Soc., *1968*, 3008
504. Sawodny, W., Dehnicke, K.: Z. anorg. allg. Chem. *349*, 169 (1967)
505. Krasnov, K. S., Timoshin, V. S., Danilova, T. G., Khandozhko, S. V.: Molekulyarnye postoyannye neorganicheskikh soedinenij, Moscow, Khimiya 1968
506. Semin, G. K., Babushkina, T. A., Yakobson, G. G.: Primenenie yadernogo kvadrupol'nogo rezonansa v khimii, Moscow, Khimiya 1972
507. Kidd, R. G., Truax, D. R.: J. Am. Chem. Soc. *90*, 6867 (1968); Canad. Spectroscopy, *14*, 1 (1969)
508. Akitt, J. W., Greenwood, N. N., Storr, A.: J. Chem. Soc. *1965*, 4410
509. Haselbach, E.: Tetrahedron Lett. *1970*, 1543

510. Almlöf, J., Haselbach, E., Jachimowicz, F., Kovalewski, J.: Helv. chim. acta *58*, 2403 (1975) (1975)
511. Koptyug, V. A., Buraev, V. I.: Zh. Org. Khim. *14*, 18 (1978)
512. Brouwer, D. M.: Rec. trav. chim. *87*, 611 (1968)
513. Hepler, L. G.: Canad. J. Chem. *49*, 2803 (1971)
514. Koptyug, V. A., Buraev, V. I., Isaev, I. S.: Zh. Org. Khim. *14*, 1922 (1978)
515. Koptyug, V. A., Buraev, V. I.: ibid. *16*, 1882 (1980)
516. Gold, V.: J. Chem. Soc. Farad. Trans. I *1972*, 1611
517. Astin, K. B.: Tetrahedron Lett. *1980*, 3713
518. Sorensen, T.: J. Am. Chem. Soc. *89*, 3782, 3794 (1967)
519. Ranganayakulu, K., Sorensen, T. S.: Canad. J. Chem. *50*, 3534 (1972)
520. Bittner, E. W., Arnett, E. M., Saunders, M.: J. Am. Chem. Soc. *98*, 3734 (1976)
521. Arnett, E. M., Petro, C.: ibid. *100*, 5402, 5408 (1978)
522. Wolf, J. F., Harch, P. G., Taft, R. W.: ibid. *97*, 2904 (1975)
523. Fry, J. L., Harris, J. M., Bingham, R. C., Schleyer, P. von R.: ibid. *92*, 2540 (1970)
524. Feldman, M. R., Flythe, W. C.: J. Org. Chem. *43*, 2596 (1978)
525. Jorgensen, W. L.: J. Am. Chem. Soc. *99*, 280 (1977)
526. Arnett, E. M., Larsen, J. W.: ibid. *91*, 1438 (1969)
527. Arnett, E. M., Larsen, J. W.: ibid. *90*, 792 (1968)
528. Arnett, E. M., Abboud, J. L. M.: ibid. *97*, 3865 (1975)
529. D'yuar, M.: Sverkhsopryazhenie, Moscow, Mir 1965
530. Ananthanarayan, K. A., Sorensen, T. S.: Canad. J. Chem. *50*, 3550 (1972)
531. Koptyug, V. A., Buraev, V. I., Isaev, I. S., Perevyazkina, O. N.: Zh. Org. Khim. *14*, 328 (1978)
532. Olah, G. A., Kobayashi, S., Tashiro, M.: J. Am. Chem. Soc. *94*, 7448 (1972)
533. McGary, G. W., Kamoto, Y. O., Brown, H. C.: ibid. *77*, 3037 (1955)
534. Ansell, H. W., LeGuen, J., Taylor, R.: Tetrahedron Lett. *1973*, 13
535. Pitzer, K. S., Scott, D. W.: J. Am. Chem. Soc. *65*, 803 (1943)
536. Allen, R. H., Yats, L. D.: ibid. *81*, 5289 (1959)
537. McCaulay, D. A., Lien, A. P.: ibid. *74*, 6246 (1952)
538. Mamedaliev, Yu. G., Topchiev, A. V., Mamedaliev, G. M., Sulejmanov, G. N.: Dokl. Akad. Nauk SSSR *106*, 1027 (1956)
539. Pearson, D. E., Buehler, C. A.: Synthesis *1971*, 471
540. Erykalov, Yu. G., Belokurova, A. P., Isaev, I. S., Koptyug, V. A.: Zh. Org. Khim. *9*, 343 (1973)
541. Wibaut, J. P., Paulis, B.: Rec. trav. chim. *77*, 769 (1958)
542. Larsen, J. W., Bouis, P. A., Riddle, Ch. A.: J. Org. Chem. *45*, 4969 (1980)
543. Shubin, V. G., Chzhu, V. P., Rezvukhin, A. I., Koptyug, V. A.: Izv. Akad. Nauk SSSR, Otd. Khim. Nauk *1966*, 2056
544. Hart, H., Swatton, D. W.: J. Am. Chem. Soc. *89*, 1874 (1967)
545. Swatton, D. W., Hart, H.: ibid. *89*, 5075 (1967)
546. Waring, A. J., Hart, H.: ibid. *86*, 1454 (1964)
547. Hart, H., Collins, P. M., Waring, A. J.: ibid. *88*, 1005 (1966)
548. Isaev, I. S., Egorova, T. G., Shlejder, I. A., Lippmaa, E. T., Pekhk, T. I., Koptyug, V. A.: Dokl. Akad. Nauk SSSR *189*, 1258 (1969)
549. Berezina, R. N., Korchagina, D. V., Shubin, V. G.: Izv. Sibirsk. Otd. Akad. Nauk SSSR, Otd. Khim. Nauk *2* (1), 57 (1980)
550. Shine, H. J.: Aromatic Rearrangements. Elsevier Pub. Company. 1967
551. Waring, A. J., in: Advances in Alicyclic Chemistry. Vol. 1, Ed. Hart, H. and Karabatsos, G. J., N.Y., Academic Press. 1966
552. Miller, B., in: Mechanisms of Molecular Migration., Ed. Thyagaragan, B. S., N.Y., Interscience 1968
553. Pople, J. A., Schneider, W. G., Bernstein, H. J.: Spektry yadernogo magnitnogo rezonansa vysokogo razreshcheniya, Moscow, IL 1962
554. Koldin, E.: Bystrye reaktsii v rastvorakh, Moscow, Mir 1966
555. Koptyug, V. A., Shubin, V. G.: Zh. Org. Khim. *16*, 1977 (1980)
556. Morozov, S. V., Shakirov, M. M., Shubin, V. G.: ibid. *17*, 154 (1981)

557. Saunders, M., Kates, M. R.: J. Am. Chem. Soc. *100*, 7082 (1978)
558. Brouwer, D. M., Hogeveen, H., in: Progress in Physical Organic Chemistry. Vol. 9. Ed. Streitwieser, A., Taft, R. W., John Wiley and Sons (1972)
559. Saunders, M., in: Magnetic Resonance in Biological Systems. Ed. Ehrenberg, A., Malmstrom, B. G., Vanngerd, T., N.Y. 1967
560. Derendyaev, B. G., Mamatyuk, V. I., Koptyug, V. A.: Tetrahedron Lett., *1969*, 5
561. Derendyaev, B. G., Mamatyuk, V. I., Koptyug, V. A.: Izv. Akad. Nauk SSSR, Otd. Khim. Nauk *1971*, 972
562. Borodkin, G. I., Derendyaev, B. G., Shubin, V. G., Koptyug, V. A.: ibid. *1974*, 235
563. Shakirov, M. M., Shubin, V. G., Rezvukhin, A. I.: Izv. Sibirsk. Otd. Akad. Nauk, Otd. Khim. Nauk *4*, 155 (1974)
564. Loktev, V. F., Korchagina, D. V., Shakirov, M. M., Shubin, V. G.: ibid. *12* (5), 146 (1978)
565. McManus, S. P., Peterson, P. E.: Tetrahedron Lett. *1975*, 2753
566. McManus, S. P., Worley, S. D.: ibid. *1977*, 555
567. D'yuar, M.: Teoriya molekulyarnykh orbitalej v organicheskoj khimii, Moscow, Mir 1972, p. 382
568. D'yuar, M., Dogerti, R.: Teoriya vozmushchenij molekulyarnykh orbitalej v organicheskoj khimii, Moscow, Mir 1977, p. 136
569. Borodkin, G. I., Korobejnicheva, I. K., Shubin, V. G.: Zh. Org. Khim. *12*, 2381 (1976)
570. Loktev, V. F., Korchagina, D. V., Shubin, V. G.: Izv. Sibirsk. Otd. Akad. Naük SSSR, Otd. Khim. Nauk *14* (6), 86 (1980)
571. Borodkin, G. I., Pletneva, L. M., Shakirov, M. M., Shubin, V. G.: Zh. Org. Khim. *15*, 652 (1979)
572. Shubin, V. G., Korchagina, D. V., Derendyaev, B. G., Borodkin, G. I., Koptyug, V. A.: ibid. *9*, 1041 (1973)
573. Shlejder, I. A.: Thesis, Novosibirsk 1972
574. Borodkin, G. I., Shakirov, M. M., Shubin, V. G., Koptyug, V. A.: Zh. Org. Khim. *12*, 2033 (1976)
575. Lathan, W. A., Hehre, W. J., Pople, J. A.: J. Am. Chem. Soc. *93*, 808 (1971)
576. Hehre, W. J.: ibid. *94*, 5919 (1972)
577. Radom, L., Hariharan, P. C., Pople, J. A., Schleyer, P. von R.: ibid. *95*, 6531 (1973)
578. Brouwer, D. M., van Doorn, A.: Rec. trav. chim. *90*, 1010 (1971)
579. Shubin, V. G., Koptyug, V. A.: Izv. Sibirsk. Otd. Akad. Nauk SSSR, Otd. Khim. Nauk *4* (2), 131 (1976)
580. Bushmelev, V. A., Shakirov, M. M., Brouwer, D. M., Koptyug, V. A.: Zh. Org. Khim. *15*, 1579 (1979)
581. Borodkin, G. I., Koptyug, V. A., Shubin, V. G.: Dokl. Akad. Nauk SSSR *255*, 587 (1980)
582. Energii razryva khimicheskikh svyazej. Potentsialy ionizatsii i srodstvo k elektronu (Ed. Kondrat'ev, V. N.), Moscow, Nauka 1974
583. Sutula, V. D., Koptyug, V. A.: Dokl. Akad. Nauk SSSR *247*, 1424 (1979)
584. Grimmer, M., Heidrich, D.: Z. Chem. *12*, 481 (1971)
585. Bernardini, F., Hehre, W. J.: J. Am. Chem. Soc. *95*, 3078 (1973)
586. Rodionov, V. I., Koptyug, V. A.: Zh. Org. Khim. *15*, 2555 (1979)
587. Pearson, D. E.: J. Tenn. Acad. Sci. *40*, 97 (1965)
588. Pearson, D. E., Buehler, C. A.: Synthesis *1971*, 455
589. Marcus, R. A.: J. Phys. Chem. *72*, 891, 4249 (1968)
590. Marcus, R. A.: J. Am. Chem. Soc. *91*, 7224 (1969)
591. Dneprovskij, A. S., in: Reaktsionnaya sposobnost' i mekhanizmy reaktsij organicheskikh soedinenij, Leningrad, Izd. LGU 1974, p. 3
592. Allen, R. H., Yats, L. D.: J. Am. Chem. Soc. *81*, 5289 (1959)
593. Fury, L. A., Pearson, D. E.: J. Org. Chem. *30*, 2301 (1964)
594. Suld, G., Stuart, A. P.: ibid. *29*, 2939 (1964)
595. Vorozhtsov, N. N., Koptyug, V. A.: Zh. Obshch. Khim. *30*, 999 (1960)
596. Oku, A., Yuzen, Y.: J. Org. Chem. *40*, 3850 (1975)
597. Packer, J., Vaughan, J., Wong, E.: J. Am. Chem. Soc. *80*, 905 (1958)
598. Mansson, M.: Acta Chem. Scand., Ser. B. *1974*, 28, 677
599. Vitullo, V. P., Logue, E. A.: J. Chem. Soc. Chem. Commun. *1974*, 228

600. Koptyug, V. A., Isaev, I. S., Vorozhtsov, N. N.: Dokl. Akad. Nauk SSSR *149*, 100 (1963)
601. Steinberg, H., Sixma, F. L.: Rec. trav. chim. *81*, 185 (1962)
602. Vitullo, V. P., Grossman, N.: J. Am. Chem. Soc. *94*, 3844 (1972)
603. Palm, V. A., Haldna, U. L., Talvik, A. J., in: The Chemistry of the Carbonyl Group. Ed. Patai, S., N.Y., Interscience Publishers 1966
604. Cook, K. L., Waring, A. J.: J. Chem. Soc. Perkin II *1973*, 88
605. Waring, A. J.: Tetrahedron Lett. *1975*, 171
606. Marx, J. N., Argyle, J. C., Norman, L. R.: J. Am. Chem. Soc. *96*, 2121 (1974)
607. Bloom, S. M.: Tetrahedron Lett. N. *21*, 7 (1959)
608. Newmann, M. S., Pawelek, D., Ramachandran, S.: J. Am. Chem. Soc. *84*, 995 (1962)
609. Vitullo, V. P., Logue, E. A.: J. Org. Chem. *38*, 2265 (1973)
610. Fischer, A., Hehderson, G. N.: J. Chem. Soc. Chem. Commun. *1979*, 279
611. Goodwin, S., Witkop, B.: J. Am. Chem. Soc. *79*, 179 (1957)
612. Palmer, J. D., Waring, A. J.: J. Chem. Soc. Perkin II *1979*, 1089
613. Acheson, R. M.: Accounts Chem. Res. *4*, 177 (1971)
614. Suehiro, T., Yamazaki, S.: Bull. Chem. Soc. Japan *48*, 3655 (1975)
615. Blackstock, D. J., Hartshorn, M. P., Lewis, A. J., Richards, K. E., Vaughan, J., Wright, G. J.: J. Chem. Soc. B. *1971*, 1212
616. Clemens, A. H., Hartshorn, M. P., Richards, K. E., Wright, G. J.: Austral. J. Chem. *30*, 113 (1977)
617. Barnes, C. E., Myhre, P. C.: J. Am. Chem. Soc. *100*, 973 (1978)
618. Coombes, R. G., Golding, J. G.: Tetrahedron Lett. *1978*, 3583
619. Gold, V., in: Friedel-Crafts and Related Reactions. Ed. Olah, G. A., N.Y., Vol. II, part II, 1964
620. Koptyug, V. A., Shubin, V. G., Korchagina, D. V.: Tetrahedron Lett. *1965*, 1535
621. Shubin, V. G., Tabatskaya, A. A., Koptyug, V. A.: Zh. Org. Khim. *6*, 2081 (1970)
622. Hyman, H. H., Garber, G. A.: J. Am. Chem. Soc. *81*, 1847 (1959)
623. Plieninger, H., Maier-Borst, W.: Angew. Chem. *75*, 1177 (1963)
624. Bailey, W. J., Baylouny, R. A.: J. Org. Chem. *27*, 3476 (1962)
625. Hunter, H., Brune, H. A.: Z. Naturforschung *23b*, 1612 (1968)
626. Deno, N. C., Groves, P. T., Saines, G.: J. Am. Chem. Soc. *81*, 5790 (1959)
627. Ginzburg, A. G., Setkina, V. N., Kursanov, D. N.: Zh. Org. Khim. *3*, 1921 (1967)
628. Setkina, V. N., Ginzburg, A. G., Fedin, E. N., Kursanov, D. N.: Dokl. Akad. Nauk SSSR *158*, 671 (1964)
629. Ginzburg, A. G., Setkina, V. N., Kursanov, D. N.: ibid. *169*, 1080 (1966)
630. Batterham, T. J., Brown, D. J., Paddon-Row, M. N.: J. Chem. Soc. B. *1967*, 171
631. Balaban, A. T., Frangopol, M., Frangopol, P. T., Gard, E.: Preparation and Bio-Medical Application of Labelled Molecules. Bruxelles, Euratom 1964
632. Gard, E., Vasilescu, A., Matescu, G. D., Balaban, A. T.: J. Label. Compounds, *3*, 196 (1967)
633. Hodges, R. J., Garnett, J. L.: J. Catalysis *13*, 83 (1969)
634. Garnett, J. L., Kenyon, R. S.: J. Chem. Soc. Chem. Commun. *1970*, 698
635. Fonken, G. J.: J. Org. Chem. *28*, 1909 (1963)
636. Koptyug, V. A., Bushmelev, V. A.: Zh. Org. Khim. *4*, 1822 (1968)
637. Bushmelev, V. A.: Thesis, Novosibirsk 1970
638. Bushmelev, V. A., Koptyug, V. A.: Zh. Org. Khim. *5*, 703 (1969)
639. Yuldashev, K. Yu., Tsukervanik, I. P.: Zh. Obshch. Khim. *34*, 2647 (1964)
640. Bushmelev, V. A., Koptyug, V. A.: Zh. Org. Khim. *6*, 1853 (1970)
641. Perrin, Ch. L.: J. Org. Chem. *36*, 420 (1971)
642. de la Mare, P. B. D., Singh, A.: J. Chem. Soc. Perkin II *1972*, 1801
643. Lyukshova, N. V., Krysin, A. P., Koptyug, V. A.: Izv. Akad. Nauk SSSR, Otd. Khim. Nauk *1976*, 936
644. Koptyug, V. A., Krysin, A. P.: Zh. Org. Khim. *6*, 526 (1970)
645. Lyukshova, N. V., Krysin, A. P., Koptyug, V. A.: Izv. Akad. Nauk SSSR, Otd. Khim. Nauk *1975*, 73
646. Karhu, M.: J. Chem. Soc. Perkin I *1980*, 1595
647. Eberson, L., Nyberg, K.: Tetrahedron Lett. *1966*, 2389; J. Am. Chem. Soc. *88*, 1686 (1966)

648. Eberson, L.: ibid. *89*, 4668 (1967)
649. Parker, V. D., Adams, R. N.: Tetrahedron Lett. *1969*, 1721
650. Parker, V. D., Burgert, B. E.: ibid. *1968*, 3341
651. Baciocchi, E., Illuminati, G.: ibid. *1962*, 637
652. Andrews, L. J., Keefer, R. M.: J. Am. Chem. Soc. *86*, 4158 (1964)
653. Baciocchi, E., Ciana, A., Illuminati, G., Pasini, C.: ibid. *87*, 3953 (1965)
654. Baciocchi, E., Casula, M., Illuminati, G., Mandolini, L.: Tetrahedron Lett. *1969*, 1275
655. Antinori, G., Baciocchi, E., Illuminati, G.: J. Chem. Soc. B. *1969*, 373
656. Suzuki, H., Tamura, Y.: J. Chem. Soc. Chem. Commun. *1969*, 244
657. Suzuki, H.: Bull. Institute Chem. Research Kyoto Univ. *50*, 407 (1972)
658. Pokhodenko, V. D., Kalibabchuk, N. N.: Zh. Org. Khim. *2*, 1397 (1966)
659. Brodskij, A. I., Pokhodenko, V. D., Kalibabchuk, N. N., Kuts, V. S.: Dokl. Akad. Nauk SSSR *172*, 122 (1967)
660. Kalibabchuk, N. N., Pokhodenko, V. D.: Zh. Org. Khim. *4*, 329 (1968)
661. Shabarov, Yu. S., Mochalov, S. S., Dajneko, V. I.: Izv. Sibirsk. Otd. Akad. Nauk SSSR, Otd. Khim. Nauk *7* (3), 43 (1981)
662. Fischer, A., Vaughan, J., Wright, G.: J. Chem. Soc. B *1967*, 368
663. Fischer, A., Wright, G.: Austral. J. Chem. *27*, 217 (1974)
664. Fischer, A., Ramsay, J. N.: Canad. J. Chem. *52*, 3960 (1974)
665. Suzuki, H.: Synthesis *1977*, 217
666. Shtejngarts, V. D., Yakobson, G. G.: Zh. Vses. Khim. Obshchestva im. D. I. Mendeleeva 15, 72 (1970)
667. Osina, O. I., Chujkova, T. V., Shtejngarts, V. D.: Zh. Org. Khim. *16*, 805 (1980)
668. Jacquesy, J. A. C., Jouannetaud, M. P.: Bull. Soc. Chim. France 267, 295, 304 (1980)
669. Koptyug, V. A.: Zh. Vses. Khim. Obshchestva im. D. I. Mendeleeva *21*, 315 (1976)
670. Repinskaya, I. B., Altykhova, L. N., Koptyug, V. A.: Zh. Org. Khim. *13*, 1442 (1977)
671. Repinskaya, I. B., Abramov, A. D., Kuklina, N. A., Koptyug, V. A.: ibid. *15*, 2178 (1979)
672. Repinskaya, I. B., Koptyug, V. A.: ibid. *16*, 1508 (1980)
673. Pozdnyakovich, Yu. V., Chujkova, T. V., Bardin, V. V., Shtejngarts, V. D.: ibid. *12*, 690 (1976)
674. Akhmetova, N. E., Bazhin, N. M., Pozdnyakovich, Yu. V., Shtejngarts, V. D., Shchegoleva, L. N.: Teor. Eksp. Khim. *10*, 613 (1974)
675. Siskin, M.: J. Am. Chem. Soc. *96*, 3640, 3641 (1974)
676. Wristers, J.: J. Chem. Soc. Chem. Commun. *1977*, 575
677. Nagakura, S., Tanaka, J.: Bull. Chem. Soc. Japan *32*, 734 (1959)
678. Nagakura, S., Tanaka, J.: J. Chem. Phys. *22*, 563 (1954)
679. Nagakura, S.: Tetrahedron *19*, Suppl. 2, 361 (1963)
680. Brown, R. D.: J. Chem. Soc. *1959*, 2224, 2232
681. Pedersen, E. B., Petersen, T. E., Torssell, K., Lawesson, S. O.: Tetrahedron *29*, 579 (1973)
682. Mulliken, R. S.: J. Phys. Chem. *56*, 801 (1952)
683. Mulliken, R. S.: J. Am. Chem. Soc. *74*, 811 (1952)
684. Endryus, L., Kifer, R.: Molekulyarnye kompleksy v organicheskoj khimii, Moscow, Mir 1967
685. Foster, R.: Organic Charge-Transfer Complexes. N.Y., Academic Press 1969
686. Epiotis, N. D.: J. Am. Chem. Soc. *95*, 3188 (1973)
687. Abronin, I. A., Zhidomirov, G. M., Gol'dfarb, Ya. L.: Dokl. Akad. Nauk SSSR *218*, 363 (1974)
688. Glowes, G. A.: J. Chem. Soc. *1968*, 2519
689. Okhlobystin, O. Yu.: Perenos elektrona v organicheskikh reaktsiyakh, Izd. Rostovskogo Universiteta 1974
690. Weisman, S. I., de Boer, E., Conrad, J. J.: J. Chem. Phys. *26*, 963 (1957)
691. Sato, Y., Kinoshita, M., Sano, M., Akamatu, H.: Bull. Chem. Soc. Japan *42*, 3051 (1969)
692. Lewis, I. C., Singer, L. S.: J. Chem. Phys. *43*, 2712 (1965)
693. Howarth, O. W., Fraenkel, G. K.: J. Am. Chem. Soc. *88*, 4514 (1966)
694. Morley, J. O.: J. Chem. Soc. Perkin II *1976*, 1554
695. Fischer, P. H. H., Zimmermann, H.: Tetrahedron Lett. *1969*, 797
696. Bazhin, N. M., Akhmetova, N. E., Orlova, L. V., Steingarts, V. D., Shchegoleva, L. N., Yakobson, G. G.: ibid. *1968*, 4449
697. Bazhin, N. M., Pozdnyakovich, Yu. V., Shtejngarts, V. D. Yakobson, G. G.: Izv. Akad. Nauk SSSR, Otd. Khim. Nauk *1969*, 2300

698. Sato, H., Aoyama, Y.: Bull. Chem. Soc. Japan *46*, 631 (1973)
699. Buck, H. M., Bloemhoff, W., Oosterhoff, L. J.: Tetrahedron Lett. N. *9*, 5 (1960)
700. Forbes, W. F., Sullivan, P. D.: J. Am. Chem. Soc. *88*, 2862 (1966)
701. Forbes, W. F., Sullivan, P. D., Wang, H. M.: ibid. *89*, 2705 (1967)
702. Hoijtink, G. J., Weijland, W. P.: Rec. trav. chim. *76*, 836 (1957)
703. Carrington, A., Dravnieks, F., Symons, M. C. R.: J. Chem. Soc. *1959*, 947
704. MacLean, C., van der Waals, J. H.: J. Chem. Phys. *27*, 827 (1957)
705. Aalbersberg, W. I., Gaaf, J., Mackor, E. L.: J. Chem. Soc. *1961*, 905
706. Zaidi, Z. H., Khanna, B. N.: J. Chem. Phys. *50*, 3291 (1969)
707. Khan, Z. H., Khanna, B. N.: Canad. J. Chem. *52*, 827 (1974)
708. Oster, G. K., Yang, N. L.: J. Phys. Chem. *77*, 2159 (1973)
709. Richardson, T. J., Bartlett, N.: J. Chem. Soc. Chem. Commun. *1974*, 427
710. Olah, G. A., Schilling, P., Gross, I. M.: J. Am. Chem. Soc. *96*, 876 (1974)
711. Buck, H. M.: Tetrahedron Lett. *1965*, 605
712. Buck, H. M., Oosterhoff, L. J., van der Sluys-van der Vlught M. J., Verhoeven, K. G.: Rec. trav. chim. *86*, 923 (1967)
713. Sep, W. J., Verhoeven, J. W., de Boer, T. J.: Tetrahedron *35*, 2161 (1979)
714. Brouwer, D. M., van Doorn, J. A.: Rec. trav. chim. *91*, 1110 (1972)
715. Olah, G. A., Liang, G.: J. Am. Chem. Soc. *99*, 6045 (1977)
716. Wasserman, E., Hutton, R. S., Kuck, V. J., Chandross, E. A.: ibid. *96*, 1965 (1974)
717. Carter, M. K., Vincow, G.: J. Chem. Phys. *47*, 292 (1967)
718. Hulme, R., Symons, M. C. R.: Proc. Chem. Soc. *1963*, 241; J. Chem. Soc., *1965*, 1120
719. Carter, M. K., Vincow, G.: J. Chem. Phys. *47*, 302 (1967)
720. Singer, L. S., Lewis, I. C.: J. Am. Chem. Soc. *87*, 4695 (1965)
721. Dessau, R. M., Shih, S., Heiba, E. I.: ibid. *92*, 41 (1970)
722. Shtejngarts, V. D., Pozdnyakovich, Yu. V., Yakobson, G. G.: Zh. Org. Khim. *7*, 2002 (1971)
723. Forbes, W. F., Sullivan, P. D.: Canad. J. Chem. *44*, 1501 (1966)
724. Nishinaga, A., Ziemek, P., Matsuura, T.: Tetrahedron Lett. *1969*, 4905
725. Nishinaga, A., Ziemek, P., Matsuura, T.: J. Chem. Soc., C *1970*, 2613
726. Sullivan, P. D.: J. Phys. Chem. *26*, 3943 (1972)
727. Pokhodenko, V. D., Khizhnij, V. A., Koshechko, V. G., Shkrebtij, O. I.: Zh. Org. Khim. *11*, 1883 (1975)
728. Kochi, K.: Tetrahedron Lett. *1975*, 41
729. Bandlich, B. K., Shine, H. J.: J. Org. Chem. *42*, 561 (1977)
730. Perrin, Ch. L.: J. Am. Chem. Soc. *99*, 5516 (1977)
731. Olah, G. A.: Accounts Chem. Res. *4*, 240 (1971)
732. Eberson, L., Jönsson, L., Radner, F.: Acta Chem. Scand. *32B*, 749 (1978)
733. Drapper, M. R., Ridd, J. H.: J. Chem. Soc. Chem. Commun. *1978*, 445
734. Bargon, J., Fisher, H., Johnson, U.: Z. Naturforsch. *22a*, 1551 (1967)
735. Ward, H. R., Lawler, R. G.: J. Am. Chem. Soc. *89*, 5518 (1967)
736. Bubnov, N. N., Bilevitch, K. A., Poljakova, L. A., Okhlobystin, O. Yu.: Chem. Commun. *1972*, 1058
737. Lippmaa, E., Pehk, T., Saluvere, T., Mägi, M.: Org. Magnetic Resonance *5*, Spec. Suppl., 441 (1973)
738. Beletskaya, I. P., Rykov, S. V., Buchachenko, A. L.: Org. Magnetic Resonance *5*, Spec. Suppl., 595 (1973)
739. White, W. N., Klink, J. R.: J. Org. Chem. *35*, 965 (1970)
740. White, W. N., White, H. S., Fentiman, A.: J. Am. Chem. Soc. *92*, 4477 (1970)
741. Stock, L. M., Brown, H. C.: In: Advances in Phys. Org. Chem., Ed. Gold V. vol. 1, N.Y.— London, Academic Press 1963
742. Hudson, R. F.: Angew. Chem. *85*, 63 (1973)
743. Perrin, Ch. L., Skinner, G. A.: J. Am. Chem. Soc. *93*, 3389 (1971)
744. Moodie, R. B., Schofield, K.: Accounts Chem. Res. *9*, 287 (1976)
745. Olah, G. A., Kuhn, S. J.: J. Am. Chem. Soc. *86*, 1067 (1964)
746. Hahn, R. C., Strack, D. L.: ibid. *96*, 4335 (1974)
747. Suzuki, H.: Bull. Inst. Chem. Res. Kyoto Univ. *50*, 407 (1972)
748. Hartshorn, S. R.: Chem. Soc. Rev. *3*, 167 (1974)

749. Koptyug, V. A., Rogozhnikova, O. Yu., Detsina, A. N.: Zh. Org. Khim. *17*, 1345 (1981)
750. Koptyug, V. A., Rogozhnikova, O. Yu., Detsina, A. N.: Izv. Akad. Nauk SSSR, Otd. Khim. Nauk *1981*, 1297
751. Banthorpe, D. V.: Chem. Rev., 76, 295 (1970)
752. von Rys, P., Skrabal, P., Zollinger, H.: Angew. Chem. *84*, 92 (1972)
753. Cristy, P. F., Ridd, J. H., Stears, N. D.: J. Chem. Soc., B *1970*, 797
754. Grastaminza, A., Ridd, J. H.: ibid. *1972*, 813
755. Hoggett, J. G., Moodie, R. B., Penton, J. R., Schofield, K.: Nitration and Aromatic Reactivity. Cambridge University Press 1971
756. Rys, P.: Accounts Chem. Res. *9*, 345 (1976); Angew. Chem. Int. Ed. *16*, 807 (1977)
757. Barnett, J. W., Moodie, R. B., Schofield, K., Weston, J. B.: J. Chem. Soc. Perkin II *1975*, 648
758. Olah, G. A., Ohnishi, R.: J. Org. Chem. *43*, 865 (1978)
759. Salem, L.: J. Am. Chem. Soc. *90*, 543 (1968)
760. Fukui, K.: Theory of Orientation and Stereoselection. Berlin, Springer Verlag 1975
761. Appelman, E. H., Bonnet, R., Matteen, B.: Tetrahedron *33*. 2119 (1977)
762. Norman, R. O. C., Radda, G. K.: J. Chem. Soc. *1961*, 3610
763. Olah, G. A., Lin, H. C.: J. Am. Chem. Soc. *96*, 2892 (1974)
764. Dubois, J. E., Aaron, J. J., Alcais, P., Doncet, J. P., Rotenberg, F., Uzan, R.: ibid. *94*, 6823 (1972)
765. Meier, B. H., Ernst, R. R.: ibid. *101*, 6441 (1979)
766. Isaev, I. S., Mamatyuk, V. I., Egorova, T. G., Kuzubova, L. I., Koptyug, V. A.: Izv. Akad. Nauk SSSR, Otd. Khim. Nauk *1969*, 2089
767. Childs, R. F., Sakai, M., Parrington, B. D., Winstein, S.: J. Am. Chem. Soc. *96*, 6403 (1974)
768. Koptyug, V. A., Kuzubova, L. I., Isaev, I. S., Mamatyuk, V. I.: Zh. Org. Khim. *6*, 1843 (1970)
769. Crilgee, R., Grüner, H., Schönleber, D., Huber, A.: Chem. Ber. *103*, 3696 (1970)
770. Rodgers, T. R., Hart, H.: Tetrahedron Lett. *1969*, 4845
771. Childs, R. F., Winstein, S.: J. Am. Chem. Soc. *96*, 6409 (1974)
772. Mamatyuk, V. I., Rezvukhin, A. I., Isaev, I. S., Buraev, V. I., Koptyug, V. A.: Zh. Org. Khim. *10*, 677 (1974)
773. Koptyug, V. A., Mamatyuk, V. I., Kuzubova, L. I., Isaev, I. S.: Izv. Akad. Nauk SSSR, Otd. Khim. Nauk *1969*, 1635
774. Koptyug, V. A., Kuzubova, L. I., Isaev, I. S., Mamatyuk, V. I.: Zh. Org. Khim. *6*, 2258 (1970)
775. Hart, H., Rodgers, T. R., Griffiths, J.: J. Am. Chem. Soc. *91*, 754 (1969)
776. Parlik, J. W., Pasteris, R. J.: ibid. *96*, 6107 (1974)
777. Hehre, W. J.: ibid. *94*, 8908 (1972); *96*, 5207 (1974)
778. Hogeveen, H., Kwant, P. W.: Tetrahedron Lett. *1972*, 3197
779. Hogeveen, H., Kwant, P. W.: ibid. *1973*, 423
780. Hogeveen, H., Volger, H. C.: Rec. trav. chim. *87*, 385 (1968)
781. Hogeveen, H., Volger, H. C.: ibid. *87*, 1042 (1968); *88*, 353 (1969)
782. Hüttel, R., Tauchner, P., Forkl, H.: Chem. Ber. *105*, 1 (1972)
783. Hogeveen, H., Kwant, P. W.: Tetrahedron Lett. *1972*, 5361
784. Burger, U., Delay, A.: Helv. chim. acta *56*, 1345 (1973)
785. Hogeveen, H., Kwant, P. W., Schudde, E. P., Wade, P. A.: J. Am. Chem. Soc. *96*, 7518 (1974)
786. Hogeveen, H., Kwant, P. W.: ibid. *95*, 7315 (1973)
787. Masamune, S., Cain, E. N., Vukov, R., Takada, S., Nakatsuka, N.: Chem. Commun. *1969*, 243
788. Lustgarten, R. K., Brookhart, M., Winstein, S., Gassman, P. G., Patton, D. S., Richey, H. G., Nichols, J. D.: Tetrahedron Lett. *1970*, 1699
789. Olah, J. A., White, A. M.: J. Am. Chem. Soc. *91*, 3956 (1969)
790. Hogeveen, H., Kwant, P. W.: Tetrahedron Lett. *1973*, 1351
791. Hogeveen, H., Kwant, P. W.: ibid. *1973*, 1665
792. Hogeveen, H., Kwant, P. W.: J. Am. Chem. Soc. *96*, 2208 (1974)
793. Jonkman, H. T., Nieuwpoort, W. C.: Tetrahedron Lett. *1973*, 1671
794. Gold, V., Laali, Kh., Morris, K. P., Zdunek, L. Z.: J. Chem. Soc., Chem. Commun. *1981*, 769
795. Farkaşiu, D., Fisk, S. L., Melchior, M. T., Rose, K. D.: J. Org. Chem. *47*, 453 (1982)
796. Farkaşiu, D.: Acc. Chem. Res. *15*, 46 (1982)
797. Bakoss, H. J., Ranson, R. J., Roberts, R. M. G., Sadri, A. R.: Tetrahedron *38*, 623 (1982)
798. Joergensen, C. K.: Naturwissenschaften *67*, 188 (1980)
799. Catalan, J., Yanez, M.: J. Am. Chem. Soc. *101*, 3490 (1979)

800. Dewar, M. J. S., Reynolds, C. H.: ibid. *104*, 3244 (1982)
801. Sycheva, I. M., Zakharov, N. A.: Izv. Sibirsk. Otd. Akad. Nauk SSSR, Otd. Khim. Nauk *12*, (5), 105 (1980)
802. Lau, Y. K., Nishizawa, K., Tse, A., Brown, R. S., Kebarle, P.: J. Am. Chem. Soc. *103*, 6291 (1981)
803. Lau, W., Huffman, J. C., Kochi, J. K.: ibid. *104*, 5515 (1982)
804. Shtejngarts, V. D.: Uspekhi khimii *50*, 1487 (1981)
805. Bardin, V. V., Furin, G. G., Yakobson, G. G.: Zh. Org. Khim. *17*, 999 (1981)
806. Dobronravov, P. N., Shtejngarts, V. D.: ibid. *17*, 1556 (1981)
807. Dobronravov, P. N., Shtejngarts, V. D.: ibid. *17*, 2622 (1981)
808. Dobronravov, P. N., Shtejngarts, V. D.: ibid. *19*, 995 (1983)
809. Borodkin, G. I., Nagi, Sh. M., Bagryanskaya, I. Yu., Gatilov, Yu. V.: ibid. *9*, 000 (1983)
810. Loktev, V. F., Shubin, V. G.: Izv. Sibirsk. Otd. Akad. Nauk SSSR, Otd. Khim. Nauk *2* (1), 108 (1981)
811. Loktev, V. F., Korchagina, D. V., Shubin, V. G.: ibid. *2* (t), 112 (1981)
812. Borodkin, G. I., Panova, E. B., Shakirov, M. M., Shubin, V. G.: Zh. Org. Khim. *18*, 2312 (1982)
813. Berezina, R. N., Akimova, S. N., Korchagina, D. V., Shubin, V. G.: Izv. Sibirsk. Otd. Akad. Nauk SSSR, Otd. Khim. Nauk *7* (3), 144 (1981)
814. Berezina, R. N., Ugryumova, L. E., Korchagina, D. V., Shubin, V. G.: Zh. Org. Khim. *18*, 592 (1982)
815. Olah, G. A., Singh, B. P.: J. Am. Chem. Soc, *104*, 5168 (1982)
816. Lammertsma, K.: ibid. *103*, 2062 (1981)
817. van de Griendt, F., Cerfontain, H.: Tetrahedron *36*, 317 (1980)
818. van de Griendt, F., Cerfontain, H.: ibid. *35*, 2563 (1979)
819. Morozov, S. V., Shakirov, M. M., Shubin, V. G.: Zh. Org. Khim. *19*, 1011 (1983)
820. Morozov, S. V., Shakirov, M. M., Shubin, V. G.: Izv. Akad. Nauk SSSR, Otd. Khim. Nauk *1983*, 204
821. Hart, H., Teuerstein, A., Babin, M. A.: J. Am. Chem. Soc. *103*, 903 (1981)
822. Alder, R. W., Goode, N. C.: J. Chem. Soc., Chem. Commun. *1976*, 108
823. Alder, R. W., Bryce, M. R., Goode, N. C.: J. Chem. Soc., Perkin Trans. II *1982*, 477
824. Brown, H. C., Periasamy, M.: J. Org. Chem. *47*, 4742 (1982)
825. Loktev, V. F., Storozhenko, V. G., Shubin, V. G.: Izv. Akad. Nauk SSSR, Otd. Khim. Nauk *1983*, 304
826. Duangthai, S., Webb, G. A.: Org. Magn. Reson. *12*, 98 (1979)
827. Brigodiot, M., Lebas, J. M., Hervieu, H.: J. Mol. Struct. *32*, 311 (1976)
828. Sycheva, I. M., Shchegoleva, L. N., Mitasov, M. M., Derendyaev, B. G.: Izv. Sibirsk. Otd. Akad. Nauk SSSR, Otd. Chim. Nauk *5*, 143 (1981)
829. Farkaşiu, D.: J. Chem. Soc., Chem. Commun. *1982*, 316
830. Sutula, V. D.: Izv. Sibirsk. Otd. Akad. Nauk SSSR, Otd. Khim. Nauk *7* (3), 103, 107 (1982)
831. Borodkin, G. I., Panova, E. B., Shakirov, M. M., Shubin, V. G.: Zh. Org. Khim. *19*, 114 (1983)
832. Borodkin, G. I., Nagi, Sh. M., Shubin, V. G.: Izv. Akad. Nauk SSSR, Otd. Khim. Nauk *1982*, 2639
833. Van de Griendt, F., Cerfontain, H.: Tetrahedron *37*, 643 (1981)
834. Childs, R. F., Zeya, M., Dain, R. P.: Can. J. Chem. *59*, 76 (1981)
835. Childs, R. F., Mika-Gibala, A.: J. Org. Chem. *47*, 4204 (1982)
836. Childs, R. F., Shaw, G. S., Varadarajan, A.: Synthesis *1982*, 198

Author Index Volumes 101–122

Contents of Vols. 50–100 see Vol. 100
Author and Subject Index Vols. 26–50 see Vol. 50

The volume numbers are printed in italics

Bei Fragen zur Produktsicherheit wenden Sie sich bitte an:
If you have any questions regarding product safety,
please contact:

Walter de Gruyter GmbH
Genthiner Straße 13
10785 Berlin
productsafety@degruyterbrill.com